Theoretical Petrology

Theoretical Petrology

Second Edition

TOM. F. W. BARTH

Professor, Mineralogisk-Geologisk Museum
of the University, Oslo, Norway

JOHN WILEY & SONS, INC., NEW YORK · LONDON

Library of Congress Catalog Card Number: 62–18349

Printed in the United States of America

Preface to the Second Edition

In the ten years since the first edition was published spectacular advances in petrology have taken place. In particular the production of quantitative data and the application of thermodynamic reasoning have yielded new and better insight into the origin of rocks.

Up to the last day of the revision of the book, new data, concepts, and theories were considered; and most chapters had to be entirely rewritten. But so fast is the progress in the field that many statements in this second edition are already superseded by new knowledge made available in the last six months.

Much of the original organization of the book is retained; but for better readability the thermodynamic data are collected at the end of the book in an Appendix which is called "Thermodynamics as a Help to the Study of Rocks." It should be emphasized that this Appendix contains definitions and concepts essential to the understanding of the arguments and the calculations given throughout the general text.

Tom. F. W. Barth

July, 1962

Preface to the First Edition

Petrology grew out of geology from the naturalist's desire to observe and interpret nature. Rocks were observed, sampled, and described. Museums were erected and textbooks were written. But the genesis of rocks and their mode of development remained obscure.

The Prometheus of petrology came at the advent of this century in the rigorous application of the quantitative laws of physics and chemistry to geological problems. Reactions are governed by the same laws in the earth's crust as in the laboratory. It is unfortunate that physics and chemistry were ever separated; they seemed wide apart only when chemistry was largely empirical and nonmathematical. Again petrology, so long as it was empirical and nonmathematical, was quite different from either, in both methods and ideas. Now that experimental methods have led to the synthesis of minerals and rocks and the determination of their thermodynamical properties, petrology has become physico-chemistry applied to the crust of the earth. The sooner we realize this the better.

The fundamentals of petrology are embedded in the physical chemistry of rock-making processes. Not until such processes are fully understood can students of geology and mining elicit truth from the existing petrological data. Comprehensive information on this subject is not readily available, however; it is rarely taught in advanced university courses, partly for lack of textbooks, partly for conventional reasons. This book is presented to students, teachers, and professional geologists with the hope that it may serve in a minor way to meliorate some of these difficulties.

TOM. F. W. BARTH

December, 1951

Contents

Part I · Physics and Chemistry of the Earth

1 · ROCKS

The humanist-mathematical bias of much of our education leaves many but little aware of their immediate physical environment. More is known of stars and atoms than of rocks.

S.W.W. Geogr. J. **91,** 1938.

Rocks are made up of definite mineral assemblages. We may use an analogy from zoological science and note that a mineral species will correspond to an animal species, whereas a rock corresponds to a fauna. It has become customary, and probably also necessary, to name the rocks according to their mineral content, taking into consideration the qualitative mineral content and relative proportions of the constituent minerals on the one hand, and the mechanical and textural relations on the other. In zoology this would mean that one had to introduce a special name for a fauna consisting of hares, foxes, and fleas. Not only this, but new names would have to be introduced as the relative proportions of these three animals were changed, according to whether the hares are big and strong with silken fur or small and miserable from hunger and flea bites. Such analogies demonstrate how difficult it is to develop a satisfactory system of rock names and explain why just in igneous geology alone more than six hundred different rock names have been introduced (not counting compound names). They explain also why petrologists are helpless and, generally speaking, unhappy about it.

A student of petrology has to know a great deal about the classification of rocks. Fortunately there are many good textbooks and handbooks on petrographic techniques, describing rock specimens, modes of occurrence, and taxonomic systems. The reader is referred to the literature given on page 2. This book attempts to give a complete and logical presentation of principles and underlying theory of the petrology of the rocks of the earth's crust.

In the past geologists did not believe that the colossal forces in the interior of the earth obeyed the ordinary physico-chemical laws as determined in the laboratories. Yet the extraordinary progress in petrology during the present century is due primarily to the theoretical and practical applications of fundamental physico-chemical principles in petrogenesis. As usual in the history of mankind, each new step has been bitterly opposed by conservative practitioners. But gradually the new school of thought has gained momentum, and, generally speaking, both academic and industrial research institutions now appreciate the need for exact physico-chemical studies of rock-making processes, and try to plan their work accordingly. In many schools, however, the curriculum lags behind sadly. It is hoped that this book will make it easier to understand and to teach this important phase of geology.

Today we understand the rock-making processes sufficiently well to undertake an interpretation and a synthesis of the existing quantitative data—the field data from all over the world, the experimental data from geophysical and geochemical laboratories, and the theoretical data of the research workers in geology, chemistry, and physics. These data, collected, organized, and interpreted, are the building bricks from which the edifice of modern petrology can be constructed.

TEXTBOOKS AND MONOGRAPHS

(For further references, see pages 226 ff. and 364 ff. under igneous and metamorphic rocks.)

General

Barth, T. F. W., C. W. Correns, and P. Eskola, *Entstehung der Gesteine,* Berlin, 1939.

Correns, C. W., *Einführung in die Mineralogie, Kristallographie und Petrographie,* Berlin, 1949.

Eskola, P., *Kristalle und Gesteine,* Wien, 1946.

Fischer, Walther, *Gesteins- und Lagerstättenbildung im Wandel der wissenschaftlichen Anschauung,* Stuttgart, 1961.

Jung, J., *Précis de Pétrographie,* Paris, 1958.

Niggli, P., and E. Niggli, *Gesteine und Minerallagerstätten,* Basel, 1948.

Rinne, F., *Gesteinskunde,* 12th ed., Leipzig, 1940.

Rosenbusch, H., *Elemente der Gesteinslehre,* 4th ed., edited by Osann, Stuttgart, 1923.

Turner, F. J., and J. Verhoogen, *Igneous and Metamorphic Petrology,* 2nd ed., New York, 1959.

Tyrrell, G. W., *The Principles of Petrology,* London, 1926.

Williams, H., F. J. Turner, and C. M. Gilbert, *Petrography,* San Francisco, 1955.

2 · ORIGIN OF THE EARTH

It is an old maxim of mine that when you have excluded the impossible, whatever remains, however improbable, must be the truth.

Arthur Conan Doyle, *The Adventures of Sherlock Holmes.*

Petrology is not primarily concerned with the origin of the earth and the planets. This problem belongs properly in the realm of astrophysics, but in so far as it has a direct bearing on geochemistry and the frequency distribution of the elements in the rocks of the earth, it must be discussed. Mechanical characteristics of all planets are given for reference in Table I-1. The terrestrial planets (Venus, Mars, Mercury) and the Moon form one group; the giant planets (Jupiter, Saturn, Uranus, and Neptune) form another group of quite different bulk properties. Likewise, the giant planets are peculiar both for their low surface temperature (less than 150°K) and the strange composition and vast depth of their atmospheres.

Ammonia (NH_3), methane (CH_4), helium, and hydrogen seem to make up the outer zone of bulk. In comparison to these giant planets, the atmospheres of the earth and the terrestrial planets are extremely thin and without hydrides. In the solar atmosphere hydrogen and

TABLE I-1

Mechanical Characteristics of the Planets

	r	M	g	V	D
Moon	0.273	0.012	0.1645	0.212	3.4
Mercury	0.39	0.054	0.2093	0.290	4.8
Venus	0.973	0.814	0.8383	0.910	4.86
Earth	1.000	1.000	1.0000	1.000	5.52
Mars	0.532	0.108	0.3717	0.447	4.1
Jupiter	11.26	317.1	2.501	5.32	1.30
Saturn	9.45	95.0	1.064	3.17	0.69
Uranus	4.19	14.7	0.840	1.88	1.56
Neptune	3.89	17.3	1.141	2.12	2.22
Pluto	0.5	0.06	...	...	...

r, equatorial radius, unit 6378.388 km; M, mass, unit $5.966 \cdot 10^{27}$ g; g, acceleration of gravity, unit 980.60 cm sec^{-2}; V, velocity of escape, unit 11.188 km sec^{-1}; D, mean density, g cm^{-3}.

helium predominate, and there is hardly 1 per cent of heavy metals; but in the interior of the sun the sum of all metals may attain 10 per cent. Mercury and the Moon have no atmosphere at all. The atmosphere of Mars, compared with that of the earth, is very thin and contains about 98 per cent nitrogen and argon, very small amounts of carbon dioxide, water, and oxygen. The density of the atmosphere of Venus is comparable to that of the earth; it contains large quantities of carbon dioxide, some water, but practically no oxygen. Its opacity is mainly due to dust stirred up from the surface by winds.

Theories on the origin of the solar system can be divided into two groups. (1) Dualistic (Chamberlain-Moulton) assuming an interaction between the sun and another celestial body, causing a catastrophic change. (2) Monistic (Kant-Laplace, the planetesimal theory; Deans, Jeffreys, the tidal theory), assuming a slow evolution of a primordial system. Now favored is the hypothesis of Weizsächer (1944) explaining the planets as remnants of a primeval tenuous solar nebula filling the space to the orbit of Pluto. According to the laws of mechanics, this nebulous shell must assume the shape of a flat disc, vortices develop, and eventually individual balls separate, which later develop into planets. Figure I-1 shows what happens to a compressible ball of gas rotating with increasing velocity. In stage 2 the equator is marked by a sharp edge along which the gas no longer is affected by gravitation but escapes into space. Eventually only a small fraction of the original gas ball remains. It condenses into a planet, the chemical composition of which is greatly different from that of the original gas ball. The heavy elements are highly concentrated, the light elements mostly lost to space. The degree of con-

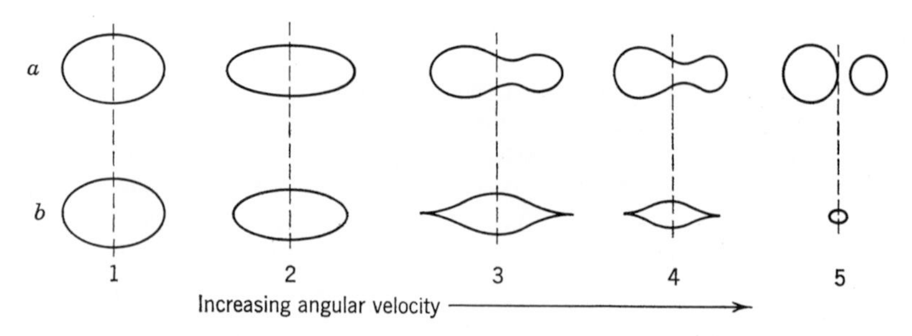

FIG. I-1. Different stages of rotating bodies. (*a*) Incompressible liquid. (*b*) Compressible gas. There is a sharp transition between (*a*) and (*b*). Physical planets must, therefore, generally speaking, behave either like (*a*) or like (*b*). (After Pahlen.)

centration depends upon the size of the original ball, the angular velocity, and the temperature. This concentration explains the peculiar chemical composition of the earth (and probably of the terrestrial planets), which seems to be unique in the solar system. The theory as developed by Weizsächer is largely qualitative. It has been discussed in a quantitative manner by Ter Haar and modified by Kuiper, but an entirely satisfactory theory which is free from objections has not yet been derived. However, it seems that a monistic theory rather than a dualistic is much more likely to succeed.

The age of the earth's crust is somewhat more than 3×10^9 years. The upper limit for the age of the earth estimated from the abundance of lead in the crust is 4×10^9 years. The age limit for meteorites (determined by the helium method) is 10×10^9 years.

The ages of the fossiliferous periods are shown in Table I-2:

TABLE I-2

Chart of Geologic Periods

(After Holmes, 1960)

	Beginning of Period, Million Years		Beginning of Period, Million Years		Beginning of Period, Million Years
Pleistocene	1	Cretaceous	135 ± 5	Devonian	400 ± 10
Pliocene	11	Jurassic	180 ± 5	Silurian	440 ± 10
Miocene	25	Triassic	225 ± 5	Ordovician	500 ± 15
Oligocene	40	Permian	270 ± 5	Cambrian	600 ± 20
Eocene	60	Carboniferous	350 ± 10		
Paleocene	70 ± 2				

3 · HEAT OF THE EARTH

The thermal history of the earth is influenced not only by the initial distribution of temperature, but also by the radioactive matter, which, if originally distributed uniformly, would keep the interior fluid below a skin a few tens of kilometres thick until the process tending to lift radioactive constituents to the top had time to produce the present concentration into the upper layers.

Jeffreys, *The Earth*, 1929.

Very little can be said about the heat distribution in the earth in pregeologic time. It is believed that the heat of accretion was sufficient to give birth to a molten globe. But some geologists today think

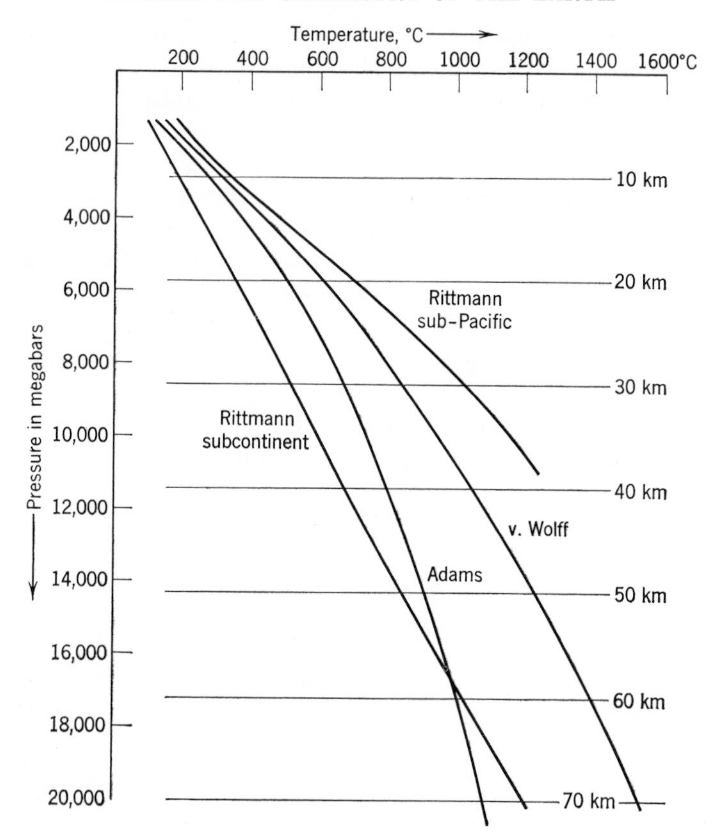

Fig. I-2. Variation of temperature with depth, according to different authors.

that the earth used to be cold and is now heating up because of radioactive processes.

The temperature of the interior of the earth today is also subject to conjecture. Direct measurements in mines and in bore holes have shown an average increase in the temperature of about 1°C per 50 m depth. However, the total variation in the (reciprocal) geothermal gradient is from 7 m (in recent lava) to 173 m per degree (lower part of a bore hole in Transvaal). Extrapolation to greater depths is very uncertain (see Fig. I-2). The temperature of the core is unknown.

The *heat conductivity* varies with the type of rock. For typical granite it is about $8 \cdot 10^{-3}$, for gabbro about $5 \cdot 10^{-3}$ cal cm^{-1} sec^{-1} $degree^{-1}$. The *specific heat* of most silicates is around 0.2 cal g^{-1}.

The *heat flow* to the earth's surface is approximately 1.2×10^{-6}

cal cm^{-2} sec^{-1}. For the whole earth it amounts to about 6×10^{12} cal sec^{-1}. It corresponds to about $2 \cdot 10^{20}$ cal per year, or a cooling of the earth of 22°C per 10^9 years, if no heat was generated. These figures are unfamiliar, but put as 40 cal cm^{-2} annually, and compared with the heat of melting of ice, 80 cal g^{-1}, we see that its amount is such as could melt an ice sheet of 0.5 cm thickness in one year.

Aside from a possible residuum of heat associated with the earth at the time of its formation, the primary source of heat in the earth is radioactivity. The important radioactive elements produce heat at an average rate which can be calculated from the measured energies of the α, β, and γ rays emitted, together with the experimentally determined decay constants. In one year:

The UI-AcU series yields 0.75 cal per gram of uranium.
The Th series yields 0.20 cal per gram of thorium.
The K series yields 0.000027 cal per gram of potassium.

The values, together with the radioactive content of terrestrial materials, make possible an estimate of the rate at which heat is now being generated by radioactive processes. For example, in one gram of granite we find the following values:

U yields	$3.0 \cdot 10^{-6}$ cal per year
Th yields	$2.7 \cdot 10^{-6}$ cal per year
K^{40} yields	$0.9 \cdot 10^{-6}$ cal per year
Rb yields	$0.04 \cdot 10^{-6}$ cal per year
Sm yields	$0.006 \cdot 10^{-6}$ cal per year
Total	*$6.646 \cdot 10^{-6}$ cal per year*

Gabbroic rocks yield $2 \cdot 10^{-6}$ cal per gram per year, whereas ultrabasic rocks yield only $0.1 \cdot 10^{-6}$ cal per gram per year. Accordingly, the lithosphere (16 km depth), which represents 2.2×10^{25} g, or less than 4 per cent of the total mass of the earth, generates radioactive heat of the order of magnitude of 10^{20} cal per year. Any radioactive heat generated in deeper regions would therefore be in excess of the observed heat flow to the earth's surface. It seems possible, therefore, that the earth may be heating, if not uniformly, at least possibly at certain depths.

Since the beginning of the earth, the radioactive isotopes have been disintegrating. The final disintegration products are shown in Table I-3. The disintegration is particularly marked in the case of U^{235} and

TABLE I-3

Properties of Natural Radioactive Isotopes

	Frequency, %	Half-Lives, T (Years)	Distribution Ratio	Disintegration Products
Re^{187}	62.9	6.2×10^{10}	...	Os^{187}
Rb^{87}	27.9	5.0×10^{10}	...	Sr^{87}
Th^{232}	100.0	1.39×10^{10}	...	Pb^{208}
U^{238}	99.27	4.51×10^{9}	...	Pb^{206}
K^{40}	0.0119	1.27×10^{9}	{89, 11}	{Ca^{40}, A^{40}}
U^{235}	0.72	7.13×10^{8}	...	Pb^{207}

K^{40}. Putting the present amount of K^{40} equal to 1 and computing the amount, N, of the isotope at a time, t, measured backwards from the present, we have:

$$\log N = \frac{t}{T} \cdot \log 2$$

The half-period for K^{40} is $T = 1.27 \times 10^9$ years.

We find that 3 billion years ago ($t = 3 \times 10^9$) the amount of K^{40} was more than five times higher than it is today, corresponding to an increase of almost 100 per cent for the total heat output in the granitic crust as compared with today. The great amount of heat thus generated in the crust of the earth would influence geologic processes such as the formation of magmas, volcanic activity, and the rise of mountain chains. It is worth noticing that the actualistic point of view in geology, a principle otherwise well established, according to these data, obviously needs qualification.

The radiant solar energy received by the earth is about 10^{-2} cal cm^{-2} sec^{-1}, or a total of about 10^{24} cal per year, overlooking the effects of the earth's atmosphere. This heat flux is several orders of magnitude greater than the internal heat flow; 0.1 per cent goes to photosynthesis, 15 per cent is absorbed by the earth, 85 per cent evaporates water.

4 · STRENGTH OF THE EARTH

If we can conceive what would happen if a 6.6 ton sphere of soft mud having a density of hematite and a diameter of 4 feet were dropped from a height of 15 feet into a concrete pavement, we will have a very good idea of what would happen if our earth were to have a similar collision while moving at its orbital velocity of 29.8 km/sec.

M. K. Hubbert, *Bull. Geol. Soc. Am.* **48,** 1937.

Some of the mechanical constants of the earth have been listed in Table I-1. It may be added that the *density* ranges from 2.7 at the surface to about 15 at the center. The over-all *modulus of rigidity* of the earth as determined in various ways is of the order of twice that of steel. The strength of the earth material increases with increase of pressure and decreases with increase of temperature. The over-all *strength* is comparable to the strength of ordinary steel, or about 4×10^9 *dynes per square centimeter.* The *viscosity* of the earth is about 10^{22} *poises,* which appear to be an average characteristic of the earth materials to great depths. It implies a material whose general physical properties differ but slightly from those of the surface rocks of the earth. These constants would seem to imply that the earth must be looked upon as a completely solid and rigid body. It represents a ball that has elastic properties rather similar to those of steel.

Numerous geophysical and geological observations, however, seem to contradict the conception of an earth as strong as steel. To explain the observed gravity anomalies in the neighborhood of mountain ranges, it has been necessary to adopt the *hypothesis of isostasy:* the crust is in floating equilibrium on the denser subcrustal stratum; the mountain ranges are made up of relatively light material that floats high and, like an iceberg in water, protrudes downward into the heavier "plastic" stratum under each mountain range. Likewise, it would seem that geological uplifts and depressions of almost any area of more than a few tens of kilometers in breadth will vary with the concurrent loading and unloading by deposition and erosion in almost perfect accordance with the dictates of the principle of isostasy.

A. A. Michelson, a renowned physicist, studied the "microtides," and found that the tides in open oceans are about 75 cm ($2\frac{1}{2}$ feet). As this is only 69 per cent of what they would be if the earth were absolutely unyielding, the theoretical total tide must be about 110 cm. The difference, or 35 cm, represents the tidal movement of the solid earth itself. Thus the ground under our feet is periodically moving up and down. It is pulled up every time the moon is high in the sky, and sinks down again as the moon approaches the horizon. Although the rigidity of our globe is higher than that of steel, in response to weak but persistent forces it behaves like a soft plastic body. Furthermore, the Wegener theory maintains that not only are the continents in a state of flotation, but they are also actually drifting about on the face of the earth.

By a study of the earth's *paleomagnetism* (remanent magnetism), it has been possible, in the first approximation, to give a record of these movements. Paleomagnetism exists, for example, in many old lava flows or in sediments. During sedimentation all small grains showing magnetic susceptibility, particularly grains of magnetite, will orient themselves parallel to the direction of the earth's magnetic field at the time. Thus measurements from England indicate that in Triassic time the North Pole was placed in what is now eastern Siberia; subsequently it migrated into the Arctic Ocean and eventually reached its present position. This corresponds to a simple migration of the pole. But measurements conducted in North America give a pole migration curve systematically displaced in relation to that determined from England; and the pole migration curve determined by measurements in Australia is again widely different from either. Such results can only be explained by a relative displacement of the several land areas. One has to conclude, therefore, that the poles did not migrate, but that different areas of the continental crust moved about approximately as the individual floes in the ice pack move and turn. In this way, pieces of the continental crust may have drifted all over the globe's surface.

King Hubbert has demonstrated most convincingly the resolution of this paradox by making use of the methods of *dimensional analysis:* Should a model of the earth on a small scale, in order to reflect the physical properties of the earth, be made of steel? Would the earth in a collision with another planet rebound like a steel ball, or would the two bodies coalesce? We shall not go into the mathematical details of the theory, but only give the answers to these queries in a dimensional analysis calculation by Hubbert for a model earth with a radius of 63.7 cm. See Table I-4.

In the light of these results, it becomes understandable that the earth, although rigid like steel when small rock specimens are examined, deforms like putty under the influence of forces extensive in space and time; for example, the Glacial depression and the post-Glacial uplift of Scandinavia are, indeed, the direct consequences of the loading and the unloading of the Glacial ice.

TABLE I-4

PROPERTIES OF THE ORIGINAL AND A MODEL EARTH

(M. K. Hubbert, Theory of scale models as applied to the study of geologic structures, *Bull. Geol. Soc. Am.*, 1937)

Property	Original	Model Ratio	Model
Mean radius	6.37×10^8 cm	10^{-7}	63.7 cm
Mean radius of orbit	1.50×10^{13} cm	10^{-7}	1.50×10^6 cm
Period of rotation (Sidereal Day)	8.6164×10^4 sec	3.16×10^{-4}	27.3 sec
Period of revolution (Sidereal Year)	3.156×10^7 sec	3.16×10^{-4}	9.98×10^3 sec (2.77 hours)
Mean orbital velocity	2.98×10^6 cm/sec	3.16×10^{-4}	9.44×10^2 cm/sec
Mean density	5.52	1	5.52
Mass	5.98×10^{27} g	10^{-21}	5.98×10^6 g
Pressure at 100 km depth	3×10^{10} dynes/cm^2	10^{-7}	3×10^3 dynes/cm^2
Shear strength	4×10^9 dynes/cm^2	10^{-7}	4×10^2 dynes/cm^2
Gravitational constant	$6.67 \times 10^{-8} \dfrac{\text{cm}^3}{\text{g sec}^2}$	10^7	$0.667 \dfrac{\text{cm}^3}{\text{g sec}^2}$

The radius of the depressed Scandinavian area is about 1000 km. The time it took the uplift to equalize one-half the initial depression was about 10,000 years and involved the filling in of a volume of almost 10^{15} m^3 (10^{21} cm^3). This enormous volume was filled through "plastic" flow in the underground. It is petrographically important to note that this and similar geological processes involve large-scale transportation of material in the solid crust.

A matter of special interest is the strength which has been taken to be the model equivalent of an earth having the strength of cold steel. A material having a shearing strength of 4×10^2 dynes cm^{-2} corresponds approximately to very soft mud. That is to say, a sphere the size of the earth, having the density of 5.52 and the strength of cold steel, does exhibit properties very similar to those of a sphere of soft mud of a radius of 63.7 cm.*

* This shows how important it is to use the right kind of material in model experiments performed to produce mountain-like structures. H. Cloos, who has made the most successful experiments, applied wet, half liquid clay and obtained striking resemblance to structures actually observed in the field.

5 · THE CONCENTRIC SHELLS OF THE EARTH

*Derart ist uns die relative Seltenheit der meisten Kulturmetalle verständlich, als Resultat einer grossartigen metallurgischen Schmelzoperation, auf deren Schlackenprodukt wir leben.**

V. M. Goldschmidt, *Der Stoffwechsel der Erde,* 1922.

Research in geochemistry has shown that some elements are characteristically concentrated in ore deposits, others in the usual rock types. They have been called metallogenic and petrogenic, respectively, by H. S. Washington.

Presuming that the earth is made up of concentric shells, (1) atmosphere and hydrosphere, (2) lithosphere, (3) chalcosphere, and (4) siderosphere or iron core, Goldschmidt has classified the elements according to their chief place of occurrence, as shown in Table I-5.

Estimates of the chemical composition of the interior of the earth are uncertain; indeed, they are based on extrapolations into the unknown. According to our best information, the outer shell, the so-called crust, is made up of the following concentric layers which imperceptibly merge into one another:

1. *Sediments,* which are partly penetrated by sial.
2. *Sial,* which extends to a depth of approximately 30 km under the continents. It is compositionally inhomogeneous, with granitic

TABLE I-5

GEOCHEMICAL CLASSIFICATION OF THE ELEMENTS

(After V. M. Goldschmidt, *Vid.-Akad. Skr. Oslo,* 1922)

Iron, Siderophile	Sulfides, Chalcophile	Silicates, Lithophile	Gases, Atmophile	Organisms; Biophile
Fe, Ni, Co	((O)), S, Se, Te	O, (S), (P), (H)	H, N, C, (O)	C, H, O, N, P
P, (As), C	Fe, Cr, (Ni), (Co)	Si, Ti, Zr, Hf, Th	Cl, Br, I	S, Cl, I, (B)
Ru, Rh, Pd	Cu, Zn, Cd, Pb	Li, Na, K, Rb, Cs		(Ca, Mg, K, Na)
Os, Ir, Pt, Au	Sn, Ge, Mo	F, Cl, Br, I	He, Ne, Ar	(V, Mn, Fe, Cu)
Ge, Sn	As, Sb, Bi	B, Al, (Ga), Sc	Kr, Xe	
Mo, (W)	Ag, (Au), Hg	Y, Rare Earths		
(Nb), Ta	Pd, Ru, (Pt)	Be, Mg, Ca, Sr, Ba		
(Se), (Te)	Ga, In, Tl	(Fe), V, Cr, Mn		
	(Cr)	Nb, Ta, W, U		

* The relative scarcity of most culture metals is declared as the result of a colossal metallurgic smelting operation, on the slag products of which we live.

rocks at the top, and gradually changes downward into sialma. Under the sial layer of the crust comes:

3. *Sialma,* which occupies the depth range of 30 to 60 km and whose average composition is that of a basalt or eclogite.

4. At still greater depth, heavy basic silicates prevail. This region has been referred to as the *dunite* or *peridotite shell.* Typically, olivine is present.

5. The composition of the deeper parts of the earth is subject to conjecture.

The favored model of the earth's layers is indicated in Fig. I-4. It was suggested by the study of meteorites, of seismic waves, and of high-pressure polymorphs.

In an earthquake two types of elastic waves are started that spread throughout the earth. One is the longitudinal P wave (irrotational, compressional, condensational, primary); the other is the transverse S wave (also called distortional, equivoluminal, secondary). Popularly they may be called the "Push" and the "Shake." The velocities of propagation of these waves in the crust are graphically shown in Fig. I-4. Comparison of these velocities with laboratory determinations of the compressibilities of rocks shows that sial is probably granite, sialma is basalt, and the lowest shell is peridotite. A first-order discontinuity within the earth at a depth of 2900 km is clearly indicated by the seismic waves. This was discovered by R. D. Oldham in 1906.

Elastic waves undergo reflection and refraction at interfaces between different media, and at free surfaces. The phenomena, however, are more complicated than in sound or optics, because each medium can in general transmit both longitudinal and transverse waves. Thus when a plane P wave is incident on a horizontal boundary between two solids, it is ordinarily broken up into four waves: a reflected and a transmitted P wave, and a reflected and a transmitted S wave.

If the velocity is less in the inner layer, all rays striking the inner layer give transmitted waves, but all are refracted down at finite angles. The result is that a definite belt of shadow is cast. The inner sphere acts like a lens, and tends to bring the rays emerging from it to a focus at the image of the origin, as illustrated in Fig. I-3.

In the earth the velocity of P increases with depth steadily from about 8 km per second at the base of the sialma to 13 km per second at a depth of 2900 km. At this depth the velocity suddenly falls to about 8.5 km per second, and then again proceeds to increase slowly with depth. The velocity of S similarly increases in the outer shells

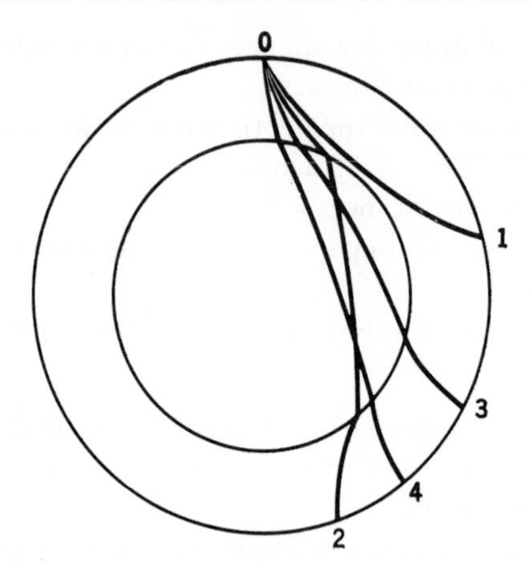

FIG. I-3. Earthquake waves are generated at 0 and follow the paths marked 1, 2, 3, and 4. The effect of a discontinuity of the first order, with smaller velocity of propagation in the inner sphere, is shown graphically. Ray 1 just grazes the discontinuity; ray 2 is the critical one, just steep enough to enter the inner sphere; 3 gives minimum deviation; 4 is even steeper. (After Jeffreys, *The Earth,* 1929.)

from 4.3 to 7.2 km per second. But S does not penetrate the core, which therefore must be fluid or an imperfectly elastic solid. Jeffreys has said: "The most striking feature revealed by seismology is, indeed, the presence of a central core, apparently liquid, with a radius rather more than half that of the earth as a whole, and with a sharp boundary."

Detailed studies of earthquake wave reflections have revealed various other breaks in the structure of the globe. Geologically important is the Mohorovičič (M) discontinuity: at an average depth of 30 km below the continents and 5 km below the ocean floors the velocity of P earthquake waves increases suddenly to about 8 km per second. This velocity essentially restricts the type of rocks in this region to peridotites and eclogites. The deeper regions are as follows (see Fig. I-4):

From the M-discontinuity to 200–400 km: dunite-peridotite grading downward into garnet peridotite.

From 400 to 900 km: transition zone dominated by inversion of olivine to spinel. Pyroxene breaks down to spinel plus coesite (page 237).

From 900 to 2900 km: practically all elements present dissolve in a homogeneous solid solution of disordered spinel structure (state of oligophase, page 139).

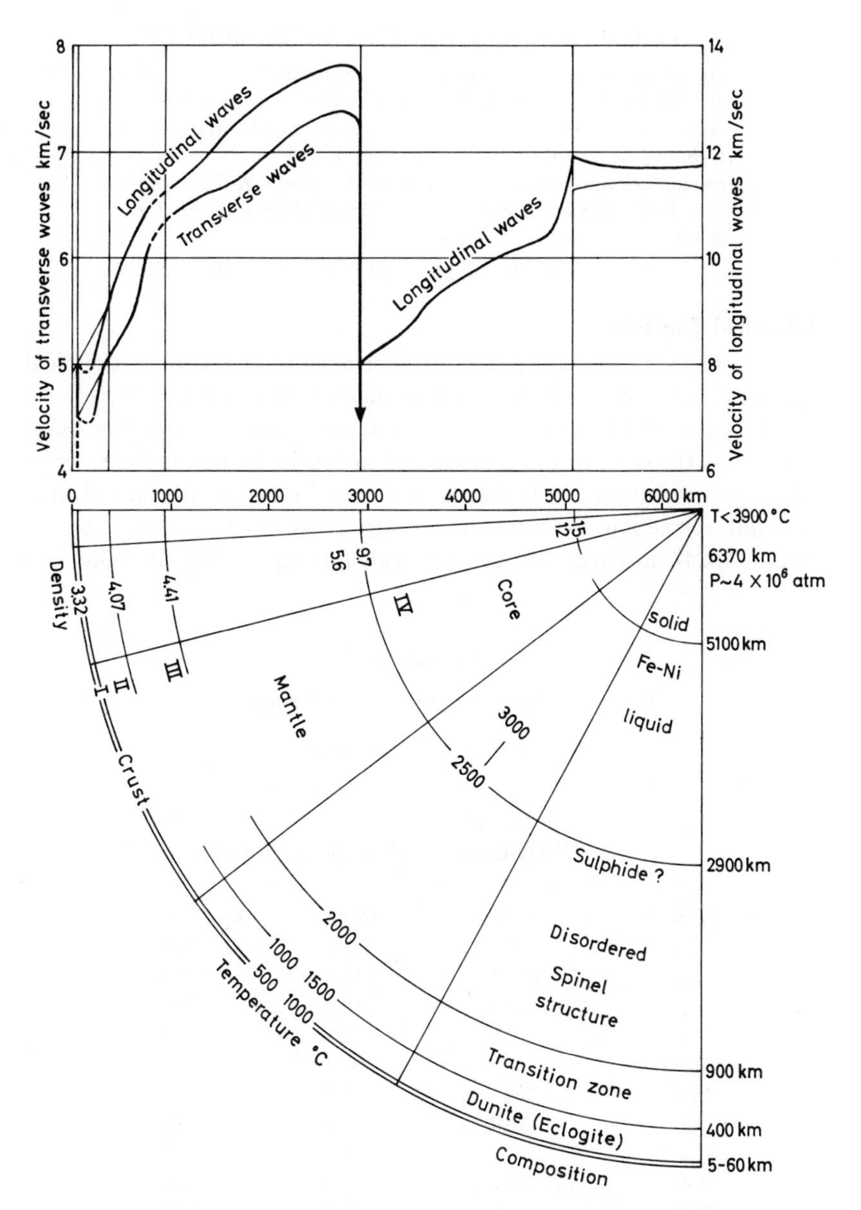

FIG. I-4. *Upper half:* Velocity of propagation of earthquake waves from the surface to the center of the earth. *Lower half:* The concentric structure of the earth, the various compositional discontinuities are reflected in discontinuities or in sudden changes in the wave velocities.

6 · COMPOSITION OF THE LITHOSPHERE

Die Ansammlung dieses ungeheuren Sauerstoffvolumens wird ermöglicht durch die Bindung mittels der zwischengelagerten Kationen, deren Ladungen den Zusammenhalt des ganzen Gebildes bewirken, die jedoch in Bezug auf wirkliche Raumbeanspruchung nur eine ganz untergeordnete Rolle spielen. Mit Recht können wir die Lithosphäre deshalb auch als Oxysphäre bezeichnen.

V. M. Goldschmidt, *N. Jahrb. Min.*, 1928.*

Chemical Elements

A rapid survey of the relative importance of the various chemical elements in the crust of the earth is made in Table I-6. Clarke and Washington (1924) determined the average composition of the *lithosphere* (or the crust of the earth to 16 km depth) by taking the average of all good analyses of the igneous rocks and subsequently introducing a small correction for the sedimentary rocks (see page 62). Sederholm (1924) insisted that in such calculations we should take into

TABLE I-6

CHEMICAL COMPOSITION OF THE LITHOSPHERE

(In Weight Percentage)

	(1) Clarke and Washington	(2) Sederholm	(3) Goldschmidt
SiO_2	59.07	67.45	59.19
TiO_2	1.03	0.41	0.79
Al_2O_3	15.22	14.63	15.82
Fe_2O_3	3.10	1.27	3.41
FeO	3.71	3.28	3.58
MnO	0.11	0.04	0.11
MgO	3.45	1.69	3.30
CaO	5.10	3.40	3.07
Na_2O	3.71	3.06	2.05
K_2O	3.11	3.55	3.93
H_2O	1.30	0.79	3.02
P_2O_5	0.30	0.11	0.22
CO_2	0.35	0.12	0.54

* Translated on page 17.

consideration the relative abundance of each rock type; according to that method he computed the weighted average composition of the rocks of Finland. Vogt (1931) likewise criticized the figures of Clarke and Washington. He calculated the average composition from the relative frequency distribution of the individual types of igneous rocks as given by Daly, taking granite to make up 60 per cent of the total. Goldschmidt (1933) pointed out that the average composition of glacial clay in Norway closely corresponds to the average rock composition, the reason being that the clay was formed simply by the mechanical grinding action of the ice and, without suffering chemical alterations, was mixed and moved by the ice to the then seashore and deposited as clay.

No special significance is attached to the weight of an element. More informative for many purposes are figures giving the numbers of each atom, that is, atom percentages. Volume percentages are also useful. The results of such calculations are presented in Table I-7.

The dominant role of oxygen in the lithosphere is demonstrated in Table I-7. Goldschmidt has called it oxysphere and regarded it as essentially a packing of oxygen ions. "The accumulation of this huge volume of oxygen is made possible through the cations, which occupy the interstices and, with their electrical charges, keep the whole structure together although their volume is comparatively insignificant."

TABLE I-7

ELEMENTAL COMPOSITION OF THE LITHOSPHERE

(Recalculated from Clarke and Washington)

	Weight, %	Atom, %	Volume, %	Atomic Radii, Å
O	46.71	60.5	94.24	1.40
Si	27.69	20.5	0.51	0.39
Ti	0.62	0.3	0.03	0.68
Al	8.07	6.2	0.44	0.50
Fe	5.05	1.9	0.37	0.70
Mg	2.08	1.8	0.28	0.65
Ca	3.65	1.9	1.04	0.99
Na	2.75	2.5	1.21	0.95
K	2.58	1.4	1.88	1.33
H	0.14	3.0	...	...

The concentration of the minor elements in the lithosphere is shown in Table I-8.

The composition of the "lithosphere" can by no means be taken to represent that of the whole earth. (See page 12, where the concentric structure of the earth is discussed.) The figures given by Clarke and Washington for the lithosphere embrace only 0.4 per cent of the total mass of the earth; this percentage is insignificant, perhaps, when compared to the entire globe, yet of magnificent extent in relation to man.

TABLE I-8

CONCENTRATION OF THE MINOR ELEMENTS IN THE LITHOSPHERE

(In Parts per Million)

Ti	4400	Ga	15	Hg	0.5
H	1300	Th	10	I	0.3
P	1180	Sm	7	Bi	0.2
Mn	1000	Gd	6	Sb	0.2
F	800	Pr	6	Tm	0.2
S	520	Sc	5	Cd	0.2
Sr	450	Hf	5	Ag	0.1
Ba	400	Dy	5	In	0.1
C	320	B	3	Se	0.09
Cl	200	Sn	3	A	0.04
Cr	200	Yb	3	Pd	0.010
Zr	160	Er	3	Pt	0.005
Rb	120	Br	3	Au	0.005
V	110	Ge	2	He	0.003
Ni	80	Be	2	Te	(0.0018)?
Zn	65	As	2	Rh	0.001
N	46	U	2	Re	0.001
Ce	46	Ta	2	Ir	0.001
Cu	45	Tl	1	Os	0.001
Y	40	W	1	Ru	0.001
Li	30	Mo	1	Ra	1.3×10^{-6}
Nd	24	Eu	1	Pa	5×10^{-7}
Nb	24	Ho	1	Po	3×10^{-10}
Co	23	Cs	1	Ac	3×10^{-10}
La	18	Tb	0.9	Ne	7.0×10^{-10}
Pb	15	Lu	0.7	Kr	1.6×10^{-11}
				Xe	2.0×10^{-12}

7 · SELECTED REFERENCES

Bullen, K. E., *An Introduction to the Theory of Seismology,* Cambridge, 1953.

Clarke, F. W., The data of geochemistry, 5th end., *U.S. Geol. Survey Bull.* **770,** 1924.

Clarke, F. W., and H. S. Washington, The composition of the earth's crust, *U.S. Geol. Survey, Profess. Paper* **127,** 1924.

Goldschmidt, V. M., *Geochemistry,* London, 1954.

Hubbert, M. K., Theory of scale and models as applied to the study of geologic structures, *Bull. Geol. Soc. Am.* **48,** 1937.

Jacobs, J. A., R. D. Russell, and J. T. Wilson, *Physics and Geology,* New York, 1959.

Jeffreys, H., *The Earth,* 3rd ed., Cambridge, 1952.

Physics and Chemistry of the Earth, I, II, III, and IV, edited by L. H. Ahrens, Frank Press, Kalervo Rankama, and S. K. Runcorn, London, 1956–1961.

Rankama, K., *Isotope Geology,* London, 1954.

Rankama, K., and T. Sahama, *Geochemistry,* Chicago, 2nd ed., 1952.

Researches in Geochemistry, edited by P. H. Abelson, Wiley, New York, 1959.

Ringwood, A. E., On the chemical evolution and densities of the planets, *Geochim. et Cosmochim. Acta* **15,** 257, 1959.

Sederholm, J. J., The average composition of the earth's crust, *Bull. comm. geol. Finlande* **70,** 1924.

Ter Haar, D., Further studies on the origin of the solar system, *Astrophys. J.* **111,** 179, 1950.

The Earth Today, edited by A. H. Cook and T. F. Gaskell, London, 1961.

The Earth as a Planet, edited by G. P. Kuiper, Chicago, 1954.

Urey, H. C., The origin and development of the earth and other terrestrial planets, *Geochim. et Cosmochim. Acta* **1,** 209, 1951.

———, *The Planets; their origin and development,* Yale University Press, 1952.

von Weizsäcker, C. V., Über die Entstehung des Planeten-systems, *Z. Astrophys.* **22,** 319, 1944.

Part II · Sedimentary Rocks

According to available statistical data, about 85–90% of the annual yield of mineral products comes from sedimentary mineral and ore deposits; this fact emphasizes the great practical importance of the geochemistry of sediments.

V. M. Goldschmidt, 1937.

No attempt will be made to treat fully the many interesting problems in sedimentary petrography. These are special problems and are described in books on sedimentology. Only a small part of the lithosphere is made up of sedimentary rocks, and many of the sedimentological problems are of no general interest for the understanding of the rock-forming processes on the earth.

This part of the book will be short, therefore, and will be limited to subjects of fundamental petrological and geochemical significance. In particular, the special petrology of salt deposits, of coal, and of petroleum will not be treated in this book. References to special articles and books are listed on page 49.

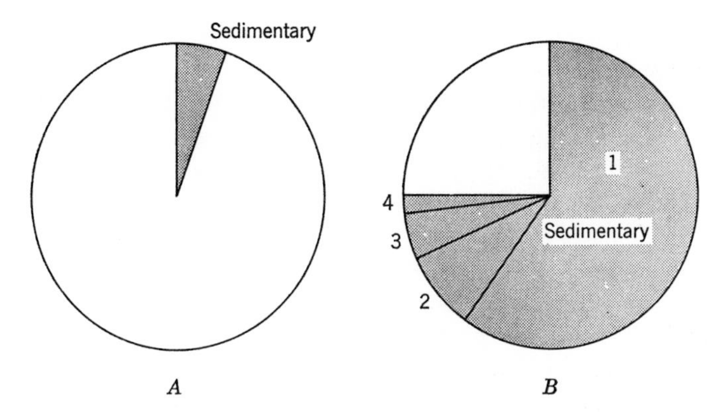

Fig. II-1. Relative abundance of sedimentary and non-sedimentary rocks. *A,* by volume of the lithosphere (outer shell 16 km thick); *B,* by surface area. Shale (1) 80%. Sandstone (2) 12%. Limestone (3) 6%. Dolomite (4) 2%.

1 · SEDIMENTARY DIFFERENTIATION

The primary constituents of the silicate crust are unstable in contact with the atmosphere. They are attacked by oxygen, carbon dioxide, and water. The rocks, thus decomposed, yield to the mechanical action of the erosion. The material is transported by wind, water, or ice, and is redeposited in other places as sediments, or remains in solution in the sea.

The first impression might be that this process would lead to a complete intermingling of the various rock components. But this is not true. The processes of weathering, erosion, and sedimentation, that is, the process of the *external geochemical migration* actually brings about an extreme differentiation of most of the chemical components.

The external geochemical migration may be compared to a colossal chemical separation of the several elements—a separation which can be compared to that being performed in a chemical laboratory in the course of a quantitative chemical analysis of rocks. The reason for this regularity is, at least partly, that in the waters carrying the minerals in solution the concentrations and the pH values will increase, and the oxidation potential will decrease with time and distance of transport. The result of this natural analysis is not always excellent, for nature often uses poor methods, but by and large the results are not bad.

Separation of Silica

Chemically resistant minerals, particularly quartz, are left intact by the weathering agencies; they form disintegrates and residual sediments (sand, gravel, etc.) high in SiO_2. This is the first step in the quantitative decomposition of the silicate rocks.

Separation of the Sesquioxides

The second step is, as in the laboratory, the separation of aluminous products. The greater part of the alumina of the primary rocks is contained in the feldspar minerals which are less resistant than quartz. Through the action of external forces, the feldspars are decomposed to give colloidal particles. This process is complicated, and recent experiments have shown that both silica and alumina are thrown into solution as ions. The solutions are unstable, however, and in short time colloidal products are formed. Depending upon climatic conditions, this process may take place in a variety of ways and with dif-

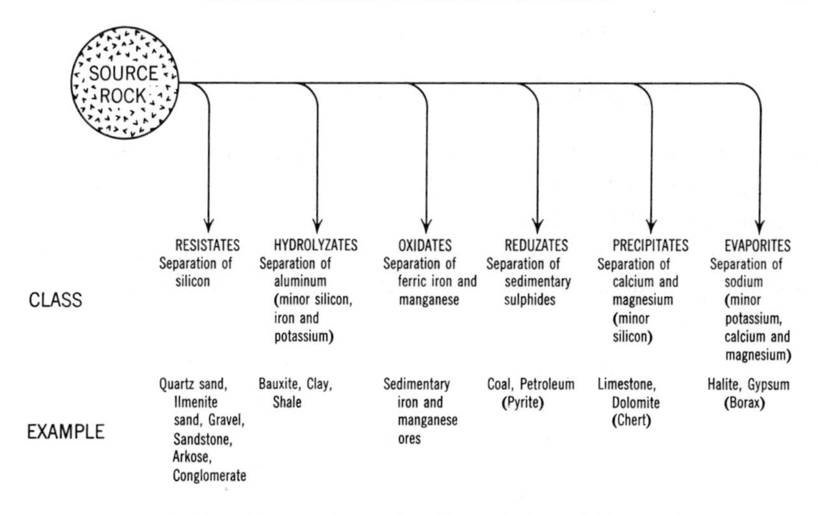

FIG. II-2. The geochemical differentiation of sediments.

ferent velocities. All these decomposition products of feldspar are highly disperse, and exhibit a clayey consistency; they form watery suspensions and are capable of long transport. They are, however, effectively sedimented by the addition of electrolytes; in nature this sedimentation occurs when the suspensions reach the sea. Clay sediments are deposited that contain the greater part of the alumina of the weathered rocks. Iron and manganese are likewise contained in the same sediments. Certain biochemical processes may cause a special separation of iron in individual deposits.

In the laboratory a certain fraction of silica, usually called x, passes into the filtrate from the silica determination and is recovered by the chemist in the sesquioxide precipitate. Corresponding to this x, some silica remains in solution in natural waters. By special biochemical processes it is precipitated in fresh water (diatomaceous earth) or sea water (marine radiolarian deposits, jasper). See page 45.

Separation of Lime and Magnesia

The order of succession again follows that of a chemical analysis. Calcium and magnesium remain in ionic solution after the precipitation of the sesquioxides and are thus delivered into the sea. From the sea lime is extracted through organisms and precipitated as calcium carbonate, forming more or less pure limestone. Magnesium remains longer in the sea (see page 40) but is slowly extracted through metasomatic reaction with limestone according to the schematic equation:

$$2CaCO_3 + Mg^{2+} \rightarrow CaMg(CO_3)_2 + Ca^{2+}$$

Many dolomite rocks appear to have this mode of formation. The dolomitization of many coral reefs can be followed step by step (borings in the atoll Funafuti, Ellice Islands, Pacific).

Ground water likewise is a factor in dolomitization. The importance of time in these metasomatic processes is indicated in the following table.

Sedimentary Carbonates	$CaCO_3 : MgCO_3$
Pre-Devonian	2.39:1
Devonian	4.49:1
Carboniferous	8.89:1
Cretaceous	40.23:1
Tertiary	37.92:1
Quaternary and recent	25.00:1

Separation of Alkalies

Ions of alkali metals and of magnesium remain longest in solution and move around in the ocean. By evaporation of shallow inland seas they crystallize out, forming salt deposits famous for their great varieties of minerals.

However, an appreciable fraction of the original potassium does not remain long in the ocean in solution. Nature, like the inexperienced student in the laboratory who forgets to wash his sesquioxide precipitate, has committed an error. Potassium is lost at an early stage of nature's analysis through adsorption and coprecipitation with the colloidal clayey sediments. Marine clay is always high in potash, a fact of great importance in agriculture. Potash fertilizes the clay, and the high adsorption power of the soil prevents loss of potash through leaching.

A rapid survey of the main sediment-forming processes is given in Fig. II-2.

Thus we find that the external geochemical migrations, the processes of weathering, erosion, and sedimentation, conspicuously separate the chemical rock components and give birth to products of extreme chemical character, like sandstone (SiO_2), clay (rich in $Al_2O_3 \pm SiO_2$), calcium carbonate rocks, and alkali salt deposits. Biochemical processes give rise to some of the rarer highly differentiated rocks: certain iron ores, sulfur deposits, phosphorites, coal, and lignite deposits. In all these rocks, the chemical differentiation has attained a maximum.

Their mode of occurrence is at the surface of the earth; their mineral assemblages are more or less completely adjusted to low temperature and pressure and the presence of oxygen, water, and carbon dioxide. The relative content of oxygen attains a maximum in the sedimentary rocks.

2 · THE PHYSICAL CHEMISTRY OF SEDIMENTATION

The sedimentation of clastic matter follows the rather simple rules of mechanical precipitation. By transport in water the material is sorted according to grain size and, to some extent, according to specific gravity. Wind likewise sorts the material according to grain size and specific gravity. The finest pelitic mud of disintegrates and residua frequently comes to rest together with the colloidal hydrolyzates which are precipitated under the same hydrologic conditions.

It is always true in petrography that the various types of rock exhibit gradual transitions. Rocks may consist of mechanically disintegrated fragments and residual matter in varying proportions. The products of hydrolysis may form individual deposits, or they may occur, in smaller or greater amounts, as components in other types of rock. For example, an ordinary graywacke is made up of (1) disintegrated rock fragments (clastic sediments), (2) clay substance (hydrolyzate sediments), and (3) calcite (carbonate sediments).

Contrary to earlier belief, all rock-making elements, including silicon and aluminum, go at least temporarily into true ionic solution. All sediments except disintegrates and residua are formed from low-temperature aqueous solutions by fractional precipitation following the laws of physical chemistry. By this process an extreme petrochemical differentiation is effected; new minerals and new rocks are formed that are radically different from the igneous rocks. Recognition and appreciation of this fact make it obviously wrong to describe the relation between sedimentary and igneous rocks as that of sawdust to a living tree.

Goldschmidt has pointed out that factors of special significance in sedimentary petrogenesis are (1) temperature, (2) hydrogen ion concentration, (3) reduction-oxidation potential, and (4) ionic adsorption phenomena.

The low temperature of the aqueous solutions favors the chemical differentiation by rather effectively preventing the formation of mixed crystals. At elevated temperatures, ions (or atoms) of different size may freely replace each other in the crystal lattices, with formation of extensive series of solid solutions. Thus we find that the pair NaCl—KCl at about 800°C forms a continuous series of mixed crystals, whereas the solid miscibility at 20°C is practically zero. The decrease in miscibility results in a geochemical differentiation of greater intensity than that encountered in magmatic melts. As an example, individual minerals of strontium or barium are never formed from magmatic melts, for strontium and barium ions enter into the crystal lattices of calcium-bearing and potash-bearing minerals at elevated temperatures. But individual minerals of both strontium and barium frequently separate from aqueous solutions, simply because the possibility of forming mixed crystals is limited.

The hydrogen ion concentration of the mineral waters is of importance, because in most cases the mineral formation takes place through interaction of ions in solution. In pure water at 20°C the hydrogen ion concentration is about 10^{-7}. If the hydrogen ion concentration of a solution is greater than that of pure water at the same temperature, it is called acid; otherwise it is called alkaline. The point of neutrality is displaced by increasing temperature toward a higher concentration of the hydrogen ions. The negative logarithm of the concentration is usually denoted by pH, and used as an inverted measure for the hydrogen ion concentration.

The chemical elements may be divided into three groups: (1) those which remain in true ionic solution even up to high pH values; (2) those precipitated by hydrolysis; (3) those forming complex anions containing oxygen and usually giving true ionic solutions.

Which group an element belongs to depends upon the function of ionic potential (Cartlege, 1928) given by the ratio between ionic charge, z, and ionic radius, r (Table II-1). The diagram in Fig. II-3

TABLE II-1

IONIC POTENTIALS, z/r

Ion	z/r	Ion	z/r	Ion	z/r
Cs^{1+}	0.61	Sm^{3+}	3.0	Ti^{4+}	6.3
Rb^{1+}	0.67	Y^{3+}	3.4	Nb^{5+}	7.3
K^{1+}	0.71	Cp^{3+}	3.5	Mo^{6+}	9.7
Na^{1+}	1.0	Sc^{3+}	3.7	Si^{4+}	10
Li^{1+}	1.3	Th^{4+}	4.2	B^{3+}	15
Ba^{2+}	1.4	Ce^{4+}	4.2	P^{5+}	15
Sr^{2+}	1.6	Zr^{4+}	4.6	S^{6+}	20
Ca^{2+}	1.9	Al^{3+}	5.3	C^{4+}	27
Mg^{2+}	2.6	Be^{2+}	5.9	N^{5+}	45

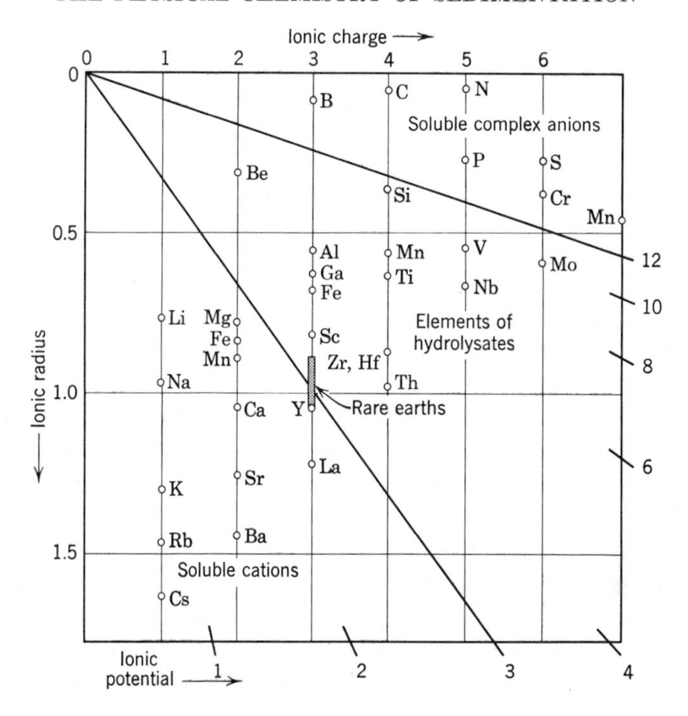

Fig. II-3. Properties of chemical elements in relation to ionic potentials.

illustrates the regular division of the elements into three fields with respect to ionic potential.

The theoretical explanation of this threefold division (Wickman, 1944) lies in the actualization of different kinds of chemical bonds in the several hydroxides. From the point of view of crystal chemistry the various hydroxides may show ionic bonds, in which case they are soluble; or they may show hydroxyl bonds, which are characteristic of elements susceptible of hydrolyzation; or, eventually, they may show hydrogen bonds, again giving soluble crystals.

The function of hydrogen in intermolecular forces is unique, since, possessing no inner electrons, it will occupy no space in combination with oxygen. The oxygen atom, under the influence of the electric field of the cations, may break up into four concentrations of negative charge, each of the value of one-half electrostatic valence, tetrahedrally arranged in space. One of these concentrations is occupied by the hydrogen ion, giving a net charge of $\frac{1}{2}+$ there. Now it can be shown that the remaining unattached negative charges are potential bonds, the nature of which is determined by the electrostatic valence, V, from the central cation. The electrostatic valence is defined as the ratio between ionic charge, z, and the number of surrounding oxygens, n; thus $V = z/n$. It follows from quantum mechanics that variations in V will influence the bonding of hydrogen in hydroxides in the following way:

$$\begin{aligned} &(1)\ \text{If}\quad V < \tfrac{1}{2} \rightarrow \text{ionic bonds}\\ &(2)\ \text{If}\ \tfrac{1}{2} \leq V \leq 1 \rightarrow \text{hydroxyl bonds}\\ &(3)\ \text{If}\quad V > 1 \rightarrow \text{hydrogen (+ hydroxyl) bonds} \end{aligned}$$

For each cation its V-value can be calculated from its valence and coordination number in its hydroxides. The calculations show the following relations to the diagram: (1) The soluble cations of ionic potential below 3 show $V < 1/2$; (2) the hydrolyzates of intermediate ionic potential which are precipitated by hydrolysis have V between 1 and $1/2$; (3) the soluble complex anions of ionic potential above 12 have $V > 1$.

A possible discrepancy exists in boron. $B(OH)_3$ has 3-coordination, and therefore hydroxyl bonds (the electrostatic valence from B is 1); consequently it should be grouped with the elements of the hydrolyzates. Its geochemical behavior supports this assumption.

Table II-2 gives the precipitation point of some hydroxides from aqueous solutions. See also Fig. II-4.

TABLE II-2

HYDROGEN ION CONCENTRATION OF NATURAL WATERS AND POINT OF PRECIPITATION OF SOME HYDROXIDES

pH	Natural Waters	Precipitates of Hydroxide
11		
10	Alkaline soils	Mg
9	Alkaline hot springs	Ag, Mn
8	Sea water	Ni, Co, Y, Ce
7	River water	Fe^{2+}, Pb
6	Rain water	Cu, Cr, Zn, Be
5	Bog water	Al
4		U, Th
3	Acid hot springs (pH 2–4)	Fe^{3+}, Sn^{2+}, Zr
2		
1		
0		
−1	Sulfuric acid in deserts	

Most natural waters go through an evolution of increasing pH, until they eventually empty into the sea, which is slightly alkaline. Silica becomes more soluble with increasing pH, and is therefore often delivered into the sea. But aluminum hydroxide is precipitated in mildly acid solutions, near the point of neutrality. Many valuable deposits of this metal are probably thus produced. The difference in behavior of ferric and ferrous iron is of interest. Ferric iron is soluble only in rather strongly acid solutions; it is therefore precipitated before alumina, but the separation is usually not clean. Ferrous iron remains longer in solution in equilibrium with carbon dioxide in oxygen-free waters. Similarly, tervalent and quadrivalent manganese ions are precipitated before bivalent manganese.

The reduction-oxidation potential is of importance, for instance, in the separation of native metals from aqueous solutions. In Table II-3

TABLE II-3

NORMAL POTENTIALS

Gold	+1.5
Silver	+0.81
Copper	+0.52
Antimony	+0.2
Lead	−0.13
Nickel	−0.25
(Iron	−0.44)
Zinc	−0.76
Sodium	−2.71
Calcium	−2.76
Lithium	−3.02

are listed the normal potentials of some metals compared to a normal hydrogen electrode. In cases of positive values the corresponding native metal is able to separate from natural waters, as, for example, gold, silver, copper. In cases of larger negative values the native metal is unable to form. In border cases an element of very low negative potential still may separate as metal if the aqueous solution has a favorable composition (for example, lead). Elements above the horizontal line in Table II-3 may occur as native metals in nature; those below the line do not. (Some terrestrial native iron is known, formed under very unusual conditions.)

Ionic adsorption phenomena take place at low temperature in colloidal phases or phase complexes that are capable of capturing and binding certain ions through adsorption. One example is the binding of potassium ions by the clayey products of the hydrolysis (see page 24). In a geochemical adsorption process the binding of the ion to the colloidal surface takes place in competition with the over-all hydration of the ion in the solvent. It can be shown that the degree of adsorption in an ion is a function of radius, charge, polarizability, and normal potential, as well as the nature of the chemical compounds formed at the phase boundary.

Through the processes of adsorption the natural waters are deprived of many of the rarer elements. Most of the ions of the heavy metals, such as ions of lead, zinc, and copper, as well as complex ions of arsenic and molybdenum, are captured by, and coprecipitated with, the colloidal particles, usually hydrolyzates, and thus supplied to the sediments.

The amounts of poisonous metals and metalloids which potentially have been delivered into the ocean from the primary rocks throughout geologic times are so considerable that a serious poisoning of the ocean would have been caused if this process of elimination of poisonous substances had not been in action. Or the evolution of life would have taken a different course, developing organisms not susceptible to our poisonous metals. This statement applies, for instance, to copper, lead, arsenic, selenium, mercury, antimony, and bismuth. In many cases these metals have been removed from aqueous solutions by a means also known in practical medicine, that is, adsorption on freshly precipitated hydroxides of iron. We find considerable concentrations of selenium, arsenic, and lead in the sedimentary iron ores. The arsenic content of these ores in most cases is so high that it brings a very notable quantity of arsenic even into iron or steel, from which this element is difficult to eliminate by the usual technical processes of refining. Molybdenum is concentrated in manganiferous sedimentary ore deposits.

3 · PETROLOGY AND MINERALOGY OF THE SEDIMENTARY ROCKS

Estimates of the average composition of sediments are listed in Table II-4. In this average the salt deposits are not included, nor are the salt brines of the pore spaces that, even at great depths, take considerable volume.

Disintegrates and Residua—Clastic Sediments

Chemical and mineralogical compositions are shown in Table II-5. The names to be applied to the various types of clastic sediments are

TABLE II-4

COMPOSITION OF THE SEDIMENTS

(After Correns)

SiO_2	55.64	Quartz	30
TiO_2	0.69	Feldspars	9
Al_2O_3	14.44	Muscovite	11
Fe_2O_3	6.87	Biotite, glauconite	8
MgO	2.93	Paragonite	4
CaO	4.69	Kaolinite, halloysite	12
Na_2O	1.21	Montmorillonite	5.5
K_2O	2.87	Chlorite	2
H_2O	5.54	Calcite	6
P_2O_5	0.17	Dolomite	2.5
CO_2	3.80	Fe_2O_3	5.5
S	0.32	H_2O^-	2
C	0.65	Inclusive	2.5

by no means standardized. The following groups are generally recognized:

Sandstone, predominantly composed of one mineral, quartz, with minor amounts of fragmental carbonate, feldspar (up to 10 per cent), mica, and stable heavy minerals.

Graywacke, made up of angular and/or rounded quartz and rock fragments with feldspar from 20–60 per cent, and with a fine-grained matrix which may be micaceous and/or chloritic.

Arkose, is characterized by quartz and feldspar (25 per cent or more), often derived from granite and having the appearance of granite; micas are common though subordinate.

Sparagmite, is used in Scandinavia for extensive Eocambrian deposits of arkoses.

The difference in the mode of formation between typical disintegrates and typical residual sediments depends upon the ratio between the rate at which the mechanical breaking up of the parental rock takes place and the rate at which the chemical weathering takes place.

In arctic and subarctic regions, where mechanical action attains its maximum and chemical action its minimum, even finely divided rock powder escapes chemical weathering and is deposited as glacial clay.

TABLE II-5

AVERAGE COMPOSITION OF DISINTEGRATES AND RESIDUA

	Gray-wacke (Pettijohn, 1957)	Arkose (Pettijohn, 1957)	Spa-ragmite (Barth, 1938)	Quartzose Sandstone (Pettijohn, 1957)
SiO_2	64.2	75.5	80.9	92.3
TiO_2	0.5		0.4	0.0
Al_2O_3	14.1	11.4	7.6	1.4
Fe_2O_3	1.0	2.4	2.9	0.2
FeO	1.0		1.3	0.3
MnO	0.1	0.2		
MgO	2.9	0.1	0.1	0.1
CaO	3.5	1.6	0.1	3.0
Na_2O	3.4	2.0	0.7	0.1
K_2O	2.0	5.6	4.7	0.1
H_2O	2.2	0.6	1.2	0.2
P_2O_5	0.1			
CO_2	1.6	0.4		2.3
Quartz	45.6	40	60	90
Chert	1.1	...	...	...
Microcline perthite Plagioclase	} 16.7	} 53	34 ...	} 2
Chlorite and muscovite	25.0	...	6	2
Carbonate Ore, apatite, etc.	4.6 +	} 7	... +	} 6
Rock fragments	6.7	...	...	...

As distinct from other clay deposits, glacial clay is a typical disintegrate sediment.

Residual products are the "clay minerals" and the hydrated sesquioxides of aluminum and iron. These end products of weathering are associated with concentrations of those minerals (especially quartz) that are not notably subjected to alterations. Particularly pure silica residua (glass-sand attaining 99.99 SiO_2) are formed by dissolution of all other minerals, for example, through the action of dense plant growth.

All minerals of high chemical and mechanical resistance as well as most minerals of high specific gravity may be present in the residua. There are residual sediments rich in zircon, monazite, magnetite, ilmenite. Often hard minerals are concentrated: garnet, tourmaline, corundum, diamond. Noble metals are also concentrated: gold, platinum, osmiridium. The most famous of the gold-bearing quartz conglomerates, perhaps, is at Witwatersrand, Johannesburg, South Africa.

Hydrolyzates and Oxidates

Rocks of this class are high in sesquioxides and show a clayey consistency. Chemically they are strongly differentiated, and are made up of one or more of the clay minerals and/or lateritic minerals. Clay minerals are pseudo-hexagonal phyllosilicates with grain size <0.02 mm, viz.: mica, hydromica, montmorillonite, and vermiculite, as well as chlorite with montmorillonite-chlorite and vermiculite-chlorite, and last but not least the series of kaolinite-antigorite inclusive of halloysite. (See Table II-6.) For further description of the X-ray crystal structure, see under phyllosilicates, page 244.

Porcelain earth is pure kaolinite; *bentonite* consists mostly of montmorillonite; *bole* (bolus) is a fine, earthy mixture of halloysite and ferric hydroxides. *Allophan, hisingerite* are amorphous, colloidal "clay" with Al:Si = 1:1, respectively Fe:Si = $1 \sim 1$.

TABLE II-6

MINERALS FORMED BY HYDROLYSIS

(Clay Minerals)

Kaolinite, dickite, nackrite	$Al_4Si_4O_{10}(OH)_8$
Montmorillonite, beidellite, nontronite	$Al_4Si_8O_{20}(OH)_4 \cdot n \cdot H_2O$, with some variation in the SiO_2 to Al_2O_3 ratio and Fe, Mg, Ca, and alkalies replacing Al
Illite (Hydromica)	$K_{0-2}Y_{4-6}Z_8O_{20}(OH)_4$, where Y = Al, Fe, Mg; Z is mainly Si with some Al. If the index for K reaches 2 the formula becomes representative of mica
Halloysite	$Al_4Si_4O_{10}(OH)_8 \cdot 4H_2O$
Vermiculite	$(Mg,Fe,Al)_3(Al,Si)_4O_{10}(OH)_2 \cdot Mg_x(H_2O_4)$
Saponite	$Mg_3(Al,Si)_4O_{10}(OH)_2 \cdot Na_{1/3}(H_2O)_4$
Chlorite (see pages 248–249)	$Y_6Z_4O_{10}(OH)_8$
Antigorite (see page 247)	$Mg_6Si_4O_{10}(OH)_8$

Hydromicas are very common; they form by leaching of mica whereby,

(1) K^+ is replaced by a hydroxonium ion $(H_3O)^+$

$$(K)^+ + 2H_2O = (H_3O)^+ + KOH$$

or

(2) K^+ and $(OH)^-$ may be replaced simultaneously by two water molecules:

$$(K^+ + OH^-) + 2H_2O = (H_2O + H_2O) + KOH$$

(3) in some hydromuscovites $(SiO_4)^{4-}$ may be partly replaced by $(H_4O_4)^{4-}$.

Thus there are three possible structural formulas for hydromuscovite. In biotite the leaching is followed by an oxidation of Fe^{2+} to Fe^{3+}.

Montmorillonite-vermiculite structurally represents talc lattice layers and intercalated cation-water layers exhibiting, respectively, a small negative and a small positive overcharge (expanded talc lattices). The weakly charged negative silicate layers repulse each other, thereby facilitating the entry of the weakly positive cation-water layers. Vermiculites take their name from *vermis*—worm—because they absorb moisture and swell and curl like worms.

Chlorites represent talc lattices expanded by (OH)-brucite layers.

Kaolinite, dickite, nackrite, and *anauxite* have the same chemical composition and are made up of single sheets of the formula $Al_4Si_4O_{10}(OH)_8$, but they differ in the stacking pattern of these sheets.

Halloysite is a kaolinite expanded by intercalated layers of H_2O. A kaolinite expanded by (OH)-cation layers is not known.

In marine sediments, kaolinite is partly lost during diagenesis and probably transferred into a mica-type clay mineral, that is, illite. Montmorillonite is also probably unstable under marine conditions. Potash and probably magnesia are taken up by the sediments of the ocean. However, the geological distribution of the various clay minerals is not completely known. Most of them reside in both ancient and recent marine sediments as well as in soils. In the kaolinite group, kaolinite itself is by far the most common mineral; the others have special modes of occurrence and have never been found in soils.

Halloysite is also rare, and has not been found in soils or recent marine sediments. Metahalloysite $Al_4Si_4O_{10}(OH)_8$ has only been reported from one locality in clay in Belgium.

The formation of the clay minerals takes place under temperate or subtropical conditions by hydrolyses of the feldspars, aided by waters rich in carbon dioxide, for example:

$$\underset{\text{(albite)}}{Na_2Al_2Si_6O_{16}} + nH_2O = \underset{\text{(kaolinite)}}{Al_2Si_2O_5(OH)_4} + \text{solution of } Na_2SiO_3$$

$$\underset{\text{(anorthite)}}{CaAl_2Si_2O_8} + H_2CO_3 = \underset{\text{(kaolinite)}}{Al_2Si_2O_5(OH)_4} + \text{solution of } CaCO_3$$

Potash feldspar behaves analogously to albite.

Sodium silicate and calcium carbonate are carried away in solution and a *clay* is formed. The actual mineral content of clays is extremely irregular, though, and depends upon a variety of factors—climate, degree of weathering, composition of the original rock, pH values, etc. Quite generally, clays consist of four groups of minerals: (1) clay minerals formed by hydrolyses, (2) residual minerals, (3) biogenic constituent, (4) authigenic minerals (formed after sedimentation).

Lateritic minerals are hydroxides of iron or aluminum (they are listed in Table II-7), and represent a further stage of hydrolysis, for example:

$$\underset{\text{(kaolinite)}}{Al_2Si_2(OH)_4} + nH_2O = \underset{\text{(gibbsite)}}{2Al(OH)_3} + \text{solution of } 2SiO_2$$

TABLE II-7

MINERALS IN LATERITE

Hydroxides of Al and Fe Commonly Present	
Gibbsite (hydrargillite)	$Al(OH)_3$
Diaspore	α-$Al_2O_3 \cdot H_2O$
Boehmite	γ-$Al_2O_3 \cdot H_2O$
Turgite	$Fe_2O_3 \cdot \frac{1}{2}H_2O$
Goethite and *Lepidocrocite*	$Fe_2O_3 \cdot 1H_2O$
Limonite	$Fe_2O_3 \cdot 1\frac{1}{2}H_2O$
Xanthosiderite	$Fe_2O_3 \cdot 2H_2O$
Limnite	$Fe_2O_3 \cdot 3H_2O$

Ocher, common name for various hydrated oxides of iron

Wad, various hydrated oxides of manganese. δ-MnO_2 and ϵ-MnO_2 have also been described. The δ-form occurs in manganese nodules of the ocean floor.

Hydrated Manganese Minerals Occasionally Present	
Ramsdellite	γ-MnO_2, isotype with groutite (diaspore series)
Polianite	β-MnO_2, or *pyrolusite* as secondary after manganite
Synthetic	α-MnO_2, tetragonal and related to:
Cryptomelane	$K_2Mn_8O_{16}$ } Form solid solutions in all proportions with MnO_2, and with:
Hollandite	$Ba_2Mn_8O_{16}$ }
Psilomelane	$(Ba,H_2O)_2Mn_5O_{10}$
Ranciéite	$(Ca, Mn^{\cdot\cdot})Mn_4^{\cdot\cdot\cdot\cdot}O_9 \cdot 3\,H_2O$
Pyrochroite	$MnO \cdot H_2O$, brucite type
Patridgeite	Mn_2O_3 and bixbyite $(Mn,Fe)_2O_3$ (fluorite type)
Manganite	γ-$Mn_2O_3H_2O$, groutite α-form, diaspore type
Hausmannite	Mn_3O_4, Hydrohausmannite $Mn_3(O,OH)_4$
Braunite	$Mn_4^{\cdot\cdot}Mn_3^{\cdot\cdot\cdot\cdot}SiO_{12}$, tetragonal, related to bixbyite structure
Långbanite	of same formula, but hexagonal

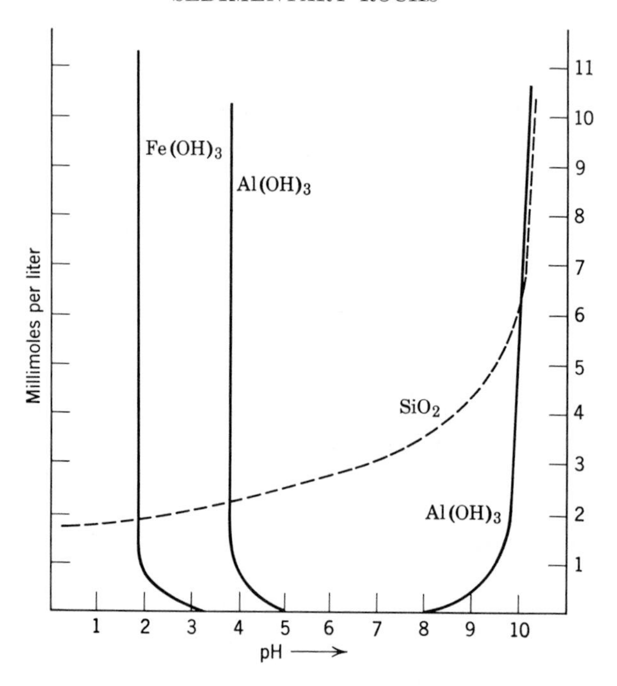

FIG. II-4. The effect of pH on the solubility of $Fe(OH)_3$, $Al(OH)_3$, and SiO_2. (Okamoto, *Geochim. Acta,* 1957.)

Laterite is a rock or soil of varying composition made up of mixtures of hydrated oxides of iron and aluminum, some titanium, and, occasionally, maganese plus minerals that resist weathering.

Bauxite is a laterite particularly high in alumina (low in iron).

Laterization is the process by which hydroxides of ferric iron and aluminum are deposited within the pore space of rocks near the surface, but the specific chemical and physical conditions for the process to occur are not known. Neither is it known where deposits of laterite and bauxite are most likely to be found—although we know that laterization is active throughout the humid tropics and subtropics and seems to be essentially independent of the parent rock. With reference to Fig. II-4, we can see that if rocks are leached in mildly alkalic solutions, SiO_2 may move away in solution while $Al(OH)_3$ and $Fe(OH)_3$ are precipitated. If the system is moderately acid, pH 4.5 to 6.5, alumina, iron, and silica are all retained as hydrolyzates (clay minerals) plus quartz. But under tropical conditions there is removal of silica, lime, magnesia, and alkalies, leaving an earthy residue of essentially gibbsite and limonite. Kaolinite is the dominant clay mineral in laterite (plus some vermiculite (?) and illite). In order to distinguish laterite from clay, the proportion between silica and sesquioxides has been used. With reference to the ratio $SiO_2:(Fe_2O_3 + Al_2O_3)$, we have

$$\text{Laterite} < 1{:}3 < \text{Lateritic soil} < 2{:}3 < \text{Non-lateritic clay}$$

Fossil deposits are known in France (at Baux), parts of Italy, Hungary, and Yugoslavia; these regions supply about 60 per cent of the bauxite of the world. In North Africa (Atlas Mountains, Morocco) there are also large deposits. In North America the more important deposits are confined to Arkansas, with minor ones in Georgia and Alabama.

TABLE II-8

CHEMICAL COMPOSITION OF CLAY SEDIMENTS

	Average Shale (Clarke) (*1*)	Average Shale from the Middle West, USA Prof. Paper, 1960, 400-A (*2*)	Abyssal Ocean Clay off Lower California (Grim) (*3*)	Clay Mean of 712, USGS Prof. Paper, 1960, 400-A (*4*)	Laterite, India (Niggli) (*5*)	Bauxite, Little Rock, Arkansas (Mead) (*6*)
SiO_2	58.1	56.95	53.8	65.10	0.90	10.64
TiO_2	0.7	0.68	1.2	0.67	1.59	1.20
Al_2O_3	15.4	15.90	19.0	15.33	26.27	57.48
Fe_2O_3	4.0	4.36	10.1	3.53	56.01	2.56
FeO	2.5	1.56	...	0.55		
MnO	...	...	...	...		
MgO	2.4	2.15	3.6	1.18	0.20	...
CaO	3.1	4.58	0.9	2.85	0.64	...
Na_2O	1.3	0.56	1.0	1.03		
K_2O	3.2	1.86	2.8	1.23		
H_2O	5.0	2.64	7.0	3.05	14.39	28.36
P_2O_5	0.2	0.34	...	0.83		
CO_2	2.6	3.96	...	3.10		
SO_3	0.6	0.49	...	0.47		
Organic	0.8	3.97	...	1.08		

(*1*) Average shale (Clarke, 1924).
(*2*) Average shale from the Middle West, U.S.A.
(*3*) Abyssal ocean clay off Lower California (Grim, 1949).
(*4*) Average clay from the Middle West, U.S.A.
(*5*) Laterite, India (Niggli, 1952).
(*6*) Bauxite, Little Rock, Arkansas (Mead, 1915).

By weathering, iron and manganese go into solution as bivalent ions (for example, as bicarbonate or sulfate). They oxidize in surface waters (bogs, lakes) and are precipitated as hydroxides. Algae and bacteria may take part in the process. Iron and manganese may occur together in varying proportions, or they may form individual deposits. Since manganese has complicated chemical properties and can assume various stages of oxidations, the mineralogy of these deposits becomes very complex. (See manganese minerals in Table II-7.) The conditions for the deposition of bog ores are widely present in glacial regions. In Canada and Sweden the ore is dredged from existing lake bottoms.

Analyses of various clay sediments are listed in Table II-8.

In shallow sea basins, commercially important sediments are limonite, goethite, hematite, iron-rich chlorite, glauconite, and occasionally siderite. The structure is frequently oolitic. Examples are the Mesozoic oolitic sedimentary iron ores in Lorraine (minette ore) and corresponding deposits in Central Germany and in Scania, Sweden. The Silurian Clinton ores in the United States, from New York to Alabama, and at Wabana, Newfoundland, are also directly sedimentary in origin. The famous iron ores in the Lake Superior Region represent weathered portions of pre-Cambrian sedimentary beds. Similar deposits are encountered at Krivoi Rog, Russia, and at other places.

Carbonate Sediments

By rock weathering, magnesium and especially calcium easily go into ionic solution, usually as bicarbonate. They remain in solution and are delivered into the sea. Investigations by Garrels and others show that the total concentration of carbonate ion in sea water can be calculated by assuming that sea water is in equilibrium with the partial pressure of carbon dioxide of the air, and that the only contributions to its concentration are $m\ CO_3^{2-}$, $m\ NaCO_3^-$ and $m\ MgCO_3^\circ$.*

Complexing (i.e., formation of $CaHCO_3$, $MgHCO_3$) of carbonate ions has a marked effect on the solubility of carbonate minerals. The high stability of $MgCO_3^\circ$ in pairs indicates that magnesium exerts an important indirect control on $CaCO_3$ solubility.

Na^+ and Mg^{2+} interact with CO_3^{2-} and HCO_3^-, thus affecting

* (m = calculated molality = moles per 1000 g H_2O of dissolved species, i.e., "free" or "uncomplexed" species; ° is used to distinguish the dissolved species from the solid, i.e., ° represents the ion pairs.)

the total dissolved carbonate in sea water. If the ocean had NaCl substituted for $MgCl_2$, the total concentration of CO_3^{2-} would diminish to 20 per cent of its present value, and total HCO_3^- would drop to about 80 per cent. The result would be a major redistribution of CO_2 between atmosphere and ocean.

Of the total CO_3^{2-} in sea water (not including HCO_3^-): 75 per cent is $MgCO_3°$, 15 per cent is $NaCO_3^-$, 10 per cent is CO_3^{2-}. Although the total concentration CO_3^{2-} can be predicted from these data, this is not possible in the case of Ca^{2+}. But the surface layers of the sea usually become supersaturated in $CaCO_3$; this facilitates the production of $CaCO_3$ in pelagic organisms, although precipitation through organic agencies does not necessarily require saturation. At greater depths the water is usually unsaturated in $CaCO_3$. Separation of solid $CaCO_3$ is therefore confined to the upper layer. From this layer the remnants of the organisms sink to the bottom. On the way down, some $CaCO_3$ may again go into solution if the water at this depth is unsaturated. In this way, sediments rich in $CaCO_3$ are produced on the sea bottom of vast areas, either pure or mixed with red deep-sea clay or with radiolarian ooze. See Table II-9.

Quantitatively most important is the *globigerina ooze* which covers 37.4 per cent of all ocean bottom, corresponding to one fourth of the

TABLE II-9

ANALYSES OF MARINE OOZE

(After Clarke)

	Radiolarian Ooze	Diatom Ooze	Globigerina Ooze
Ignition	7.41	5.30	7.90
SiO_2	56.02	67.92	31.71
Al_2O_3	10.52	0.55	11.10
Fe_2O_3	14.99	.39	7.03
MnO_2	3.23	...	trace
CaO	0.39	...	0.41
MgO	0.25	...	0.12
$CaCO_3$	3.89	19.29	37.51
$Ca_3P_2O_8$	1.39	0.41	2.80
$CaSO_4$	0.41	0.29	0.29
$MgCO_3$	1.50	1.13	1.13
Insoluble	...	4.72	...

TABLE II-10

DISTRIBUTION OF SEDIMENTS

(Gmelins Handbuch: Sauerstoff, 1952)

Hemipelagic Deposits	Area 10^6 km^2	%	Eupelagic Deposits	Area 10^6 km^2	%
Blue mud	37.5	10.8	Red deep-sea clay	133.0	38.8
Volcanic sand	1.5	0.4	Radiolarian ooze	5.9	1.6
Red mud	0.26	0.07	Diatom ooze	28.2	8.2
Green sand and clay	2.2	0.6	Globigerina ooze	128.0	37.3
Calcareous sand	6.6	1.9	Pteropod ooze	1.0	0.3

total surface of the earth. Globigerina ooze is a deep-water sediment, usually deposited below 2000 m. See Table II-10.

In more shallow waters (250–2000 m), $CaCO_3$-rich sediments are less abundant, the organic remnants rich in $CaCO_3$ are here diluted with terrestrial clay material. But at very shallow depths (0–250 m) again sediments rich in $CaCO_3$ are formed through the action of algae, corals, shellfish, etc. Under these conditions inorganic precipitation of $CaCO_3$ also may occur if the water is supersaturated in $CaCO_3$.

A type of Mesozoic foraminiferal limestone is the chalk (Denmark, North France, England) which is nearly pure carbonate of lime in exceedingly fine subdivision and, for some mysterious reason, not altered by processes of diagenesis. In America the Niobrara and Selma chalks are analogous deposits.

Very fine-grained, somewhat porous limestones are of use in lithography. The "lithographic limestone" from Solenhofen, Bavaria, is famous for its fossils.

Through the action of organisms all three modifications of $CaCO_3$ may form: calcite, aragonite, and vaterite.* But in older limestone sediments all $CaCO_3$ is converted into calcite, which is the stable, and therefore the least soluble, modification.

Dolomite ($CaMgC_2O_6$) is an important carbonate rock. In contact with sea water it is formed metasomatically after calcite as de-

* Calcite occurs in all foraminifera, tetracorals, bryozoa, echinoderms, crustaceans, in most algae and lamellibranchiates. Aragonite occurs in most hexacorals and cephalopods, in many lamellibranchiates and gastropods, and in some algae. Aragonite and calcite together occur in some lamellibranchiates and gastropods. Vaterite is said to occur in some gastropods. Dolomite and calcite occur in some algae, corals, and echinoderms.

scribed on page 24. Likewise in salt deposits, dolomite is often formed by the action of magnesium-rich salt solutions. Dolomitization of limestone through magnesium-rich aqueous solutions is very common, especially in orogenic mountain chains, and belongs to the general rock metamorphism.

Siderite ($FeCO_3$) is found as a component in oolitic sedimentary iron ores. Pure siderite rocks also occur. Usually they are formed by metasomatic precipitation of siderite in limestone and should, therefore, be classified with metamorphic rocks.

Magnesite ($MgCO_3$) usually represents a metasomatic alteration product of other carbonate rocks. But, as a rarity, primary sedimentation of magnesite occurs in some inland lakes in the deserts of the Southwestern United States.

Evaporites

The soluble matter in the ocean and in lakes without outlets (desert lakes), and in the ground water as well, is precipitated if the solutions are evaporated beyond the saturation point of the several salts.

Crystallization starts in sea water when the salinity has reached 72 per cent, with calcite as the first solid phase; at 202 per cent gypsum, and at 388 per cent halite, crystallize. Calcite is only precipitated until the volume is reduced by half, then gypsum alone until the volume is reduced by 91.7 per cent. Then, from 12 to 63 times normal concentration NaCl is precipitated with a little gypsum. The potash salts with magnesium are the last to come. Such high salinities are attained by evaporation in inland lakes and in sequestered bays of the ocean; for example, a strong current flows into the Karabogas Bay on the east shore of the Caspian Sea carrying 8 million tons of salt per year into the bay. The salinity in the bay will in places attain 285 per cent; salts are therefore constantly being precipitated and are accumulating on the sea bottom.

Ocean salt deposits are very special rocks. The minerals are listed in Table II-11. Their mode of formation through evaporation of sea water, their thermometamorphism, and their mechanical "flow" (salt domes, diapirs) have long interested geologists and are now rather well understood. To be sure, at the turn of the century, the famous Dutch scientist, Van't Hoff, was one of the first to study rocks on a physico-chemical basis; these "rocks" were the ocean salt deposits at Stassfurt, Germany. By experiments and by physico-chemical reasoning he was able to sort out certain mineral associations characteristic of certain temperatures. Each characteristic association corresponds to what we now call a mineral facies. Thus the mineral

TABLE II-11

THE MOST IMPORTANT MINERALS OF OCEANIC SALT DEPOSITS

Halite	$NaCl$ } "Hartsalz"
Sylvite	KCl } "Hartsalz"
Kieserite	$MgSO_4 \cdot H_2O$ } "Hartsalz"
Carnallite	$KCl \cdot MgCl_2 \cdot 6H_2O$
Tachydrite	$CaCl_2 \cdot 2MgCl_2 \cdot 12H_2O$
Kainite	$KCl \cdot MgSO_4 \cdot 3H_2O$
Bischofite	$MgCl_2 \cdot 6H_2O$
Hexahydrite	$MgSO_4 \cdot 6H_2O$
Epsomite (reichardtite)	$MgSO_4 \cdot 7H_2O$
Bloedite (astrakhanite)	$Na_2Mg(SO_4)_2 \cdot 4H_2O$
Langbeinite	$K_2Mg_2(SO_4)_3$
Loeweite	$Na_2Mg(SO_4)_2 \cdot 2H_2O$
Polyhalite	$2CaSO_4 \cdot MgSO_4 \cdot K_2SO_4 \cdot 2H_2O$
Glauberite	$Na_2SO_4 \cdot CaSO_4$
Vanthoffite	$3Na_2SO_4 \cdot MgSO_4$
Boracite	$Mg_7B_{16}O_{30}Cl_2$
Gypsum, anhydrite, calcite, dolomite	

THE MOST IMPORTANT MINERALS OF SODA SEDIMENTS

Trona (Urao)	$Na_2CO_3 \cdot HNaCO_3 \cdot 2H_2O$
Soda	$Na_2CO_3 \cdot 10H_2O$
Thermonatrite	$Na_2CO_3 \cdot H_2O$
Pirssonite	$CaCO_3 \cdot Na_2CO_3 \cdot 2H_2O$
Gay-Lussite	$CaCO_3 \cdot Na_2CO_3 \cdot 5H_2O$

THE MOST IMPORTANT MINERALS OF BORATE SEDIMENTS

Borax	$Na_2B_4O_7 \cdot 10H_2O$
Kernite	$Na_2B_4O_7 \cdot 4H_2O$
Ulexite	$NaCaB_5O_9 \cdot 8H_2O$
Probertite	$Na_2CaB_6O_{11} \cdot 6H_2O$
Colemanite	$Ca_2B_6O_{11} \cdot 5H_2O$
Pandermite	$Ca_8B_{20}O_{38} \cdot 15H_2O$
Thenardite	Na_2SO_4
Glaserite	$NaKSO_4$
Hanksite	$9Na_2SO_4 \cdot 2Na_2CO_3 \cdot KCl$
Gypsum, halite, and carbonates	

THE MOST IMPORTANT MINERALS OF NITER DEPOSITS

Soda niter	$NaNO_3$
Niter	KNO_3
Thenardite	Na_2SO_4
Lautarite	$Ca(IO_3)_2$
Dietzeite	$Ca(IO_3)_2CaCrO_4$
Tarapacaite	K_2CrO_4
Gypsum, halite, usually also borates	

facies principle was discovered and already developed for salt rocks by Van't Hoff twenty years before Eskola used the term (see page 307). It is a tribute to the genius of Van't Hoff that he, long before any ordinary geologist had even thought of such possibilities, was able to work out a complete quantitative "facies" system for the Stassfurt deposits. It is a regrettable fact that it took geologists twenty years to catch up with this development in physical chemistry, and that we still but reluctantly admit the necessity of physico-chemical reasoning in geological problems.

In the system ($K^{\cdot}$,$Mg^{\cdot\cdot}$–Cl',SO_4'') several reactions occur in the range 0° to 100°C. Sylvite (KCl) and halite (NaCl) are stable in the whole range and take part in the associations, but most other salt minerals have a more restricted range of stability. Magnesium sulfate crystallizes at low temperature as reichardite, $MgSO_4 \cdot 7H_2O$, at somewhat higher temperatuere as hexahydrite, $MgSO_4 \cdot 6H_2O$, and above 72°C as kieserite, $MgSO_4 \cdot H_2O$. At 72°C the following reaction also occurs:

$$\underset{\text{kainite}}{2KClMgSO_4 \cdot 3H_2O} + \underset{\text{carnallite}}{KCl \cdot MgCl_2 \cdot 6H_2O} \xrightarrow{72^\circ} \underset{\text{kieserite}}{2MgSO_4 \cdot H_2O} + \underset{\text{sylvite}}{3KCl}$$

At 83°C kainite reaches the upper limit of its stability range and is replaced by langbeinite:

$$\underset{\text{kainite}}{2KCl \cdot MgSO_4 \cdot 3H_2O} + \underset{\text{kieserite}}{MgSO_4 \cdot H_2O} \xrightarrow{83^\circ} \underset{\text{langbeinite}}{K_2SO_4 \cdot 2MgSO_4} + \underset{\text{in solution}}{MgCl_2}$$

The minerals on the right side of the first equation plus a possible excess of either kainite or carnallite, plus admixed halite, are stable above 72°, and known as the so-called *Hartsalz* of the Central European salt deposits. The fact that this paragenesis is stable at relatively high temperatures has led to the conclusion that there is a secondary development through recrystallization at somewhat elevated temperature. The necessary heat was presumably supplied by deep burial (geothermal metamorphism). At normal geothermal gradient the depth of recrystallization would be several kilometers.

If sea water is evaporated at 25°C the dissolved salts will precipitate in a certain sequence, yielding a theoretical primary profile which in Table II-12 is compared with the profile actually observed in salt deposits at Stassfurt, Germany. Sodium chloride is present in all zones. Polyhalite is present in zones 3, 4, and 5. Anhydrite is interesting. It appears first intermittently in zone 6, and thenceforth continues to crystallize through all subsequent zones. The difference

TABLE II-12

Salt Profiles

Theoretical Profile	Natural Products
8. Zone of bischofite	Removed in solution
7. Zone of carnallite	Zone of carnallite Zone of sylvite-kieserite (Hartsalz)
6. Zone of kainite	Zone of sylvite-kieserite (Hartsalz)
5. Zone of K-Mg sulfates	Zone of langbeinite
4. Zone of K-free Mg sulfates	Zone of vanthoffite
3. Zone of polyhalite	Zone of polyhalite
2. Zone of anhydrite }	{ Zone of glauberite
1. Zone of gypsum }	{ Zone of anhydrite

between the theoretical and the natural profile indicates a thermo-metamorphism of the natural salts of about 80°C.

The metamorphism proceeded in three steps characterized by the following events:

1. Recrystallization into normal Hartsalz: sylvite-kieserite-halite. Concretions of boracite and of rinneite were formed at this time.

2. Local mobilization of solutions of NaCl and $CaSO_4$, which through retrograde metamorphism produced either langbeinite-sylvite-kieserite-halite, or in other places, according to the amount of the intruding solutions, produced so-called "Reichsalz" (kieserite-sylvite-halite), normal Hartsalz (sylvite-kieserite-halite), or langbeinitic Hartsalz (sylvite-kieserite-langbeinite-halite).

3. Intrusion of solutions of $MgCl_2$ into the association of step 1, whereby the Hartsalz was recarnallitized. These solutions came from the main mass of anhydrite. During step 1 of the metamorphism they had ascended from the salt layers into the anhydrite rocks, in which they became temporarily stored; they were then pressed back into the salt layers in conjunction with the intensified diapiric rise of the salt masses during Senonian time.

It has been possible to follow the metamorphism of salt deposits in some detail—giving us a good picture of the mobility of the various ions and the entangled interrelations of time-temperature-pressure-chemical composition, and indicating thereby the complexities involved in the much less understood reactions of metamorphism of silicate rocks.

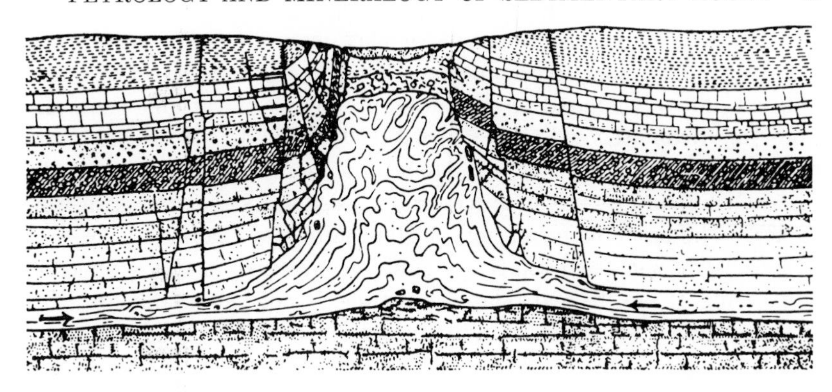

FIG. II-5. Scheme of diapiric movement. A salt eczema has dislocated its roof and flows upward, while the surrounding rock faults and founders into the open space left by the ascending salt. (After F. Rinne, *Gesteinskunde,* 1928.)

During the recrystallization, the salt deposits rose to the surface through an interesting process of flow. The metamorphism was not static, therefore, but accompanied by movement, squeezing, and differentiation. Being lighter than the surrounding rocks, the salt bodies, driven by their own buoyancy, would rise through the overlying strata as domes or diapirs, a schematic cross section of which is shown in Fig. II-5. (It is of more than ordinary interest that granites likewise are able to rise in the crust in a diapir-like fashion.)

Other evaporite sediments are, for instance, sodium carbonate minerals, occurring in deserts as precipitates from "soda lakes," and likewise borate sediments. (See Table II-11.)

An interesting and special group of evaporite sediments are represented by the Chilean niter deposits, containing some highly oxidized salts, for example, chromates, perchlorates, and iodates, in addition to nitrates, sodium chloride, and sulfates. (See Table II-11.)

Brief Survey of Some Special Sediments

Silica may be precipitated as opal, as chalcedony, or as quartz. Chalcedony is the dominant constituent in most chert and flint. The sequence of changes is often (1) original hydrous gel (opal), which subsequently crystallizes to (2) chalcedony and then to (3) quartz.

Silicates, other than clay minerals previously mentioned, are rare as sedimentary precipitates. These may in some places form a large or important part of the deposits in which they occur. (1) *Glauconite,* is approximately $K(Mg,Fe)(Fe,Al)_4Si_8O_{20}(OH)_6$, forming individual deposits and also occurring in scattered grains in some sandstones and limestones. Glauconitization of clay minerals occurs at the sea bottom in some modern sediments. (2) *Chamosite* is a chlorite-like mineral probably related to the kaolin group and of variable composition [approximately $(Fe,Al)_{\gtrless 5}(Si,Al)_4O_{10}(OH)_8$]. It occurs as a constituent in marine clay iron stones and mudstones, or mixed with calcite, clay, and quartz in shales and sandstones associated with iron stones. The mode of formation of glauconite and chamosite, whether by reconstruction from relatively simple compounds or by submarine weathering of biotite and related minerals, is not yet established.

Manganese may occur in individual sedimentary deposits, as mentioned above. The minerals are very diversified (see Table II-7), manganite ($Mn_2O_3 \cdot H_2O$) pyrolusite (MnO_2) and its hydrates being the most common. The most important manganese deposits in the world are formed in this way. For example, the Tertiary brackish water deposits of carbonates and oxides of manganese in clay shale at Nikopol, South Russia, and at Tshiaturi, Caucasus, together making up one-third of the world yield.

Highly oxidized manganese hydroxides are also precipitated at great depths on the floor of the ocean, forming concretions rich in MnO_2, pelagite. In these nodules is stored an enormous amount of manganese, but they are so disseminated that they are not now minable. They parallel former shore lines and indicate that low oxidation potentials obtained in the outlying portions of their depositional basins.

Barium sulfate (barite) concretions are found as great rarities. The interaction of hot spring water rich in barium with the sulfate present in the sea water is considered responsible for the formation.

Radium and other radioactive elements are markedly concentrated in marine sediments. See Table II-13.

TABLE II-13

AVERAGE RADIUM CONTENT OF PACIFIC OCEANIC SEDIMENTS

(After Sverdrup et al.)

Radiolarian ooze	14.1×10^{-12} g per gram
Red clay	8.7×10^{-12} g per gram
Globigerina ooze (siliceous)	7.2×10^{-12} g per gram
Diatom ooze	5.0×10^{-12} g per gram
Globigerina ooze	3.7×10^{-12} g per gram
Terrigenous mud	2.5×10^{-12} g per gram
Granite (for comparison)	1.6×10^{-12} g per gram
Basalt (for comparison)	0.6×10^{-12} g per gram

The distribution with depth of radioactivity in drill core samples presents us with a highly important quantitative method in Pleistocene chronology, *the ionium method.* This method is based on the fact that uranium in sea water produces ionium, which is immediately precipitated as ionium fluoride, and then sediments uniformly when overall sedimentation is uniform. In the sediments, uranium, ionium, and radium then decay at different rates, so that the proportions of the three elements in a sample will change with time. In long cores from the sea bottom the age of the sediments has been traced back to 300,000 years.

The unique uranium-vanadium deposits in sandstone of the western Colorado Plateau are probably hydrothermal in origin and not reworked sedimentary deposits.

Phosphate Sediments. The supply of dissolved phosphorus delivered into the sea is balanced by the removal in skeletal structures, some of which are extremely high in phosphorus, and by the precipitation as phosphorite (see below).

Phosphorite nodules are found in near-shore localities at depths less than 1000 m, commonly associated with calcareous remains and glauconite.

Larger sedimentary phosphate deposits may form in various ways:

1. Direct sedimentation of shells or bones rich in phosphate (respectively from brachiopods and crustaceans, or vertebrates). *Example:* Lower Ordovician phosphate deposits formed through accumulation of phosphate shell of *Obolus spollinis,* in Estonia.

2. Direct sedimentation of excrements of carnivorous animals, but especially sea birds. *Example:* Peruvian sea bird guano formed on islands west of South America.

3. Metasomatic action of phosphate-rich solutions from (2) on underlying limestone. *Example:* Various phosphate islands in the Pacific (Christmas Island, Nauru, and others). Solutions from guano layers have reacted with the underlying coral limestone, with formation of much calcium phosphate. Similar occurrences are known in the West Indies (Curaçao and Aruba). Guano deposited on laterite may give rise to aluminum phosphate instead of calcium phosphate.

4. Concentration of relatively insoluble phosphate concretions as residua after weathering and erosion of phosphate-rich limestone. *Example:* Some of the most important phosphate occurrences in Florida are formed as residua from limestone originally poor in phosphate.

5. Marine sedimentation of phosphorite possibly formed through precipitation of phosphorite from deep-sea water rich in phosphate and fluorine. *Example:* Thick layers of marine phosphorite extending over

hundreds of kilometers occur in North Africa (Morocco, Algeria, Tunis) and in the Rocky Mountains in Canada and the United States.

The most important, commercially, are those of 3, 4, and 5.

The most important minerals of the phosphate sediments are *apatite* $Ca_5(PO_4)_3(Cl,OH,F)$ and hydroxyapatite $\rightarrow$ carbonate apatite or *francolite* (varieties are dahlite, staffelite, podolite). *Vivianite* ($Fe_3P_2O_8 \cdot 8H_2O$) is often formed through recrystallization in old bones. *Struvite* (guanite) $NH_4MgPO_4 \cdot 6H_2O$ and *wavellite* $[Al_3(PO_4)_2(OH)_3 \cdot 5H_2O]$ are present in many occurrences but they are seldom found in quantities.

Phosphorite is the name used for the common phosphate rock. It is composed of "amorphous" *collophanite* and varying amounts of hydroxy apatite and/or carbonate apatite. Most phosphorites contain an excess of fluorine and carbonate (compared to the apatite formula); this can be explained by the fact that collophanite is made up of minute crystallites of apatite structure, each crystallite containing only a few unit cells. The CO_3-groups (and the F-groups) attach themselves to each unit cell in greater numbers than demanded by the formula unit, some of the groups actually belonging to the neighboring cells which, however, are not, or only incompletely, developed. This accounts for the excess fluorine and carbonate in collophanite. When apatite (francolite) appears in phosphorite it is always associated with a small amount of fluorite and calcite; both are formed secondarily by recrystallization of collophanite. By this process the excess fluorine (and CO_2) is wrung out and must appear in an independent crystalline phase, viz.: fluorite or calcite.

Biosediments rich in carbon are always formed in reducing environments. They consist of peat, lignite, anthracite, oil, etc., and will not be described further here. Their study is specialized; their properties and occurrences are described in books on coal petrography, petroleum geology, etc.

In most fine-grained sediments deposited in water, we find small amounts (usually around 1 per cent) of organic substances rich in carbon and related to humus (alteration product of lignine). Nearshore sediments generally contain somewhat larger amounts, averaging around 2.5 per cent. The extreme range is from less than 0.5 to more than 10 per cent. Slime and mud with large amounts of organic substance are often deposited at the bottom of stagnant waters poor in oxygen. Through the action of anaerobic bacteria, such deposits develop into sapropel or rotten ooze rich in bitumen, which again gives rise to bituminous shales ("black shales") often rich in uranium and vanadium, oil schists, or stinky limestone. Organic matter in sedimentary formations is considered to have been the source of petroleum.

Reduzate sediments other than those rich in carbon may form under reducing conditions, as, for example, by the deposition of sapropel sediments.

In sulfate-bearing waters (sea water, for example) the presence of methane (formed through fermentation of cellulose) may cause a reduction of the sulfate ion:

$$CaSO_4 + CH_4 = CaCO_3 + H_2S + H_2O$$

Iron sulfides are formed from the hydrogen sulfide thus generated. Marcasite or pyrite is typically present in bituminous sediments, for example, black schists and limestones. In this way large sedimentary sulfide ore deposits are formed (Rammelsberg and Meggen, Germany, and many other places).

Other metal sulfides are also formed in this way in sediments. Siderite ($FeCO_3$) is also typically associated with reduzate sediments.

Native sulfur may form through oxidation of hydrogen sulfide (often effected by bacteria consuming the energy of combustion):

$$2H_2S + O_2 = 2H_2O + S_2 + 122 \text{ Cal}$$

Rich sedimentary sulfur deposits are formed in this way.

Sulfur (and gypsum) deposits not associated with any reducing environment are formed by oxidation (through the air) of hydrogen sulfide emanations from volcanoes, fumaroles, and hot springs.

4 · SELECTED REFERENCES

Balk, R., Structure of Grand Saline salt dome. Texas, *Am. Assoc. Petrol. Geologists* **33,** 1944.

Cartlege, G. H., The ionic potential, *J. Am. Chem. Soc.* **50,** 1928.

Clarke, F. W., Data of geochemistry, *U.S. Geol. Survey Bull.* **770,** 1924.

Engelhardt, v. W., *Der Porenraum der Sedimente,* Berlin, 1960.

Garrels, R. M., *Mineral Equilibria: At Low Temperature and Pressure,* Harvard University, 1960.

Garrels, R. M., M. E. Thompson, and R. Siever, Control of carbonate solubility by carbonate complexes, *Am. J. Sci.* **259,** 24–45, 1961.

Goldschmidt, V. M., The principles of the distribution of the chemical elements in minerals and rocks, *J. Chem. Soc.* (London), 1937.

Hill, Warner, and Horton, Chemical composition of sedimentary rocks (preliminary paper), *U.S. Geol. Survey, Profess. Paper* **400A,** 62, 1960.

Kokorsch, R., Genesis, Metamorphose und Faziesverhältnisse des Stasfurtlagers, *Geol. Jahrb. Landesamt Bodenforschung Beiheft* (Hannover), **41,** 1960.

Krumbein, W. C., and F. J. Pettijohn, *Manual of Sedimentary Petrography,* New York and London, 1938.

Kuenen, Ph. H., *Marine Geology,* New York, 1950.

Lotze, F., *Steinsalze und Kalisalze,* 2nd ed., Berlin, 1957.

Miller, H. P., The problems of coal geochemistry, *Econ. Geol.,* 1949.

Milner, H. B., *Sedimentary Petrography,* 3rd ed., London, 1940.

Petrascheck, W., Die Metamorphose der Kohle, etc., *Sitzungber. österreich. Akad. Wiss.* **156,** 1948.

Pettijohn, F. J., *Sedimentary Rocks,* New York, 2nd ed., 1957.

Phillips, F. C., Oceanic salt deposits, *Quart. Rev.* **1,** 1947.

Pia, J., Die rezenten Kalksteine, *Tschermaks Min. petr. Mitt.* (Ergänzungsband), Leipzig, 1933.

Stutser, O., and A. C. Noe, Geology of coal, University Chicago Press, 1940.

Sverdrup, H., M. W. Johnson, and R. H. Fleming, *The Oceans* (especially chapter on marine sedimentation, pp. 946–1049), New York, 1942.

Recent Marine Sediments, edited by P. D. Trask, American Association Petroleum Geologists, Tulsa, Oklahoma, 1939.

Twenhofel, W. H., *Principles of Sedimentation,* 2nd ed., New York, 1950.

Twenhofel, W. H., and S. A. Tyler, *Methods of Study of Sediments,* New York and London, 1941.

Van't Hoff, J. H., *Zur Bildung der ozeanischen Salzablagerungen,* I and II, Braunschweig, 1905 and 1909.

X-ray Identification and Crystal Structures of Clay Minerals, edited by G. Brown, Mineralogical Society, London, 1961.

Part III · Igneous Rocks

"Igneous rock" in the wide sense comprises the crystalline or glassy rocks which, belonging neither to the sedimentary nor clearly metamorphic classes, look as if they might have cooled from a magma.

A Dictionary of Geology, 1961.

Geology recognizes three major divisions of rock according to their mode of formation: igneous, sedimentary, and metamorphic. But the century-old idea that all so-called igneous rocks had congealed from a molten magma has, in the light of modern knowledge, met with severe criticism. The composition of many "igneous" rocks is such that a corresponding magma becomes a physico-chemical impossibility. The fact that 95 per cent of the deep-seated rocks are granitic in composition is contrary to the hypothesis of a magmatic evolution. Petrology is, indeed, in a state of doubt and confusion.

Most petrologists today—each according to his special experience and knowledge—want to narrow the group of "truly" igneous rocks. Bowen's experiments have been construed to mean that peridotite magmas are non-existent. Many granites show clear evidence of being metasomatic rocks. Most radical is the French school of Perrin and Roubault, claiming categorically that magmatic granites do not exist. These and other hypotheses will be discussed in later chapters. Let us first consider the historical development.

The adjective *igneous* is derived from the Latin *ignis,* fire. *Magma* is a Greek word meaning a paste; in geology it is now used to designate larger intracrustal bodies of molten rocks, replacing the older term *subterraneous lava. Eruptive rocks* (*roches éruptives, Eruptivgesteine*) imply a breaking-out from lower to higher levels in the earth's crust. The adjectives *igneous, magmatic,* and *eruptive* are currently applied to rocks as if they were synonymous.

Nobody has seen a magma in a subterranean position. And no instrument has as yet registered its presence or measured any of its properties. But of course we have seen molten lavas, hot silicate melts, extruded from craters or fissures of some of the 622 active volcanoes in the world. There is no doubt that a lava rock is igneous and eruptive. By definition it is also magmatic, since the lava evidently rose from deeper parts of the earth's crust (subterraneous lava = magma).

A ready conclusion and extrapolation is that much magma, unaffected by extrusion processes, might remain buried in the depth of the earth. It would cool and congeal under a more or less thick cover of supracrustal rocks, and thus become those intrusive and plutonic igneous rocks which by subsequent upfolding and/or denudation became exposed to the surface and accessible to investigation. This explanation obviously holds for some of the coarse-grained rock bodies, but it does not necessarily hold for all. Bodies of batholithic dimensions would require enormous magma reservoirs, and the existence of such large subterraneous bodies of molten silicates is hypothetical and not an established fact.

The generalization that all holocrystalline silicate rocks must be of magmatic origin has resulted in the copius development of the school of "magmatists," who inter alia have constructed a great number of "magma types," some of which do not correspond to any existing lava (Niggli, 1936). It is thought best today to exercise more caution and require positive proofs before a rock is regarded as igneous. Rocks of dubious origin should be critically discussed before they are admitted or rejected (page 220). The crux of the definition is the congealing of magma. With Read we regard the dictum, "no magma, no igneous rock," as a fair statement of a reasonable position.

A. DESCRIPTIVE CLASSIFICATIONS

We can classify rocks, for petrological purposes, exactly, definitely, and strictly only by creating arbitrary divisions, cutting them up by sharp planes and putting them into man-devised pigeon-holes. Such a classification is a pis-aller, a makeshift, a classification of convenience; it may or may not correspond to the evolution of igneous rocks as it really is.

H. S. Washington, *Bull. Geol. Soc. Am.*, **33,** 1922.

A descriptive classification is one that describes without taking the genetic relations into consideration. Therefore the genetic implications of the name "igneous" will not be discussed here. Igneous rocks are in this classification because they have been called so by previous authorities; they are the massive rocks which show no sign of a "secondary" mode of origin.

1 · THE MINERALOGICAL CLASSIFICATION OF IGNEOUS ROCKS

Geologically, the igneous rocks may be divided into three groups: (1) *effusive* or *extrusive rocks* [*lavas*], (2) *massive, deep-seated rocks* [*plutonites*], and (3) *hypabyssic rocks* [*dikes*].

The mode of formation is usually reflected in the texture. The effusives are fine-grained, dyscrystalline, or even glassy; the deep-seated rocks are coarse-grained and eucrystalline.

In further classification, it is of great advantage to distinguish between a mineralogical classification and a chemical classification. Several attempts have been made to combine both; and for certain rocks, and for certain purposes, special chemico-mineralogical systems have proved themselves very useful. But in the great majority of cases, greater clarity and more accurate definitions are obtained by using a twofold classification: on the one hand a mineralogical classification of the natural rocks according to their actual mineral composition, and on the other a purely chemical classification which can be worked out with great accuracy.

Mineralogically, a large number of rock types have been established, but in the following we shall be concerned with only a few leading principles in the mineralogical classification.

The rock-making minerals may be divided into two groups:

Light or Leucocrate Minerals	*Colored or Melanocrate Minerals*
Quartz, feldspar, feldspathoid	Mica, amphibole, pyroxene, olivine

The color index of a rock is the sum of the colored minerals expressed in percentages. According to this index the rocks are divided into (*A*) *leucocratic* (color index, 0–30), (*B*) *mesotype* (color index, 30–60), (*C*) *melanocratic* (color index, 60–100).

The ratio of silica to metal oxides is the chief factor in determining the *degree of saturation.* We distinguish among *oversaturated, saturated* (*neutral*), and *undersaturated.* Oversaturation (excess of SiO_2) produces free quartz, whereas deficiency of SiO_2 results in minerals of a low degree of silification (for example, olivine, feldspathoids). Hornblende also is frequently undersaturated.

Quartz and feldspathoids are chemically incompatible; consequently, the leucocrate minerals may combine into but four mineral assemblages.

1. Quartz and feldspar: oversaturated rocks.
2. Feldspar: saturated rocks.
3. Feldspar and feldspathoids: undersaturated rocks.
4. Only feldspathoid.

Each of these types is again divided into two groups, according to whether alkali feldspar or plagioclase dominates.

All criteria so far mentioned are combined in the scheme of classification given in Tables III-1 and III-2.

Leucite in a deep-seated rock is very rare. We shall see (page 87) that pressure hinders the formation of leucite, and that other mineral combinations (for example, orthoclase + nepheline = pseudoleucite) form instead. There are no good examples of leucite-bearing plutonic rocks. (*Italite* and *missourite* are probably not truly plu-

TABLE III-1

LEUCOCRATIC ROCKS

Lavas

		Alkali Feldspar ⟵	⟶	Plagioclase
SiO_2 →	1. Quartz and feldspar	*Rhyolite*	*Rhyodacite*	*Dacite*
	2. Feldspar	*Trachyte*	*Latite and trachyandesite*	*Andesite*
	3. Feldspar and feldspathoid	*Phonolite*	*Feldspathoidal latite*	
	4. Only feldspathoids	*Nephelinite, Leucitite*		

Plutonites

		Alkali Feldspar ⟵	⟶	Plagioclase
SiO_2 →	1. Quartz and feldspar	*Granite*	*Granodiorite*	*Quartz diorite, tonalite*
	2. Feldspar	*Syenite*	*Monzonite*	*Diorite* *
	3. Feldspar and feldspathoid	*Nepheline syenite*	*Nepheline monzonite*	
	4. Only feldspathoids	*Urtite*		

Dikes

Felsite, aplite, pegmatite exhibit a granite-like composition.

Chemically the *quartz porphyry* and the *feldspar porphyry*—which also may occur as lavas—correspond to granite and syenite respectively.

* *Anorthosite* is exclusively composed of plagioclase.

tonic. *Fergusite* contains no true leucite, only pseudoleucite.)

Melanocratic and hypermelanic rocks are rare. Some dikes of this type are listed in Table III-3.

Many of the dark rocks approach a monomineralic composition: *Peridotite-dunite* constitutes a series from almost pure to very pure

TABLE III-2

MESOTYPE ROCKS

Lavas

		Alkali Feldspar ⟷ Plagioclase	
$SiO_2 \rightarrow$	1. Quartz and feldspar	*Obsidians* and *pitchstones*	*Quartz basalt*
	2. Feldspar	*Mugearite—Trachybasalt* *Trachydolerite*	*Basaltic rocks* *
	3. Feldspar and feldspathoid	*Tephrite and basanite*	
	4. Only feldspathoids	*Nepheline basalt, Leucite basalt*	

Plutonites

		Alkali Feldspar ⟷ Plagioclase	
$SiO_2 \rightarrow$	1. Quartz and feldspar		*Quartz gabbro*
	2. Feldspar	*Shonkinite* (*kentallenite*)	*Gabbroic rocks* †
	3. Feldspar and feldspathoid	*Nepheline syenite*	*Theralite and essexite* ‡
	4. Only feldspathoids	*Ijolite, Jacupirangite*	

Dikes

Dikes of gabbroic composition are called *diabase* or *dolerite*. Other dike rocks are listed in Table III–3.

---

* *Basaltic rocks* are subdivided into alkali basalt and tholeiite. Addition of olivine makes for the series *olivine basalt* → *oceanite* → *picrite basalt*, with increasing amounts of olivine. *Ankaramite* and *limburgite* (with glassy groundmass) are rich in olivine and augite.

Spilite has albite or oligoclase as the principal feldspar. It usually shows evidence of hydrothermal alteration.

† *Gabbroic rocks* are generally subdivided according to the nature of the dark minerals: *gabbro* = plagioclase and monoclinic pyroxene; *norite* = plagioclase and rhombic pyroxene; hyperite has an intermediate position. Addition of olivine makes for *olivine gabbro* (*olivine norite*); with much olivine it is called *picrite*. *Troctolite* is pyroxene-free and contains only plagioclase and olivine.

‡ *Teschenite* has analcite instead of feldspathoid.

TABLE III-3

MELANOCRATIC AND HYPERMELANIC DIKES

	Biotite	Pyroxene and/or Hornblende	Alkali Pyroxene and/or Hornblende
Alkalifeldspar	Minette	Vogesite	Soda minette
Plagioclase	Kersantite	Spessartite	Camptonite
No feldspar	Alnöite ⟵⟶ Monchiquite		

TABLE III-4

APPROXIMATE MINERAL COMPOSITIONS OF SOME PLUTONIC ROCK TYPES

(After E. S. Larsen, *Handbook of Physical Constants*, 1942)

	Granite	Syenite	Granodiorite	Quartz Diorite	Diorite	Gabbro	Olivine Diabase	Diabase	Dunite
Quartz	25	..	21	20	2	..	..	..	..
Orthoclase and microperthite	40	72	15	6	3	..	..	..	..
Oligoclase	26	12	..	..	..	..	..	..	..
Andesine	..	..	46	56	64	..	..	..	..
Labradorite	..	..	..	..	..	65	63	62	..
Biotite	5	2	3	4	5	1	..	1	..
Amphibole	1	7	13	8	12	3	..	1	..
Orthopyroxene	..	..	..	1	3	6	..	..	2
Clinopyroxene	..	4	..	3	8	14	21	29	..
Olivine	..	..	..	..	..	7	12	3	95
Magnetite	2	2	1	2	2	2	2	2	3
Ilmenite	1	1	..	..	..	2	2	2	..
Apatite	trace	trace	trace	trace	trace	..	..	..	..
Sphene	trace	trace	1	trace	trace	..	..	..	..
Color Index	*9*	*16*	*18*	*18*	*30*	*35*	*37*	*38*	*98*

olivine rock. *Perknites* are rocks composed exclusively of pyroxene and hornblende. They are subdivided into *pyroxenites* and *hornblendites*. Corresponding lavas are not known. Dikes of similar composition are listed in Table III-3. The genesis of these rocks presents today a highly debatable subject in petrology. *"Magmatic" iron ore* and *sulfide ores* are often genetically associated with melanocratic deep-seated rocks.

The approximate mineral compositions of some of the principal rock types mentioned above are listed in Table III-4.

2 · THE CHEMICAL COMPOSITION OF IGNEOUS ROCKS

Everyday metals like zinc, lead, tin, mercury, silver, gold, platinum, arsenic, antimony, and bismuth—metals that are essential to our civilization and our daily needs—are found in igneous rocks, if at all, in scarcely detectable amounts. Although they are derived ultimately from igneous rocks they are made available for our use only by natural processes of concentration into ore bodies.

Clarke and Washington,
The Composition of the Earth's Crust, 1924.

Although a certain mineral assemblage uniquely determines the chemical composition of a rock, the opposite is not true; the mineral assemblage is not a one-valued function of the chemical composition (heteromorphism of rocks).

Every named rock type shows variation in chemical composition. Daly (1933) established the average composition of a great number of igneous-rock types by collecting all good analyses and computing the several averages according to the actual number of determinations. New averages were computed by Nockolds (1954). See Table III-5.

The existing petrochemical data show that each of the rock-making oxides has a characteristic frequency distribution. See Figs. III-1, III-2, and III-3.

Silica exhibits great variations but is usually limited to the range of 35 to 80 per cent. Even for rocks of more than 90 per cent SiO_2 a magmatic mode of origin has been claimed. But the origin of such rocks is rather uncertain. It is genetically significant that the distribution curve of silica shows two maxima. (See Fig. III-1.) We shall have occasion to return to this fact in later chapters. Because of its great variability, silica is often used as a reference substance, and all other oxides are plotted as dependent variables of the silica content.

Alumina, soda, and *ferric oxide* display a regular distribution, each

TABLE III-5

Average Chemical Compositions

(After Nockolds)

	Plutonic Alkali Granite	Effusive Alkali Rhyolite	Plutonic Grano-diorite	Effusive Rhyoda-cite	Plutonic Tonalite	Effusive Dacite
SiO_2	73.86	74.57	66.88	66.27	66.15	63.58
TiO_2	0.20	0.17	0.57	0.66	0.62	0.64
Al_2O_3	13.75	12.58	15.66	15.39	15.56	16.67
Fe_2O_3	0.78	1.30	1.33	2.14	1.36	2.24
FeO	1.13	1.02	2.59	2.23	3.42	3.00
MnO	0.05	0.05	0.07	0.07	0.08	0.11
MgO	0.26	0.11	1.57	1.57	1.94	2.12
CaO	0.72	0.61	3.56	3.68	4.65	5.35
Na_2O	3.51	4.13	3.84	4.13	3.90	3.98
K_2O	5.13	4.73	3.07	3.01	1.42	1.40
H_2O	0.47	0.66	0.65	0.68	0.69	0.56
P_2O_5	0.14	0.07	0.21	0.17	0.21	0.17

	Plutonic Alkali Syenite	Effusive Alkali Trachyte	Plutonic Monzo-nite	Effusive Trachy-andesite	Plutonic Diorite	Effusive Ande-site
SiO_2	61.86	61.95	55.36	54.02	51.86	54.20
TiO_2	0.58	0.73	1.12	1.18	1.50	1.31
Al_2O_3	16.91	18.03	16.58	17.22	16.40	17.17
Fe_2O_3	2.32	2.33	2.57	3.82	2.73	3.48
FeO	2.63	1.51	4.58	3.98	6.97	5.49
MnO	0.11	0.13	0.13	0.12	0.18	0.15
MgO	0.96	0.63	3.67	3.87	6.12	4.36
CaO	2.54	1.89	6.76	6.76	8.40	7.92
Na_2O	5.46	6.55	3.51	3.32	3.36	3.67
K_2O	5.91	5.53	4.68	4.43	1.33	1.11
H_2O	0.53	0.54	0.60	0.78	0.80	0.86
P_2O_5	0.19	0.18	0.44	0.49	0.35	0.28

	Plutonic Nepheline Syenite	Effusive Phonolite	Plutonic Nepheline Monzonite	Effusive Nepheline Latite	Plutonic Essexite	Effusive Teph-rite
SiO_2	55.38	56.90	50.38	52.95	46.88	44.82
TiO_2	0.66	0.59	2.49	1.43	2.81	2.65
Al_2O_3	21.30	20.17	19.97	19.14	17.07	15.42
Fe_2O_3	2.42	2.26	2.77	3.25	3.62	4.28
FeO	2.00	1.85	3.96	2.86	5.94	6.61
MnO	0.19	0.19	0.13	0.20	0.16	0.16
MgO	0.57	0.58	2.15	2.02	4.85	7.27
CaO	1.98	1.88	6.01	5.33	9.49	10.32
Na_2O	8.84	8.72	6.35	6.55	5.09	5.30
K_2O	5.34	5.42	3.97	4.37	2.64	1.26
H_2O	0.96	0.96	1.37	1.12	0.97	1.56
P_2O_5	0.19	0.17	0.45	0.37	0.48	0.35
	(x)	(1)	—	(2)		

(x) Contains CO_2 = 0.17
(1) Contains Cl = 0.23, SO_3 = 0.13
(2) Contains Cl = 0.09, SO_3 = 0.34

TABLE III-5 (*Continued*)

AVERAGE CHEMICAL COMPOSITIONS

(After Nockolds)

	Plutonic Quartz Gabbro	Plutonic Gabbro	Plutonic Norite	Plutonic Olivine Gabbro	Plutonic Tracto-lite	Plutonic Anortho-site
SiO_2	54.39	48.36	50.28	46.83	43.84	54.54
TiO_2	1.29	1.32	0.89	0.97	0.67	0.52
Al_2O_3	16.72	16.84	17.67	17.38	13.46	25.72
Fe_2O_3	2.49	2.55	1.30	1.91	2.20	0.83
FeO	7.15	7.92	7.46	8.20	9.24	1.46
MnO	0.20	0.18	0.14	0.14	0.13	0.02
MgO	4.15	8.06	9.27	10.03	19.71	0.83
CaO	6.68	11.07	9.72	11.36	8.10	9.62
Na_2O	3.15	2.26	1.96	2.03	1.33	4.66
K_2O	1.58	0.56	0.63	0.40	0.58	1.06
H_2O	1.85	0.64	0.47	0.63	0.50	0.63
P_2O_5	0.35	0.24	0.21	0.21	0.18	0.11

	Effusive Quartz Basalt	Effusive Tholeiite	Effusive Alkali Basalt	Effusive Olivine-rich Basalt	Effusive Nepheli-nite and Nepheline Basalt	Plutonic Ijolite
SiO_2	55.46	50.83	46.77	43.69	39.07	42.58
TiO_2	0.88	2.03	3.00	2.12	3.86	1.41
Al_2O_3	16.85	14.07	14.65	9.06	12.82	18.46
Fe_2O_3	2.13	2.88	3.71	3.46	8.75	4.01
FeO	4.86	9.06	7.94	9.43	6.39	4.19
MnO	0.22	0.18	0.15	0.16	0.26	0.20
MgO	6.31	6.34	6.82	19.68	6.14	3.22
CaO	7.86	10.42	12.42	9.18	14.20	11.83
Na_2O	3.30	2.23	2.59	1.49	4.09	9.55
K_2O	1.40	0.82	1.07	0.69	2.07	2.55
H_2O	0.58	0.91	0.51	0.74	1.59	0.55
P_2O_5	0.15	0.23	0.37	0.30	0.76	1.52

	Spes-sartite	Minette	Kersan-tite	Voge-site	Campto-nite	Monchi-quite	Alnöite
SiO_2	53.52	49.45	50.79	52.62	40.70	45.17	32.31
TiO_2	1.24	1.23	1.02	0.54	3.86	1.90	1.41
Al_2O_3	14.57	14.41	15.26	14.86	16.02	14.78	9.50
Fe_2O_3	3.52	3.39	3.29	3.60	5.43	5.10	5.42
FeO	5.29	5.01	5.54	4.18	7.84	5.05	6.34
MnO	0.38	0.13	0.07	0.84	0.16	0.35	0.01
MgO	6.60	8.26	6.33	8.55	5.43	6.26	17.43
CaO	7.03	6.73	5.73	5.86	9.36	11.06	13.58
Na_2O	3.48	2.54	3.12	3.21	3.23	3.69	1.42
K_2O	2.28	4.69	2.79	2.83	1.76	2.73	2.70
H_2O	1.75	3.04	5.71	2.70	5.59	3.40	7.50
P_2O_5	0.34	1.12	0.35	0.21	0.62	0.51	2.38

* Contains CO_2 = 0.38

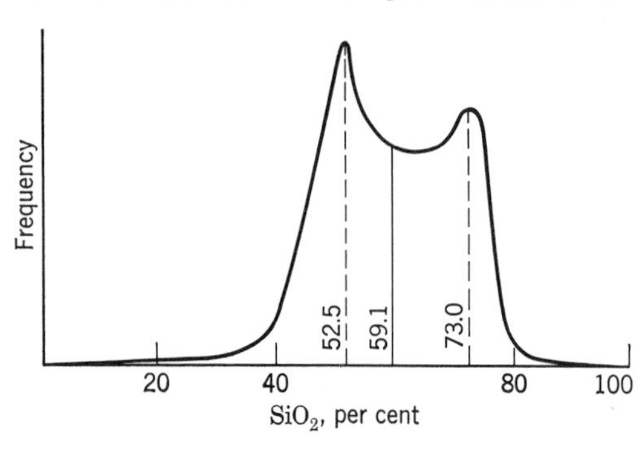

Fig. III-1. Frequency distribution of silica. (After Richardson and Sneesby, 1922.)

possessing a well-defined maximum. In most rocks alumina is between 10 and 20 per cent; certain corundum-bearing rocks with much higher alumina content are probably not igneous. Soda seldom exceeds 15 per cent, although in certain nephelinites it may attain 18 per cent.

Lime, potash, magnesia, and *ferrous oxide* show a less regular distribution. Lime may attain 29 per cent in certain pyroxenites, but usually it stays in the range of 0 to 15 per cent. Potash, like soda, seldom exceeds 15 per cent, but in italites it may attain 18 per cent. Magnesia is usually encountered in amounts from 0 to 25 per cent; but in dunites it may attain 50 per cent. Iron oxides are present in some rocks only in traces and rarely pass 15 per cent, except in some "magmatic" ore bodies. The four curves of Fig. III-3, although somewhat irregular, are exponential in character, with the highest value being at 0 per cent, indicating that these elements exist at the surface of the earth only through the courtesy of statistical fluctuations.

These data show clearly that no chance distribution exists in the various oxides of the igneous rocks. The proportions are statistically governed according to strict rules. For example, rocks with high silica content are usually high in alkalies, but low in lime and magnesia; whereas rocks with low silica content are high in lime and magnesia, and low in alkalies. Rocks that are distinctly or highly sodic are most likely to contain the rarer elements and to have the largest number of these elements typically associated with them. Thus we find lithium, cesium, yttrium, titanium, zirconium, cerium, and fluorine closely and generally, if not exclusively, bound up with sodic rocks. The association of iron with sodium is also of great interest. Such relations greatly facilitate the grouping of the rocks into a comparatively small number of types.

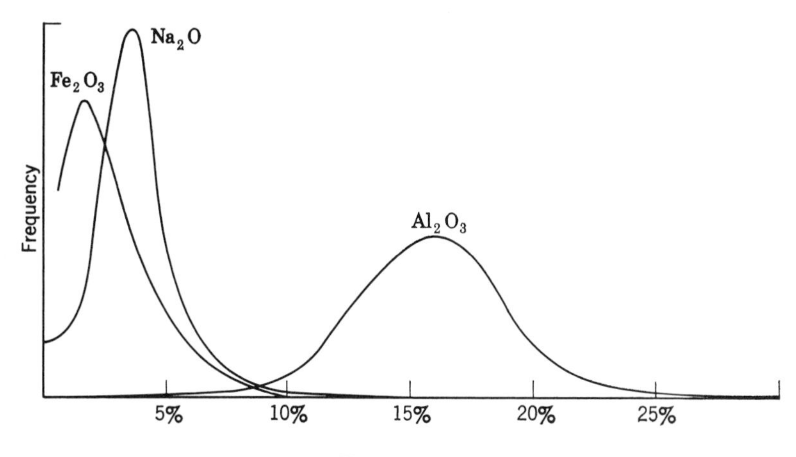

FIGURE III-2

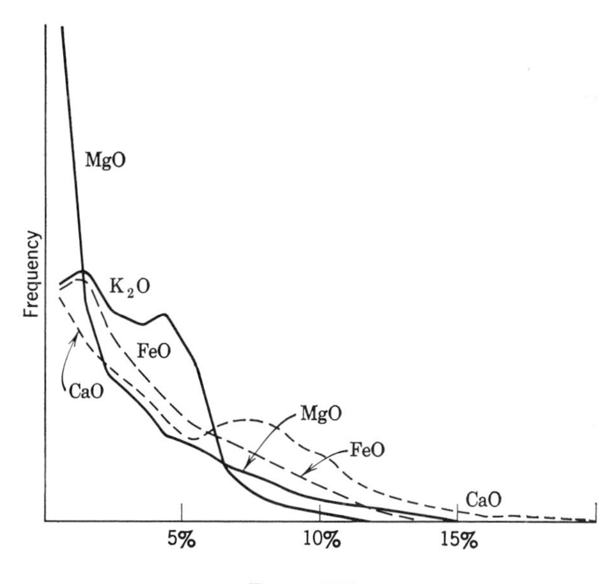

FIGURE III-3

FIGS. III-2 and III-3. Frequency distribution of major oxides. (After Richardson and Sneesby, *Min. Mag.*, 1922.)

Again these data may serve to compute the chemical composition of the most abundant rock types. We first have to distinguish between the most frequent type of rock and the average type of rock; in Table III-6 computations are listed for these types.

All these peculiar statistical facts are simple consequences of special

TABLE III-6

CHEMICAL COMPOSITION OF THE MOST FREQUENT ROCK TYPES (I AND II) AND THE AVERAGE ROCK TYPE (III)

(I and II according to Richardson and Sneesby, III after Washington)

	I	II	III
SiO_2	52.5	73.0	59.1
Al_2O_3	16.0	14.0	15.3
Fe_2O_3 FeO	10.0	1.5	6.9
MgO	6.0	0.2	3.5
CaO	9.0	0.5	5.1
Na_2O	3.0	4.0	3.8
K_2O	1.0	3.5	3.1
Sum	97.5	96.7	96.8

I. The most frequent rock type; basaltic composition.
II. The next most frequent rock type; granitic composition.
III. The average rock type, showing a granodioritic composition. (It has a lower frequency than types I and II.)

rock-making processes governing the chemical development of the several rocks. Any theory of origin must take cognizance of these relations. We shall see that the theory of crystal fractionation (the interaction crystal $\rightleftharpoons$ silicate melt) is capable of explaining many of the facts.

3 · PETROCHEMICAL CALCULATIONS AND CLASSIFICATIONS

. . . for it is obvious that the ever increasing accumulation of knowledge of the chemical and mineral properties of rocks, and of their relation to one another and to their texture, mode of occurrence, and their origin impose a greater task on modern petrographers than was borne by those of a former generation. And since it is not to be assumed that the present state of petrographical knowledge is complete, provisions should be made for expansion and adjustment along lines which seem to be those in which future development will take place.

Conclusion in Cross, Iddings, Pirsson, and Washington's "A quantitative chemico-mineralogical classification and nomenclature of igneous rocks," *J. Geol.*, **1902**.

Niggli Values

The twofold way of classifying rocks—mineralogical and chemical—has been advocated and its advantages have been convincingly demonstrated by Niggli in several papers of outstanding importance to systematic petrology. In the system that Niggli has contrived to build, the chemical classification becomes identical with definition and classification of what he calls the *magma types* (see page 181).

The chemical analysis of a rock is the basis for its classification. The procedure of calculation is as follows. The figures giving the weight percentages are divided by the molecular weight of the corresponding oxide; thus we arrive at a series of relative figures—molecular numbers—which indicate the molecular proportions of the several oxides in the rock.

The molecular number for Fe_2O_3, which in terms of ferrous oxide corresponds to 2FeO, must be multiplied by 2 and then added to FeO. The remaining molecular numbers are grouped together and added in the following way:

$$\begin{aligned} &Al_2O_3 \text{ (including } Cr_2O_3 \text{ and rare earths)} \\ &+ FeO + MgO\ (+ MnO + CoO + NiO) \\ &+ CaO \text{ (including } BaO + SrO) \\ &+ K_2O + Na_2O \text{ (including rare alkali earths)} \end{aligned}$$

The sum is recalculated to 100, and the numbers designated *al*, *fm*, *c*, and *alk*, respectively. Thus the sum $al + fm + c + alk = 100$ specifies one group of chemical substances.

The molecular numbers for SiO_2 (if desirable also for TiO_2, ZrO_2, P_2O_5, H_2O, CO_2) are recalculated according to the equation:

$$\text{Molecular number for } SiO_2 : \text{molecular number for } Al_2O_3 = x : al$$

The resulting figure for x is called *si* (respectively, *ti*, *zr*, *p*, *h*, co_2).

The two groups, $al + fm + c + alk = 100$, and *si* (and *ti*, etc.), include the most important chemical constituents of the rock. But two additional parameters are of great value: the molecular proportion of K_2O compared to the sum of the alkalies in *alk*, and the molecular proportion of MgO compared to the sum of the divalent elements in *fm*. Separately, therefore, the following values are computed:

$$\frac{K_2O}{K_2O + Na_2O + (Li_2O)} \quad \text{and} \quad \frac{MgO}{FeO + MnO + MgO}$$

and respectively designated k and mg. For many purposes in the discussion of the chemical relations in primary and metamorphic rocks, it is sufficient to set down the values si, al, fm, c, alk, as well as k and mg. In a general way they characterize a rock.

The *quartz index*, qz (Quarzzahl), is a derived quantity and is obtained as follows. The most highly silicified minerals in the rocks are feldspars and pyroxenes. In the alkali feldspars and the lime feldspars the molecular proportions are $alk:al:si = 1:1:6$ and $c:al:si = 1:1:2$, respectively. In the pyroxenes the theoretical proportions are $fm:si = 1$ and $c:si = 1$. In a rock composed of these minerals, the following relation holds:

$$si' = 6alk + 2(al - alk) + 1[c - (al - alk)] + fm \cdots \quad (1)$$

The expression $(al - alk)$ corresponds to the fraction of c that is bound to alumina in feldspar, provided $alk < al$; and $c - (al - alk)$ is the fraction of c that is associated only with si. If $al > alk + c$, the excess of al is supposed to combine with an equal amount of si, with formation of sillimanite.

Remembering that $al + fm + c + alk = 100$, and substituting in equation 1, we obtain:

$$si' = (100 + 4alk)$$

With $alk > al$, the excess of alkalies will usually form aegirite, and the corresponding si-value becomes $si' = (100 + 3al + 1alk)$.

In a rock showing a higher si-value than the calculated si'-value, free quartz will appear. If the rock shows a lower si-value, some minerals of low degree of silification (for example, olivine, biotite, feldspathoids, ore) are to be expected. The difference $si - si'$ is the quartz index, qz, which may be either positive or negative, and is a valuable indicator of the minerals to be expected.

In the table below the most important Niggli values are listed:

al from Al_2O_3	si from SiO_2
fm from FeO, Fe_2O_3, MnO, MgO	$k = \dfrac{K_2O}{K_2O + Na_2O}$
c from CaO	$mg = \dfrac{MgO}{FeO + MnO + MgO}$
alk from $K_2O + Na_2O$	$si' = 100 + 4alk$ (usually)
$al + fm + c + alk = 100$	$qz = si - si'$

In the example of Table III-7 the analysis of a granodiorite from California is recalculated after Niggli.

TABLE III-7

GRANODIORITE, PLACER COUNTY, CALIFORNIA

	Weight, %	Approximate Molecular Weight	Molecular Proportions × 1000		Calculations	
SiO_2	65.54	60	1092			
Al_2O_3	16.52	102	162	162	$al = (100 \cdot 162):453 =$	36
Fe_2O_3	1.40	160	9 as FeO 18			
FeO	2.49	72	35	117	$fm = (100 \cdot 117):453 =$	26
MnO	0.06	71	1			
MgO	2.53	40	63			
CaO	4.88	56	87	87	$c = (100 \cdot 87):453 =$	19
Na_2O	4.09	62	66	87	$alk = (100 \cdot 87):453 =$	19
K_2O	1.95	94	21			
TiO_2	0.39	80	5			
P_2O_5	0.18	142	1	453		100
H_2O	0.71	...				
	100.74					

$$si = \frac{1092 \cdot 100}{453} = 241$$

$$k = {}^{21}\!/_{87} = 0.24$$

$$mg = {}^{63}\!/_{117} = 0.54$$

$$qz = 241 - (100 + 4 \cdot 19) = +65$$

CIPW Norm Calculations and Niggli's Molecular Norm

It is about sixty years since the principles of the norm system were published by the well-known authors Cross, Iddings, Pirsson, and Washington. This method offers a taxonomic system, according to which rock analyses can be quantitatively studied and compared from a petrological point of view. The leading idea is to recalculate the rock analyses—which are usually given in terms of oxides—into a standard set of mineral molecules. The standard minerals of the norm were chosen with these purposes in mind: (1) the calculations should be simple; and (2) the normative minerals and mineral associations should correspond to natural conditions. In Table III-8 the most important normative minerals are listed.

For the actual calculation, again the chemical analysis forms the basis. In the usual way the molecular proportions of the several oxides were ascertained, and the molecular numbers combined according to rules set down in the original papers of the four authors—with minor amendments proposed later by Barth and Washington.

The most important change, introduced by Niggli (1936), rests on the following considerations. In the original calculations the molecular proportions were employed to form the several mineral mole-

TABLE III-8

NORMATIVE MINERALS

Mineral	Symbol	Formula	Molecular Weight
Salic Group			
Quartz	q	SiO_2	60
Corundum	c	Al_2O_3	102
Orthoclase	or	$K_2O \cdot Al_2O_3 \cdot 6SiO_2$	556
Albite	ab	$Na_2O \cdot Al_2O_3 \cdot 6SiO_2$	524
Anorthite	an	$CaO \cdot Al_2O_3 \cdot 2SiO_2$	278
Leucite	lc	$K_2O \cdot Al_2O_3 \cdot 4SiO_2$	436
Nepheline	ne	$Na_2O \cdot Al_2O_3 \cdot 2SiO_2$	284
Kaliophilite	kp	$K_2O \cdot Al_2O_3 \cdot 2SiO_2$	316
Femic Group			
Wollastonite	wo	$CaO \cdot SiO_2$	116
Enstatite	en	$MgO \cdot SiO_2$	100
Ferrosilite	fs	$FeO \cdot SiO_2$	132
Forsterite	fo	$2MgO \cdot SiO_2$	140
Fayalite	fa	$2FeO \cdot SiO_2$	204
Acmite	ac	$Na_2O \cdot Fe_2O_3 \cdot 4SiO_2$	462
Magnetite	mt	$FeO \cdot Fe_2O_3$	232
Hematite	hm	Fe_2O_3	160
Ilmenite	il	$FeO \cdot TiO_2$	152
Apatite	ap	$3(3CaO \cdot P_2O_5) \cdot CaF_2$	336
Pyrite	pr	FeS_2	120
Calcite	cc	$CaO \cdot CO_2$	100

The following derived normative minerals are used:

hypersthene (hy) = $(Mg,Fe)O \cdot SiO_2$

diopside (di) = $CaO \cdot (Mg,Fe)O \cdot 2SiO_2$

olivine (ol) = $2(Mg,Fe)O \cdot SiO_2$

cules. Subsequently the mineral molecules were recalculated into weight percentages, and the final result stated in weight percentages (the weight norm). However, the chemical relations of a rock in terms of weight percentages obscure the comprehensive view; moreover, the computations become unnecessarily cumbersome. Niggli, therefore, rejected the weight units and introduced what he called *equivalent molecular units.* These equivalent units must be specifically defined, for if we write, for example, "1 Nepheline" it may mean $NaAlSiO_4$ or $Na_2O \cdot Al_2O_3 \cdot 2SiO_2$, or $\frac{1}{3}(NaAlSiO_4)$, etc. The equivalent formula units are therefore defined as those which contain the same number of constituent cations, and it is simplest to reduce the several equivalent units to a sum of the cations = 1. Then the sum of all cations in the formula represents the number of equivalent units. Consequently, $Na_2O \cdot Al_2O_3 \cdot 2SiO_2 = 6Ne$ (because the formula contains 6 cations), and $MgSiO_3 = 2En$ (because the formula contains 2 cations), etc.

We use the same abbreviations as in the normative minerals (Table III-8) but capitalize the symbols. Observe that Ne does not mean just the composition $NaAlSiO_4$ or $Na_2O \cdot Al_2O_3 \cdot 2SiO_2$, but it means a quantity of this composition corresponding to one cation.

In this way the mineral equations become very simple; for instance,

$$Mg_2SiO_4 + SiO_2 = 2MgSiO_3$$

$$3Fo + 1Q = 4En$$

It is important to note that the sum of the coefficients of reaction on both sides of the equation is the same, for instance,

$$\underbrace{3Ne + 2Q}_{5} = \underbrace{5Ab}_{5}$$

Now we shall return to the rules of the norm calculation in its newest development which will be easily explained by the sample calculation given on page 69. The essential feature is that the molecular proportions (shown in Table III-7) are abandoned in favor of the proportions of the cations (metals and silicon), which are also the proportions of the equivalent molecular units. Since some of the oxide molecules (for example, Al_2O_3, Na_2O, etc.) contain two cations, the proportions of such oxide molecules are converted into the proportions of the cations by multiplying the corresponding molecular number by 2. For those oxides which contain but one cation (SiO_2, CaO, etc.) the cation number is, of course, the same as the molecular number. The cation numbers, thus established, are now recalculated to a sum of 100, which gives us the cation percentage of the rock.

From what has been said, any recalculation of the cation percentages in terms of mineral molecules can now be made without affecting the sum, which remains 100. Once the rock analysis is recalculated to a sum of 100 cations, the molecular norm can be computed with great ease. We have to take care, however, to perform the calculations in the right sequence. The rules of the norm calculations are:

1. Calcite is formed from CO_2 and an equal amount of Ca.
2. Apatite is formed from P and 1.67 times this amount of Ca.
3. Pyrite is formed from S and half this amount of Fe''.
4. Ilmenite is formed from Ti and an equal amount of Fe''.
5. The alkali feldspars are formed *provisionally* from K and Na combined in the right proportions with Al and Si to form Or and Ab.

6*a*. If there is an excess of Al over K + Na, it is assigned to anorthite, one half the amount of Ca being allotted to the excess of Al.

6*b*. If there is an excess of Al over this Ca, it is calculated as corundum.

6*c*. If there is an excess of Ca over this Al of 6*a*, it is femic and reserved for wollastonite (8).

7*a*. If in 5 there is an excess of Na over Al, it is to be combined with an equal amount of Fe''' to form the acmite molecule. There is then no anorthite in the norm.

7*b*. If, as usually happens, there is an excess of Fe''' over Na, it is assigned to magnetite, one half the amount of Fe'' being allotted to it out of what remains from the formation of pyrite and ilmenite.

7*c*. If there is still an excess of Fe''', it is calculated as hematite.

8. Wollastonite is formed from the amount of Ca left over from 6*a*.
9. Enstatite and ferrosilite are formed *provisionally* from all the Mg and Fe'' remaining from the previous allotments.

All the cations except silicon have now been assigned to actual or provisional mineral molecules, and we have next to consider the distribution of silicon.

10*a*. If there is an excess of Si, it is calculated as quartz.

10*b*. If there is a deficiency of Si, minerals of a lower degree of silification have to substitute, in part or wholly, for those minerals that were formed provisionally.

In order to do that correctly, wollastonite, Wo, should be combined with an equal amount of (En + Fs) to form diopside (Di). In diopside and olivine the proportion FeO:MgO should be the same as in the provisionally formed pyroxenes.

10*c*. The necessary amount of (En + Fs) remaining from 10*b* is converted into olivine (Fo and Fa) according to the equation:

$$4\text{En} = 3\text{Fo} + 1\text{Q}$$

10*d*. If there still is not enough Si in the analysis, albite is turned into nepheline according to the equation:

$$5\text{Ab} = 3\text{Ne} + 2\text{Q}$$

10*e*. Finally, if the analysis is very low in Si, orthoclase is in part or wholly converted into leucite:

$$5\text{Or} = 4\text{Lc} + 1\text{Q}$$

10*f*. In rare cases there is not even enough Si to form leucite. Then kaliophilite is formed:

$$4\text{Lc} = 3\text{Kp} + 1\text{Q}$$

The example in Table III-9 will elucidate these calculations.

TABLE III-9

OLIVINE BASALT, GOUGH ISLAND, SOUTH ATLANTIC

	Weight, %	Equivalent Molecular Weight	Cation Proportions × 1000	Cation, %	
SiO_2	49.10	60	818	46.0	Ap = 0.5 + 1.67 × 0.5 = 1.3
TiO_2	3.59	80	45	2.5	Il = 2.5 + 2.5 = 5.0
$AlO_{1\frac{1}{2}}$	16.21	51	317	17.9	Or = 3.3 + 3.3 + 3 × 3.3 = 16.5
$FeO_{1\frac{1}{2}}$	2.87	80	36	2.0	(Ab = 6.4 + 6.4 + 3 × 6.4 = 32.0)
FeO	6.84	72	95	5.3	An = 4.1 + 8.2 + 8.2 = 20.5
MnO	0.05	71	1	0.1	Mt = 2.0 + 1.0 = 3.0
MgO	5.04	40	125	7.0	Wo = 4.1 + 4.1 } Di = 16.4
CaO	8.90	56	159	9.0	Hy = 4.1 + 4.1 }
$NaO_{\frac{1}{2}}$	3.53	31	114	6.4	Ol = 4.8 + 2.5 = 7.3
$KO_{\frac{1}{2}}$	2.76	47	59	3.3	
$PO_{2\frac{1}{2}}$	0.54	71	8	0.5	102.0
H_2O	0.47		....		
	99.90		1777	100.0	

We have now used 2.0% Si too much; some Ab must be converted into Ne:

Ab = 5.4 + 5.4 + 3 × 5.4 = 27.0

Ne = 1.0 + 1.0 + 1.0 = 3.0

Thus the molecular norm is:

Or	16.5	Di	16.4
Ab	27.0	Ol	7.3
An	20.5	Mt	3.0
Ne	3.0	Il	5.0
		Ap	1.3
Σ salic =	67.0	Σ femic =	33.0

Let us consider the relation between the molecular norm and the mode (or the actual mineral composition) of the rock. In a great many natural igneous rocks we find a rather close agreement, but we almost always find a minor discrepancy in the ore minerals. In the norm all ferric iron and all titania are used to form ore minerals (Mt and Il, respectively), whereas some Fe_2O_3 and TiO_2 enter into the natural pyroxene. This again affects the distribution of silica, usually in the way that small amounts of normative Ne are converted into Ab.

These points are brought out by a direct comparison of the normative and modal mineral composition of the olivine basalt from Gough Island, as shown below.

Norm				Mode	
Or	16.5	= Feldspar	67	Feldspar	68
Ab	27.0				
An	20.5				
Ne	3.0				
Di	16.4			Pyroxene	17
Ol	7.3			Olivine	11
Il	5.0	= Ore	8	Ore	3
Mt	3.0				
Ap	1.3			Apatite	1
	100.0				100

The proponents of the norms system used to give the norm in weight percentages. The molecular norm, as given in the example above, may, of course, be calculated in weight percentages; but nothing is gained and much is lost by doing so. The molecular norm differs but slightly from the weight norm. This arises from the fact that the constituent oxides, SiO_2, $\frac{1}{2}Al_2O_3$, CaO, etc., have approximately the same molecular weights.

Calculation of the Standard Cell

In most rocks, oxygen makes up about 94 per cent by volume; all cations taken together (silicon and metals) make up less than 6 per cent by volume. Consequently, the number of oxygen ions is of the utmost importance for the volume relations in rocks. Of secondary importance is the packing and the number of kinds of the constituent cations.

The chemical composition of most rocks is such that on 160 oxygens there are about 100 cations. The *standard cell* is defined as a unit containing 160 oxygens (Barth, 1948).

In petrographic calculations it is often important to compare rocks of equal volume: magmatic mingling, recrystallization, and most replacement processes take place without appreciable change in volume. Since no change in volume is equivalent to no change in the number of oxygens, we may compare standard cells if we want to compare isovolumetric rock units. However, for less accurate calculations it is possible to make the same comparisons and reach the same conclusions directly from the cation percentages. For, as explained above, 100 cations usually require approximately 160 oxygens.

B. THE CRYSTALLIZATION OF SILICATE MELTS

> *One of the principal aims of modern petrologic research is to develop an understanding of the varieties of mineral assemblages that occur in nature. The mineral assemblage that one finds in a particular rock is wholly a function of the bulk chemistry of the rock and of the physical conditions under which it formed. A laboratory worker active in this field of research tries, in his experiments at high temperature and pressure, to duplicate natural mineral assemblages under controlled physical conditions. Having succeeded in this attempt, he is then in a position to interpret natural assemblages in terms of known mineral compatibilities and incompatibilities and to specify, in so far as the observed phase relations make it possible, the physical conditions under which the natural assemblages formed.*
>
> P. H. Abelson, *Carnegie Institute of Washington Year Book,* No. 54, 1955, pp. 114–115.

Igneous minerals have crystallized from lava (or magma). Synthetical studies on silicate melts are, therefore, of great value. No natural magma is as simple as the experimental melts; but the composition of a melt may be so chosen that addition of very small quantities of substances makes it nearly identical to a lava. The general laws of melting and crystallization and the physico-chemical properties of the crystalline phases as studied in the artificial minerals and melts may, therefore, be applied to natural rock and minerals according to the principle of continuity.

Among earlier works on the physico-chemistry of igneous minerals, the papers by J. H. L. Vogt merit special mention. In 1884 he published on the crystallization of slags and emphasized their similarity to igneous rocks. In later papers he systematically applied the laws of physical chemistry to the crystallization and differentiation of igneous rocks and showed that silicate melts obey the laws of solutions. He eventually worked out an elaborate theory to account for most of the common rocks. His general thesis is similar to Bowen's crystal-differentiation theory.

The present and by far the most important stage in the development of modern petrology began with the establishment in 1904 of the Geophysical Laboratory by the Carnegie Institution of Washington, D. C. Practically all data presented in this chapter are extracted from papers of the Geophysical Laboratory. One of the most successful workers in the Laboratory was N. L. Bowen.

TABLE III-10

THE FREQUENCY DISTRIBUTION OF MINERALS IN IGNEOUS ROCKS

(After Barth, *J. Geol.*, 1948)

		Oslo Province	Pacific Lavas *	All Igneous Rocks
Light colored	Glass	+	7.4	+ †
	Quartz	8	1.1	12.4
	Alkali feldspar ‡	62	13	31.0
	Plagioclase	16	30	29.2
	Nepheline	0.3	2.2	0.3
Dark colored	Olivine	0.3	9.5	2.6
	Pyroxene	5	27	12.0
	Hornblende	2	0.1	1.7
	Biotite	2	+	3.8
	Muscovite	0.2	—	1.4
	Ores §	2	7.4	4.1
Accessory and secondary	Apatite	0.5	1.2	0.6
	Sphene	0.5	+	0.3
	Chlorite and serpentine	0.8	0.4	0.6
	Calcite	0.4	+	+
	Inclusive	+ ‖	0.7 ¶	+
		100.0	100.0	100.0

* Lavas from intra-Pacific Islands.

† Glass, an unstable phase in all igneous rocks, has been recalculated into 2 per cent quartz and 5 per cent alkali feldspar.

‡ The composition of alkali feldspar is supposed to be Or:Ab = 1.

§ Magnetite, hematite, ilmenite.

‖ Present in amounts <0.1 per cent: serpentine and epidote; <0.01 per cent: fluorite, sodalite, zircon, glass, zeolite, astrophyllite, elpidite.

¶ Present in amounts <0.1 per cent: biotite, tridymite, cristobalite, epidote, prehnite, calcite, melilite, sodalite, noselite, hauyne, sphene, zircon, kaolinite, sulfur, and sulfates.

Before the properties of the minerals can be adequately understood it is necessary to define and explain some of the fundamental conceptions of physical chemistry. The chapter on thermodynamics (pages 381 ff.) gives a brief treatment of these conceptions and should be consulted for equations, accurate definitions, and for all questions not fully explained in the following text.

There are but a small number of igneous rock-forming minerals (see Table III-10): quartz, feldspar, and feldspathoids are light colored; mica, hornblende, pyroxene, and olivine are dark-colored minerals.* With the limitation placed on the number of coexisting minerals by the phase rule (see page 396) it is not difficult to survey the principal types of igneous rocks.

1 · SILICA (SiO_2)

Silica is polymorphous and occurs in a great number of different modifications, some of which are unstable under all conditions. See Fig. III-5.

The silica minerals are quartz, tridymite, cristobalite, and coesite. According to the phase rule, however, only one of these is stable at any one arbitrarily chosen temperature-pressure combination.

All modifications of silica form crystal structures that are known as tectosilicates (see Fig. III-4), in which each of the four oxygens surrounding the silicon ion is held in common by similar adjacent groups conforming with the two to one ratio of oxygen to silicon. The tetrahedra thus linked together by shared oxygen ions form endless, crossing chains parallel to six directions in space. Between the chains are large interstices in which various foreign ions may be accommodated; the mineral thus becomes "stuffed," as frequently noticed in tridymite and cristobalite, and as is also found in quartz.

Figure III-5 gives a graphical survey of the polymorphic phenomena in silica at low pressure. In pure silica, quartz is the only stable modification below 867°C, and, if the pressure is increased, it will be stable at much higher temperatures (see Fig. III-6). Nevertheless, both tridymite and cristobalite are commonly encountered as products of pneumatolytic action at much lower temperatures in lavas or other fumed rocks.

The explanation is that the thermochemical properties of silica are changed by the "stuffing." The transformation temperatures in a

* In some calcium-rich rocks *melilite* may become a prominent constituent. Melilites are mixed crystals essentially of $Ca_2Al_2SiO_7$ (gehlenite) and $Ca_2MgSi_2O_7$ (akermanite). But they are scarce or absent in most rock series, so no discussion of them will be given here. For information on their crystallization see Osborn and Schairer (1941).

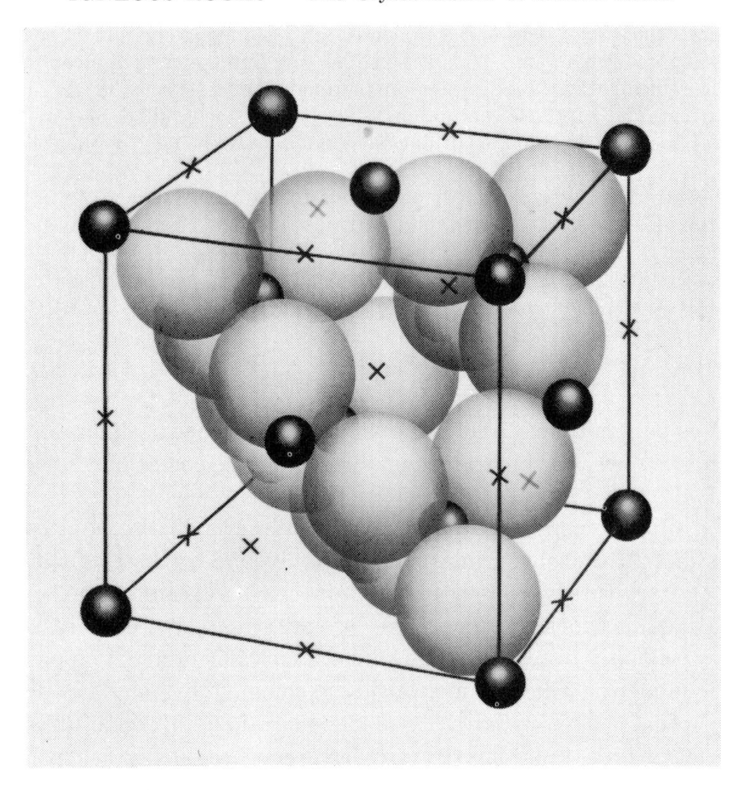

Fig. III-4. The unit of the cristobalite structure. The large transparent balls represent oxygen ions, and the small black ones silicon ions, bringing out their relative size. Crosses indicate the centers of the large interstices into which foreign ions may be stuffed.

stuffed crystal are different from those of the pure compound. It is possible, therefore, that tridymite and cristobalite in rocks are or may have been stable, impure silica. Indeed, evidence has been presented in support of the view that tridymite is always "stuffed" and never represents pure silica, but actually is a compound of SiO_2 with about 0.5 per cent of foreign oxides. Pure quartz is on this view stable up to 1025°C, where it inverts directly to cristobalite (Holmquist, 1961).

Obviously the phase rule prohibits the occurrence of tridymite and quartz in the same rock if tridymite is to be regarded as the pure silica phase, but as a contaminated phase it is permitted. The same holds true for cristobalite.

Properties of the Several Modifications

(1) *Quartz* is stable at ordinary temperatures; the symmetry is trigonal trapezohedral. On heating it changes at 573°C into (2) another modification with hexagonal trapezohedral symmetry. The inversion is spontaneous and accompanied by an abrupt change in directly observable properties: volume, refractive index, optical activity, etc. (displacive transformation). However, the changes are not great, and the name quartz is used for both modifications (distinguished as low quartz and high quartz.* On cooling, high quartz spontaneously converts into low quartz at 573°C. At temperatures above 870° quartz is no longer stable, but inverts into tridymite. (3) *Tridymite* also exhibits hexagonal symmetry; the lattice geometry is, nevertheless, very different from that of quartz. The inversion takes a long time and probably requires a catalyst to go at all; it is accompanied by a complete destruction of the quartz lattice and a slow growth of the new tridymite lattice (reconstructive transformation); it is stable in the range 870° to 1470°C. (4) *Cristobalite* is cubic and stable from 1470° to 1713°C, which is the melting point. (5) *The melt* is the stable phase above 1713°. (6) *Coesite* is a sixth modification of silica, it is monoclinic, pseudo-hexagonal and stable only at high pressure, see Fig. III-6. It has a density of 3.01 g per cm^3, which is 14 per cent higher than that of quartz, in harmony with the fact that it requires high pressures to form. (7) *Stishovite* is a very high pressure modification (above 120,000 atm) of density 4.3 g per cm^3 recently found at Meteor Crater, Arizona (v.i.).

It is worthy of note that the inversions high quartz $\rightarrow$ tridymite $\rightarrow$ cristobalite, and vice versa, are sluggish and so retarded that both tridymite and cristobalite by simple cooling can be brought down to ordinary temperatures without inverting into the stable phase (low quartz). Once cooled, both tridymite and cristobalite will, at low temperatures, where the reactions are slow, remain as metastable phases for geological periods. The same holds for the melt which undercools and remains as a glass at ordinary temperatures. The undercooled tridymite and cristobalite suffer certain changes, analogous to the change of high quartz–low quartz, but the variants are of little petrographical interest. The five modifications with real fields of stability are important, however, and may serve as geological thermometers:

* The designations β-quartz and α-quartz, respectively, are also used for low quartz and high quartz. However, because some authors use the Greek letters in the opposite way, the expressions are ambiguous.

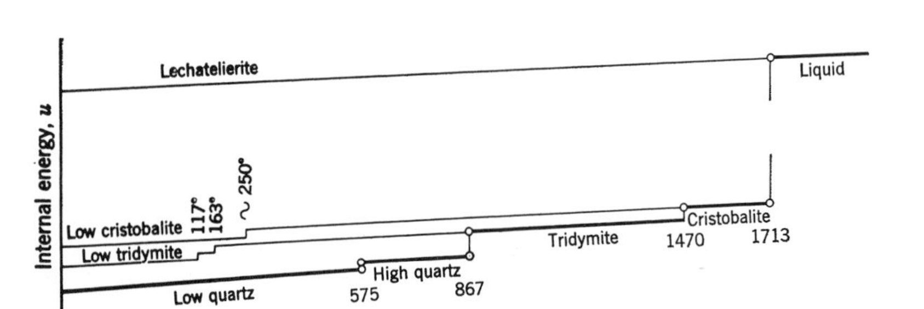

FIG. III-5. Stable and unstable forms of silica produced at room pressure. At any one temperature the form with the lowest internal energy is stable. The transformation from one form to another with increasing temperature takes place with a discontinuous rise of the internal energy. Liquid silica, cristobalite, and tridymite easily undercool and remain as metastable forms for indefinite periods (they may be stabilized by stuffing with various foreign ions, see text). At lower temperature they undergo so-called α-β inversions, but always remain as metastable forms. The diagram shows that below 867°C (and ordinary pressure) quartz is the only stable form of pure silica.

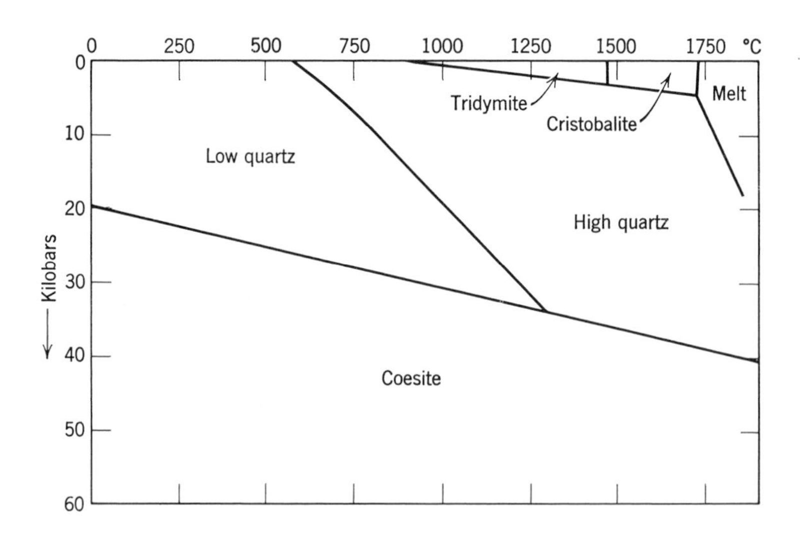

FIG. III-6. The fields of the stable modifications of SiO_2 (Boyd and England, 1960).

(1), (2) High quartz and low quartz are formed only within their respective fields of stability.

(3), (4) At lower temperatures, particularly during rapid crystallization (for instance in the presence of volatiles), both tridymite and cristobalite may form, although quartz represents the stable phase.

(5) Pure silica glass, known as the mineral lechatelierite, always forms above 1713°C. But small amounts of impurities lowers the melting point very considerably.

(6) Coesite forms at high pressure. It is of more than ordinary interest that it has been identified as an abundant mineral in sheared sandstone at Meteor Crater, Arizona (also at Canyon Diablo, Ries, S. Germany, etc.). It was obviously formed by impact of the meteor under pressures in excess of 20 kilobars. The additional presence of silica glass indicates temperatures higher than 1000°C.

(8) *Keatite* is the name given to an eighth modification of silica which can be made artificially at high vapor pressures. It is unstable at all temperatures and pressures; its density is 2.50 g per cm^3, and it has the tetragonal structure of spodumene, $LiAlSi_2O_6$.

"Silica O" has a structure similar to high quartz, and crystallizes sometimes from silica gel or other amorphous forms of silica. It is, however, a "stuffed" form of quartz; the stuffing prevents the instantaneous inversion to low quartz, but eventually it will invert into low quartz at low temperature, just as will "silica K" (keatite).

Application of the High-Low Inversion of Quartz in Petrology. Primary quartz in a lava indicates a temperature of crystallization below 870°C. But the presence of tridymite or cristobalite does not necessarily indicate anything as to the temperature conditions. Although quartz, as we now see it in a rock, is always low quartz, microscopical studies of its crystal habit and twin laws suggest that, when present in a lava, it represents pseudomorphs after high quartz or, rarely, pseudomorphs after tridymite. In pegmatites and mineral veins, however, low quartz is often encountered as a primary phase. We conclude, therefore, that the main crystallization of lavas took place in the temperature range 573° up to and beyond 870°C, whereas the formation of pegmatites and veins took place, at least partly, below 573°C.

The temperature data need, however, a modification for high pressures. Clapeyron's equation, which should be well-known and thoroughly understood by all petrologists, describes the relation between the pressure and a temperature of transition when two modifications of the same compound coexist in equilibrium. (See page 387.)

$$\frac{dT}{dt} = \frac{T}{\Delta H}(V - V') \qquad \text{(III–1)}$$

where T = absolute temperature, p = pressure, ΔH = the change of heat of the reaction, V = specific volume of the high-temperature phase, and V' = specific volume of the low-temperature phase.

As an example we shall compute the variation with pressure of the inversion temperature quartz $\rightleftharpoons$ tridymite (see Table III–11). We have

TABLE III-11

SPECIFIC WEIGHT OF DIFFERENT MODIFICATIONS OF SILICA

°C	°K	Quartz	Tridymite	Cristobalite	Glass	Coesite
0°	273	2.651	2.262	2.320	2.203	3.01
867°	1140	2.536	2.189	2.200	2.200	...

$T = 1140°K$, $V = 1/2.189 = 0.457\ cm^3/g$, $V' = 1/2.536 = 0.395\ cm^3/g$, and $V - V' = 0.062\ cm^3/g$. ΔH is approximately 2 cal/g $= 2 \times 42.7$ kg cm/g, $dT/dp = 1140 \times 0.062/854 = 0.83$ degrees/kg/cm^2, or 0.8 degrees per atmosphere, that is, the inversion point changes very rapidly with pressure. (See Fig. III-6.) For the inversion low quartz $\rightleftharpoons$ high quartz, which is a very important point in geological thermometry, the computations give $dT/dp = 0.0215$ degree per atmosphere.

These values indicate the change in the inversion temperature at one atmosphere. At higher pressures the inversion temperature is found by extrapolation. In order to place it accurately it must be checked by experiments.

2 · EUTECTIC MELTING

From the foregoing discussion and from Fig. III-5 the student should not be so confused as to think that quartz and tridymite never can crystallize directly from a melt. Although it is true that the crystallization temperature of a pure SiO_2 melt is above the stability range of quartz and tridymite, there are other melts from which these minerals separate as primary phases.

The crystallization temperature of any melt will drop if a foreign substance is dissolved in the melt and no mixed crystals are formed. In the following approximate equation

$$\Delta t = \frac{RT^2}{l} \cdot x \qquad \text{(III–2)}$$

Δt is the depression of the absolute melting temperature T, R is the gas constant, l is the heat of melting, and x is the mole fraction of the added substance. Thus, if we have two pure substances, A and B, each of them will depress the melting point of the other until the maximum depression is reached in the so-called eutectic mixture somewhere between A and B.

A good example is afforded by the system anorthite ($CaAl_2Si_2O_8$)–diopside ($CaMgSi_2O_6$). See Fig. III-7. The approximate heat of melting of anorthite and of diopside can be computed from equation III-2. Equation III-2 applies to the initial slope of the melting curve as given by the dashed lines in Fig. III-7. As diopside is added to the anorthite melt, the melting (freezing) point decreases, and the extrapolated temperature for 100 per cent addition of diopside is at 1290°C;

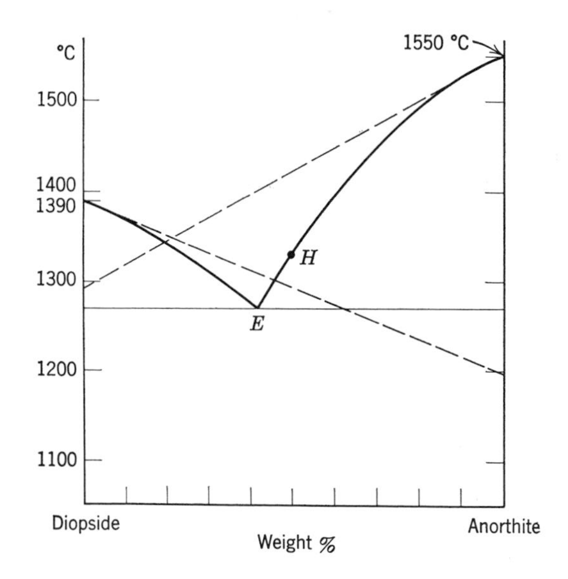

Fig. III-7. Equilibrium diagram of the system anorthite-diopside after Bowen (1951). Anorthite melts at 1550°C, diopside at 1390°C. Each lowers the melting point of the other. The curves of melting meet at 1270° and 42 per cent anorthite and 58 per cent diopside, which is therefore the eutectic point, E. A melt of anorthite and diopside in equal proportions begins to crystallize at 1330° (H) with the separation of anorthite. As the temperature falls, anorthite continues to separate, and the melt changes in composition along the curve HE, i.e., becomes richer in diopside. At 1270°, when the melt has the composition E, diopside joins, such that diopside and anorthite now separate in constant proportions, the melt remaining of constant composition and the temperature remaining at 1270° until the whole mass has crystallized. The dashed lines refer to the initial slope of the melting curves.

thus for $x = 1$, $\Delta t = 260°$. The melting point of anorthite is $T = 1823°K$, the gas constant can with sufficient accuracy be taken as $R = 2$ cal per mole. Thus $l = 2 \times 1823^2/260 = 25{,}000$ cal per mole anorthite. For diopside the corresponding values are $\Delta t = 200°$, $T = 1664$ and $l = 28{,}000$ cal per mole diopside. The actual values are 29,000 and 23,000 calories respectively. By using refined calculation methods, and by taking into consideration the heat of mixing of the two melts, more accurate values may be computed. By this refined method, Bowen has calculated the heat of melting of various silicates; his values for anorthite and albite are still regarded as the most accurate figures. He also has computed the heat of mixing of anorthite and diopside by determining the deviation of the actual from the ideal melting curves.

From the phase diagrams such as Figs. III-10 to III-27 accurate and quantitative relations can be extracted. Valuable information is thus obtained about the melting phenomena and about the physical properties of the solid and liquid phases.

3 · FELDSPARS

Among the rock-making minerals the feldspars occupy a special position, in that they make up more than 50 per cent of all igneous rocks. The physical chemistry of the feldspars, their interaction, and their reactions with other minerals are, therefore, of the utmost importance in petrology.

All feldspars are built over the general formula WZ_4O_8, where W = K, Na, Ca, Sr, Ba, and Z = Si, Al, Fe. Of petrological importance, however, are only the three molecules: Orthoclase ($KAlSi_3O_8$), Albite ($NaAlSi_3O_8$), and Anorthite ($CaAl_2Si_2O_8$). Albite-anorthite form a complete series of mixed crystals (the plagioclase series). Orthoclase and albite are completely miscible only at elevated temperatures; they form the series of the alkali feldspars. Orthoclase-anorthite are hardly miscible at any temperature. See Fig. III-8.

Plagioclase or Soda-Lime Feldspar

Any mixed crystal between albite and anorthite is called plagioclase. All plagioclases as well as the two end components are triclinic at room temperature. But above 1060°C a monoclinic albite becomes stable.

In systems of mixed crystals the law of the depression of the melting point is not valid, and equation III-2 cannot be used. The melting relations in a binary system are explained below, and applied to a typical example, the plagioclases. The melting point (liquidus

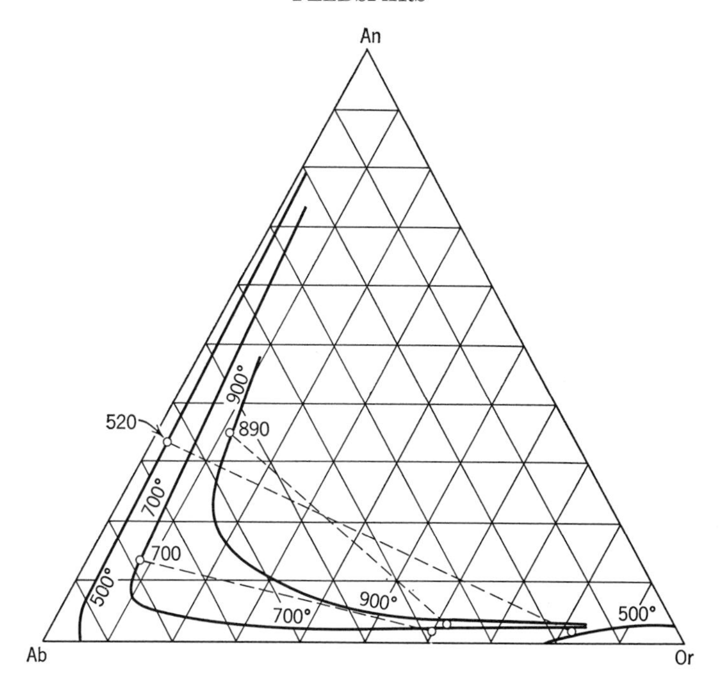

FIG. III-8. Fields of solid solubility in feldspars at various temperatures (Barth 1962).

point) of a mixture always lies between the melting points of the components, but there is no linear relation, not even in the theoretical case of ideal osmotic miscibility of the components.

The melting relations of the plagioclases are diagrammatically illustrated in Fig. III-9. The upper curve is called liquidus; the lower, solidus. Above the liquidus everything is molten, below the solidus everything is solid; between the curves, melt and crystal coexist. Pure albite melts at 1118°C, pure anorthite at 1553°C. The diagram as a whole shows the exact values of the melting points of the two feldspars and the exact melting interval of all mixtures of them. Thus we read from it that the mixture containing 50 per cent of each begins to melt at 1287° (point f), and the melting is complete at 1450° (point a); or, if we are cooling the mixture from a high temperature, such as that represented by the point x, it begins to crystallize at 1450°, and its crystallization is complete at 1287°. In addition the diagram shows us the composition of the crystals that are in equilibrium with any liquid. Thus the crystals that are in equilibrium with the liquid a have the composition b, the crystals that are in equilibrium with the

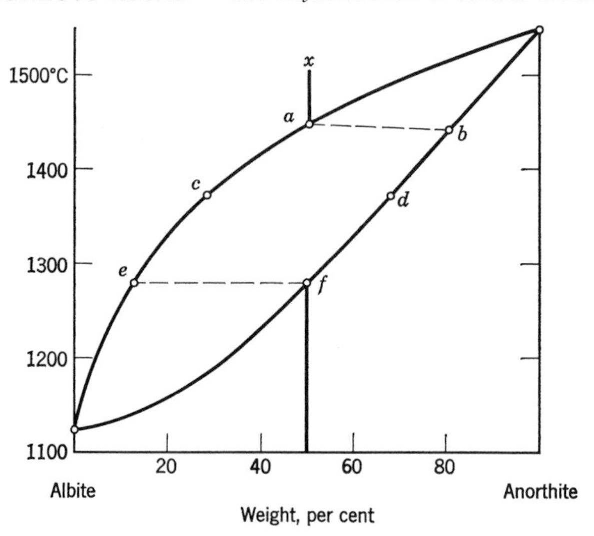

FIG. III-9. Equilibrium diagram of the system albite-anorthite. (After N. L. Bowen, *Am. J. Sci.* **33,** 1912.)

liquid *c* have the composition *d*, and so on for other compositions and temperatures. It will be noted that the crystals in equilibrium with any liquid, that is, the crystals that will form from that liquid, are always richer in lime feldspar, and their subtraction will, therefore, always cause the liquid to be enriched in soda feldspar. The composition of the liquid, therefore, moves to the left along the curve *a–c* as crystallization proceeds with falling temperature.

We may now distinguish between two extreme cases.

(1) The crystals remain suspended in the liquid, and cooling is slow enough always to allow complete reaction between crystals and liquid. The early formed crystals will, on cooling, continuously react with the liquid and thereby gradually change their composition along the solidus curve from *b* to *f*, while simultaneously the liquid changes from *a* to *e*. In such circumstances the crystals will not move beyond *f* (which is the same composition as the original liquid *x*); the last vanishing amount of liquid in equilibrium with *f* has the composition *e*. The end product of the cooling is a homogeneous mixed crystal of the same composition as the initial liquid.

(2) The crystals are continuously removed from the liquid. Reaction is thus prevented, and the composition of the liquid will continue to migrate down the liquidus curve, that is, towards soda feldspar; and there is no limit to this change of composition except the

pure soda feldspar itself. The end product of the congealing is a mixture of crystals varying in composition from *b* to pure albite. This kind of crystallization is called fractional crystallization. Obviously fractional crystallization may produce drastic effects in these feldspar mixtures. If we had a liquid mass of such character in nature, and if during its crystallization the crystals sank under the action of gravity, or were squeezed out by orogenic movements and "filter pressing," then in the part of the mass into which they moved there would be strong enrichment in lime feldspar, and the residual liquid from which they separated could be enriched in soda feldspar to the near exclusion of lime feldspar. This is the type of relation that is often seen in the members of an igneous-rock grouping in nature, in so far as their plagioclase constituents are concerned.

Zoning in plagioclases is also explained. If high viscosity or fast cooling prevents the early formed crystals from reacting with the liquid, the later crystals, which are more sodic, will be precipitated in a shell around the early crystals. As crystallization proceeds, the shell grows, successively and continuously becoming more sodic, the result being a zoned feldspar with calcic core surrounded by a series of shells of increasing sodium content towards the circumference. Most plagioclases found in lavas exhibit this kind of structure. (See also Fig. III-45.) But plagioclases of gneisses and of crystalline schists, if zoned, usually show inverse order, that is, sodic core, calcic shell. This type of zoning is probably induced by increasing metamorphic temperature.

Potash Feldspar

The composition $KAlSi_3O_8$ is frequently referred to as the orthoclase molecule (abbreviated Or). With a somewhat different meaning, orthoclase is used as a name for the common monoclinic feldspar. Four closely related minerals have the composition of potash feldspar (Table III-12). The structural relations of these four minerals are of the order-disorder type. By prolonged heat treatment all potash feldspars gradually convert into sanidine.

The melting phenomena of potash feldspar (see Fig. III-10) involve reactions between crystal and liquid of a special kind known as incongruent melting. When crystals of $KAlSi_3O_8$ are heated, melting begins at 1150°C with separation of crystals of $KAlSi_2O_6$ (leucite). The material becomes completely liquid only at 1530°C. Conversely, on cooling leucite crystals separate at this latter temperature and continue to separate until the temperature of incongruent melting

TABLE III-12

THE PRINCIPAL CRYSTALLINE FORMS OF POTASH FELDSPAR

Name	Sanidine	Orthoclase	Microcline	Adularia
Symmetry	monoclinic	monoclinic	triclinic	triclinic and pseudo-monoclinic
Axial angle	30°‖(010) → 40° ⊥ (010)		80° ⊥ (010)	variable
Temperature of stability	high	medium	low	unstable

(1150°C) is reached. At this point they react completely with the liquid of composition leucite 57.8 per cent, silica 42.2 per cent (or potash feldspar 73.7 per cent, silica 26.3 per cent) to form $KAlSi_3O_8$. During this reaction the temperature remains constant at 1150°C; the heat subtracted from the system is balanced by the heat of re-

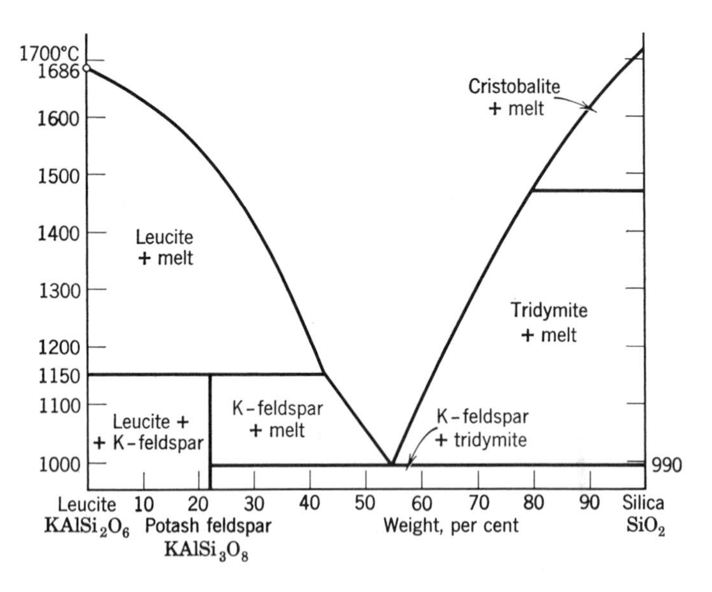

FIG. III-10. Equilibrium diagram of the binary system leucite-silica. In this system the melting phenomena of K-feldspar are illustrated. (After Schairer and Bowen, 1947.)

action. Thus we see that the liquid of composition $KAlSi_3O_8$, with perfect equilibrium, follows an involved path of crystallization, but yields only crystals of $KAlSi_3O_8$. An adequate presentation of these melting relations can be given only in a binary diagram leucite-silica.

Alkali Feldspars

Mixed crystals in the system Or-Ab represent an important mineral series in nature. At high temperatures the series is continuous; but, on cooling, exsolution sets in, that is, the homogeneous mixed crystal separates into two solid phases, one rich in potash feldspar, the other in soda feldspar. The two phases form an intergrowth of analogously oriented lamellae known as perthite. The exsolution starts at about 660°C. The exsolution area, outlined by the *solvus* curve, is shown schematically in Fig. III-11. Obviously alkali feldspars of medium composition can form only above ca. 660°C. At lower temperatures two feldspar phases will form, the composition of each phase being a function of the temperature of formation. (See Fig. IV-10.)

Because of the incongruous melting of Or the melting phenomena of the alkali feldspars cannot be accurately represented by a binary

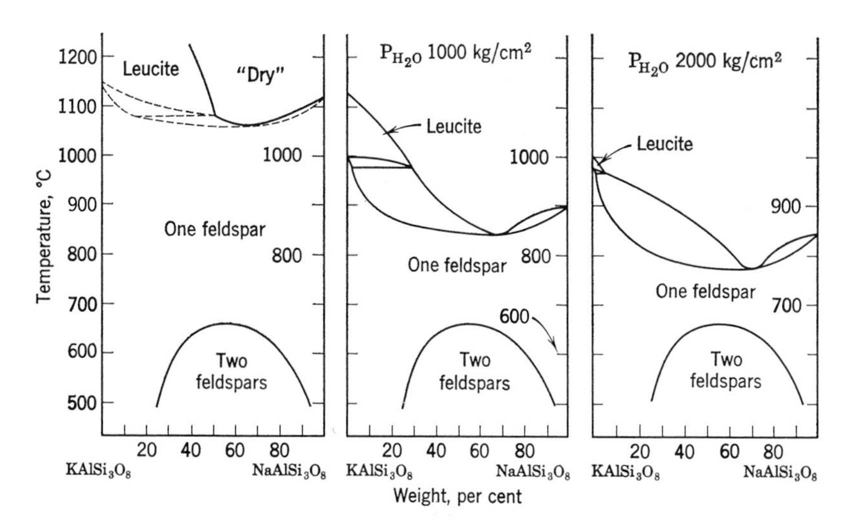

FIG. III-11. The system of the alkali feldspars. Melting relations (after Bowen and Tuttle, 1950) in dry melts, at 1000 kg per cm², and at 2000 kg per cm² pressure of H_2O. A temperature minimum occurs on the liquidus curve. In the lower part of the diagrams the exsolution area is schematically indicated. If a lava, say a trachyte, congeals at higher temperatures, alkali feldspars of any composition may form. But at lower temperatures, no feldspar of intermediate composition exists. The path of crystallization of the alkali feldspars will differ, therefore, according to the temperature of the liquid from which they crystallize.

diagram. However, under high water pressure, water will dissolve in the melt, depress the melting points and eliminate the leucite field.

Under these conditions the Ab-Or system can be treated as truly binary. There is a minimum on the liquidus curve at about 60Ab, 40Or as shown in Fig. III-11, toward which all residual liquids will move. At this composition the residual liquid will freeze to a crystal of the same composition. On each side of the minimum the mechanism of crystallization is analogous to that of the plagioclases.

The System Or-Ab-An

The composition of the rock-making feldspars can be expressed in terms of the mineral molecules $KAlSi_3O_8$, $NaAlSi_3O_8$, and $CaAl_2Si_2O_8$. But the corresponding physico-chemical system is not a true ternary system under ordinary pressure; because of the incongruent melting of Or with separation of leucite it is actually quaternary.

In Fig. III-12 a triangular diagram has been used to show the equilibrium relations in the quaternary system Or-Ab-An; the presentation is therefore by necessity inadequate, but nevertheless good enough for an approximate survey of the conditions.

At the Or-corner leucite is the primary phase. The liquidus point is at about 1530°C, and the lowest temperature on the join Or-An is approximately 1350°. (*A* in Fig. III-12; this is not a true eutectic, however—these parts of the diagram are actually being governed by quaternary equilibria.) Only in a very small field is orthoclase solid

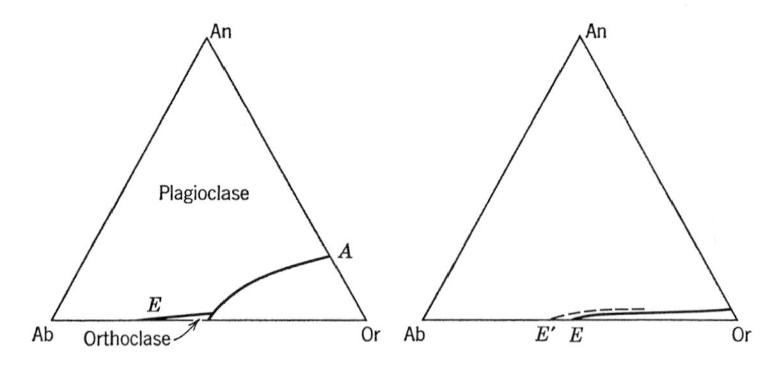

FIG. III-12. *Left:* Schematic equilibrium diagram of the system orthoclase-albite-anorthite, being part of the quaternary system silica-leucite-albite-anorthite. *Right: Full line* illustrates schematically the equilibrium conditions on a section with 30 per cent silica in the system silica-orthoclase-albite-anorthite. *Dashed line* refers to the equilibrium diagram of the truly ternary system orthoclase-albite-anorthite, at some 2600 atm.

solution a primary phase. In the whole remaining area of the diagram, plagioclase is the primary phase (including here soda-rich anorthoclase, which physico-chemically represents the same phase). The lowest point on the fusion surface towards which all residual liquids will trend is at 35Or, 65Ab, with a temperature of 1063° (E in Fig. III-12, left).

In natural magma these relations will be somewhat modified, partly by the presence of other substances, partly by increased pressures. Substances of modifying influence are, for instance, water and other mineralizers. The effect of excess silica has been shown to involve interesting consequences in igneous geology. Addition of silica causes the leucite field to shrink until it eventually disappears at about 30 per cent excess silica. The equilibrium conditions in a feldspar melt with 30 per cent excess silica are schematically shown to the right in Fig. III-12. This system, of course, is not ternary, and the triangle diagram cannot, therefore, give full information on the various equilibria. Schematically it shows, however, that no leucite field is present, and that the plagioclase field has encroached still further on the domain of Or-solid solutions, which now is restricted to a narrow area in the lower right-hand part of the diagram. The lowest point on the fusion surface is now at 48Or, 52Ab (+30 per cent excess SiO_2), where the temperature is around 1000° (point E, Fig. III-12, right).

The mode of occurrence of leucite and its small density had long been taken as indicative of a reduced stability field at higher pressure. Later Goranson demonstrated experimentally that at a pressure of some 2600 atm water will dissolve in orthoclase liquid to the extent of about 6 per cent; and from this liquid, orthoclase crystallizes directly. In other words, the incongruent relation is destroyed. The consequence is that at such pressures the system Or-Ab-An becomes truly ternary. The dashed line in Fig. III-12 indicates the approximate position of the boundary between the plagioclases and the orthoclase solid solution at pressures above 2600 atm.

In certain respects the equilibrium conditions are rather similar to those obtained by ordinary pressure with 30 per cent excess silica. Again the lowest point on the fusion surface is on the Or-Ab join (at approximately 38Or, 62Ab, that is, at E' in Fig. III-12), and the field of the orthoclase solid solutions is reduced to a narrow strip in the lower right-hand part of the diagram.

Depending upon the chemical composition, a magma may correspond to a point in the plagioclase field or in the orthoclase field of the present system. Accordingly, either plagioclase or orthoclase will crystallize first, thus in any case pushing the residual liquid towards

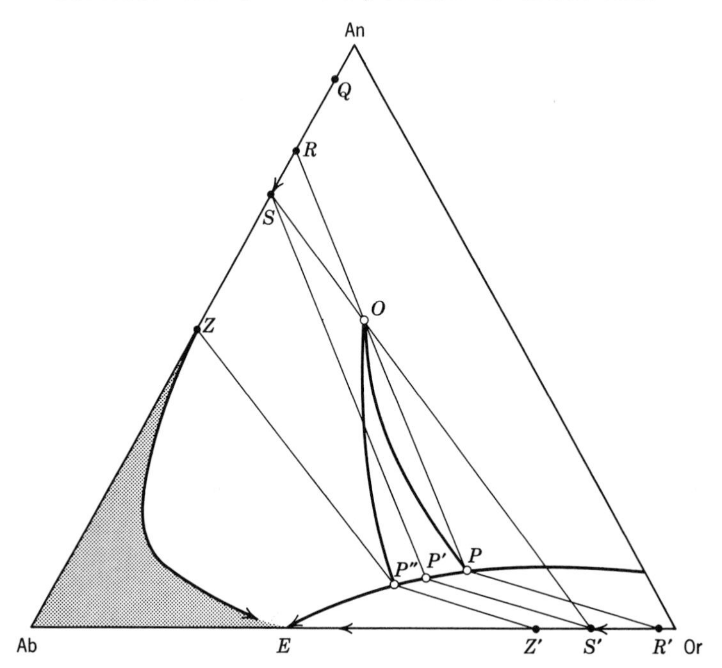

FIG. III-13. Schematic illustration of crystallization of feldspars from a deep-seated magma. (1) Equilibrium is maintained during the crystallization: From the initial liquid *O*, a plagioclase of composition *Q* begins to separate. In the course of crystallization the composition of the crystals changes from *Q* to *R*, while the melt migrates from *O* to *P* in such a way that *O* always remains on the tie-line between the crystal and the coexisting melt at any one time. At *P* the boundary curve is reached, and a simultaneous crystallization of plagioclase *R* and alkali feldspar *R′* begins. Thereby the liquid is pushed from *P* to *P′*, the crystalline phases simultaneously changing from *R* to *S* in the plagioclase series, and from *R′* to *S′* in the alkali feldspar series. The last vanishing amount of liquid will be at *P′*, and the now completely solid material will be a mixture of *S* and *S′* in the proportion (*O*—*S′*):(*O*—*S*). (2) By fractional crystallization the early formed crystals are prevented from reacting completely with the liquid: Again the first plagioclase to crystallize has the composition *Q*. As the crystallizing material changes from *Q* to *Z*, *Q* and other early crystals will not be resorbed; the melt therefore migrates from *O* to *P″* in such a way that the tie-line between crystal and melt at any one time is the tangent to *O*—*P″*. At *P″* an alkali feldspar of composition *Z′* joins, thus the liquid will be pushed to point *E* where it congeals to a crystal of the same composition. The completely solid material will now be a mixture of plagioclases ranging from *Q* over *Z* to *E*, and of alkali feldspars ranging from *Z′* to *E*. In the Ab-corner there are ternary solid solutions that complicate the path of crystallization, but do not affect the principle that by fractional crystallization point *E* is always reached regardless of the initial position of point *O*.

the boundary line. When the boundary line is reached, a simultaneous crystallization of plagioclase and orthoclase solid solution begins, and thenceforth, by strong fractionation, continues until the lowest point on the fusion surface is reached. The position of this point is always on the Or-Ab join (no An present). Thus all natural rest magmas should be enriched in an alkali feldspar containing about 40 per cent of the orthoclase molecule. The detailed crystallization history is explained in Fig. III-13.

4 · FELDSPATHOIDS AND SILICA

The feldspathoids will crystallize when there is not enough silica to make the feldspars. The most important feldspathoid is *nepheline* ($NaAlSiO_4$). Note that, although the compound $NaAlSiO_4$ can be made artificially and may be called conveniently the *nepheline molecule*, most natural nephelines are more complicated.

Consider the formulas:

$NaNaAl_2Si_2O_8$	nepheline
$Si_2\,Si_2O_8$	tridymite
$Ca\,Al_2Si_2O_8$	anorthite
$K\,K\,Al_2Si_2O_8$	kaliophilite

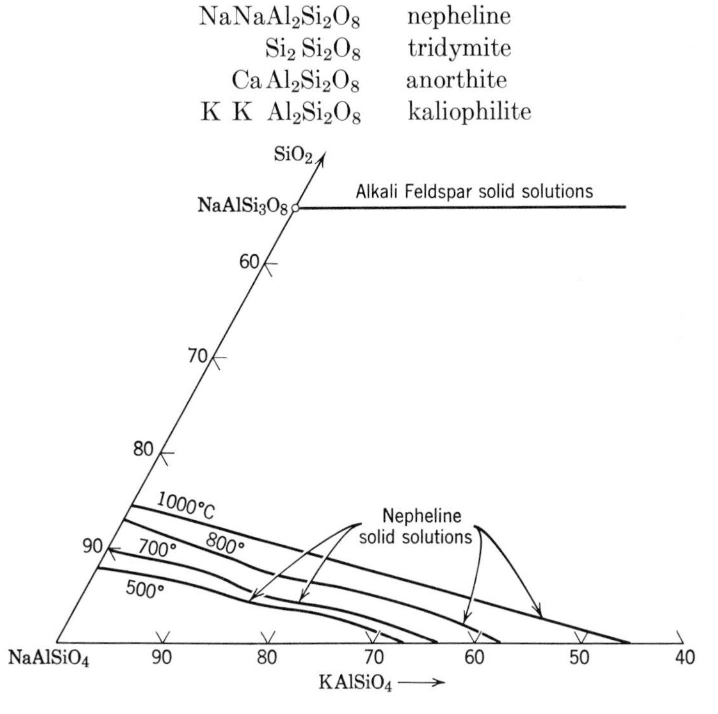

Fig. III-14. Diagram of part of the system $NaAlSiO_4$—$KAlSiO_4$—SiO_2 showing nepheline solid solutions at 500, 700, 800, and 1000°C. (D. L. Hamilton, 1961.)

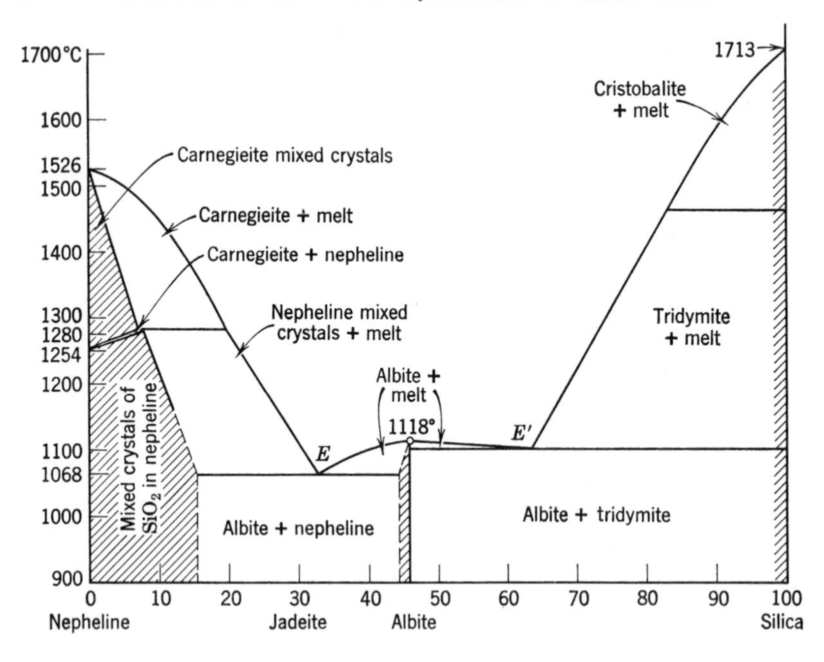

Fig. III-15. Equilibrium diagram of the system nepheline-silica (after Greig and Barth, 1938). Albite ($NaAlSi_3O_8$) is a binary compound dividing the system into two parts: (1) the albite-silica system, which is of the simple eutectic type. (2) the albite-nepheline system which is also eutectic. But both carnegieite and nepheline are able to form mixed crystals with silica (partial solid solution between $NaAlSiO_4$ and SiO_2, illustrated by the shaded fields in the diagram).

The $(Al_2Si_2)O_8$-part of the nepheline formula corresponds to the structure of tridymite except Al substitutes for Si in the proportion 1:1, exhibiting endless Al-Si-oxygen chains between which are open cavities into which Na-ions enter in order to balance the valences (see Fig. III-4). But if the proportion Al:Si < 1, less Na is required to balance the valences, and some of the cavities remain open. If Ca, which is bivalent, substitutes for Na, only half the number is necessary to saturate the valences and one set of cavities remains open. If K enters it could fill all cavities, but since only one cavity out of four is large enough for K there is a tendency for natural nepheline to approach the composition $KNa_3Al_4Si_4O_{16}$, but with some Ca and some open holes present.

The degree of solid solution is determined by the temperature (see

Fig. III-14), therefore the composition of natural nepheline can be used as a geological thermometer. (*Nepheline* inverts to carnegieite at 1260°C, but this is of no importance to petrology.)

Leucite occurs in certain lavas rich in K_2O, and it is also formed by the incongruent melting of potassium feldspar (see Fig. III-10). It has an inversion point at 603°C, so leucite, too, can be used as a geological thermometer: Above this temperature it is isometric, and shows complex twinning easily seen with the microscope that develops upon inversion; and it appears that natural leucite has always separated from a lava above this temperature.

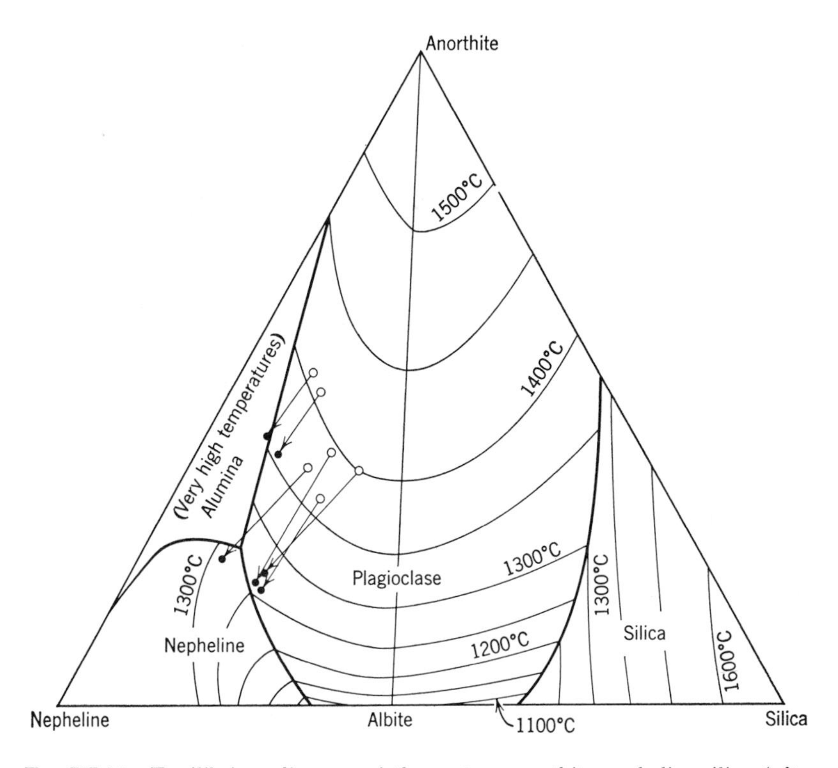

Fig. III-16. Equilibrium diagram of the system anorthite-nepheline-silica (after Schairer, 1957). Open circles representing the positions of glass-bearing lavas of the Pribilof Islands are connected with corresponding glasses by arrows. The trend of differentiation is in all rocks "down" the fusion surface, toward the cotectic curve separating the field of the feldspars from the field of the nephelines.

Of minor importance, petrographically, are *kaliophilite* and *kalsilite*, two polymorphous forms of $KAlSiO_4$.

The sodalite family minerals: *sodalite, noselite,* and *hauyne* are feldspathoids with foreign anions in their lattice. They are cubic, and

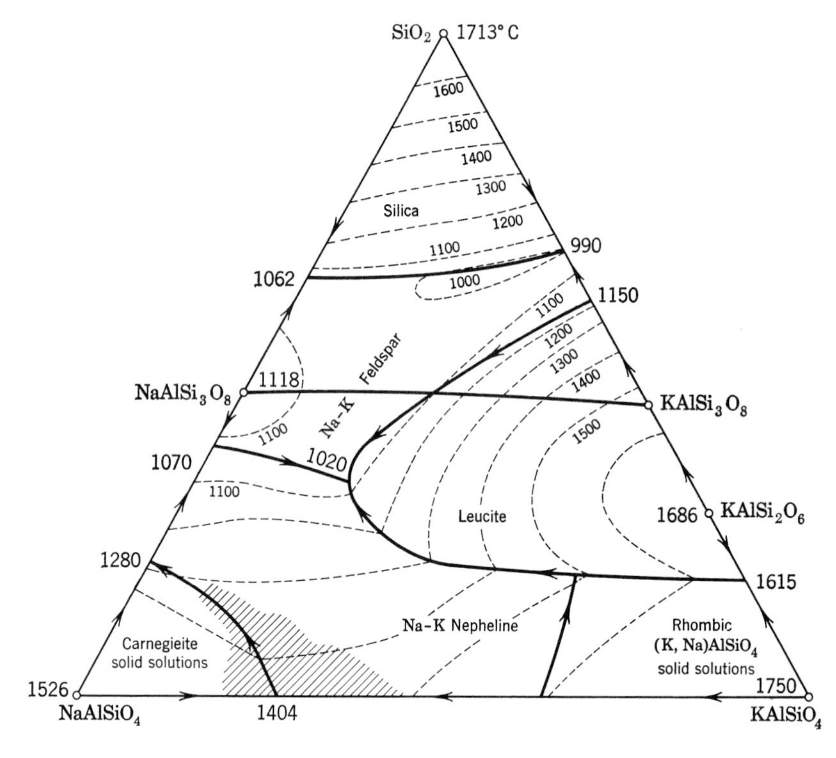

Fig. III-17. The ternary system $NaAlSiO_4$—$KAlSiO_4$—SiO_2 (after Schairer, 1950). The arrows indicate the directions of falling temperature. The primary phases are indicated, and their corresponding fields of crystallization are outlined in the diagram. Natural nephelines are not pure $NaAlSiO_4$, but include both SiO_2 and $KAlSiO_4$ in their composition.

the general formula is $(Na,Ca)_8Al_6Si_6O_{24}(Cl,SO_4)_{1-2}$. The hexagonal *cancrinite* with CO_3 as foreign anion is closely related. It occurs in rocks rich in volatiles. The cubic mineral *analcite* $NaAlSi_2O_6 \cdot H_2O$, has a similar mode of occurrence (in teschenites, etc.). It is characteristic of the feldspathoids that they are incompatible with quartz. They react according to the scheme:

$$KAlSi_2O_6 + SiO_2 = KAlSi_3O_8 \text{ (see Fig. III-10)}$$

and

$$NaAlSiO_4 + 2SiO_2 = NaAlSi_3O_8 \text{ (see Fig. III-15)}$$

The binary system nepheline-silica ($NaAlSiO_4$–SiO_2) falls into two subsystems: Nepheline-albite and albite-silica, both showing simple eutectic relations (see Fig. III-15).

The ternary system anorthite-nepheline-silica ($CaAl_2Si_2O_8$—$NaAlSiO_4$—SiO_2) sheds light on the fractional crystallization of feldspathic minerals in certain lavas poor in potash, for example, basanites from the Pribilof Islands, see Fig. III-16.

The ternary system $NaAlSiO_4$—$KAlSiO_4$—SiO_2 (Fig. III-17) includes the binary systems nepheline-silica, leucite-silica, nepheline-leucite, as well as the system of the alkali feldspars. The feature in this diagram to which attention is especially directed is the belt of low-melting temperatures in the middle of the field of the alkali feldspars. Through fractional crystallizations, any liquid in this diagram, regardless of initial composition, will migrate towards this belt. The result is a mother liquor (or rest magma), again yielding alkali feldspars of about 40 per cent Or as the final crystallization product in addition to nepheline or silica.

Natural rest magmas usually have some dissolved water, for, as crystallization proceeds, water will concentrate in the residual solutions. Thereby the melting temperatures will be considerably lowered, and the leucite field will decrease and eventually, at a content of about 7 per cent H_2O, will completely disappear.

The behavior of some alkalifeldspar-silica melts under these conditions merits special consideration. A liquid at F in Fig. III-18 will begin to precipitate feldspar having the composition F' at 850°C, and the liquid will move along its fractionation curve to the boundary $E-E'$ at G, at which time the crystal will be continuously zoned from F' to G'. As the liquid at G crystallizes further and quartz and feldspar are both being precipitated, the crystals now change composition from G' toward N, and the final outermost zone will have the composition N. These crystals will now have two zones of the composition N, one formed when the temperature was at H, and the last, outer zone at the temperature of the ternary minimum M.

The preceding survey shows that the relations of the light-colored constituents are such that alkali feldspar approaching the composition 40Or, 60Ab will be continually concentrated in residual liquids during fractional crystallization. The last liquid to crystallize will therefore yield this feldspar plus either silica or nepheline. And we shall see that the other rock constituents—the dark minerals—do not change this relation, but follow separate paths of crystallization without interfering seriously with the light constituents.

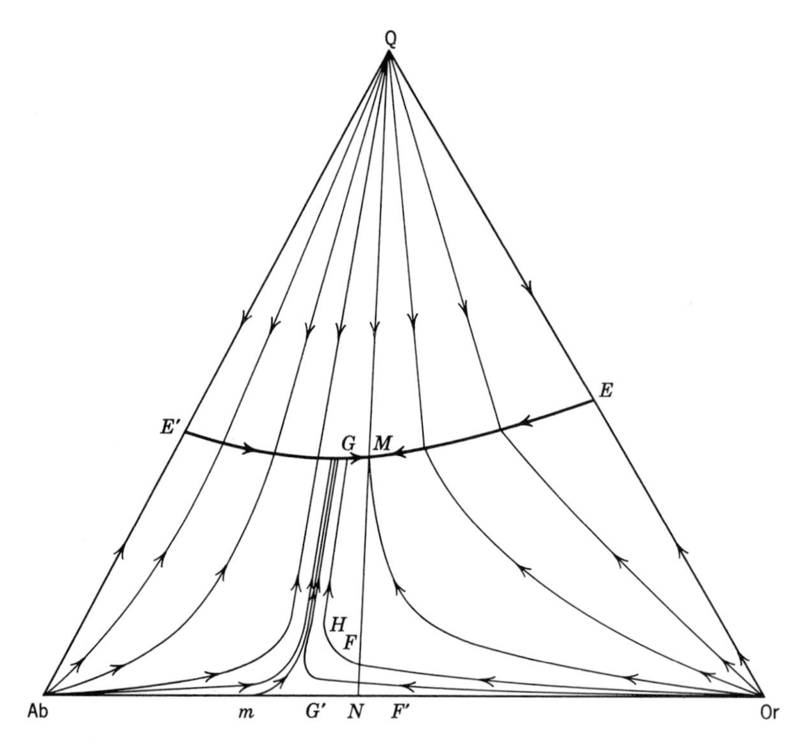

Fig. III-18. Isobaric fractionation curves for water pressure of 1000 atm projected onto the system $NaAlSi_3O_8$—$KAlSi_3O_8$—SiO_2. (Tuttle and Bowen, 1958.) Disregarding water, the line *E*—*E′* divides the field into two parts: in the upper part quartz crystallizes as primary phase; in lower part an alkali feldspar solid solution is the primary phase. Between Ab and Or there is a minimum at *m*. The lowest point on the fusion surface, or rather the temperature saturation surface, is the ternary minimum at *M*, where quartz and feldspar solid solutions crystallize together. The fractionation curves simply give the paths that must be taken by each liquid during *fractional* crystallization.

5 · OLIVINE AND PYROXENE

The *rock-forming olivines* are simple mixed crystals between Mg_2SiO_4 (forsterite) and Fe_2SiO_4 (fayalite). Their melting relations are shown in Fig. III-19.

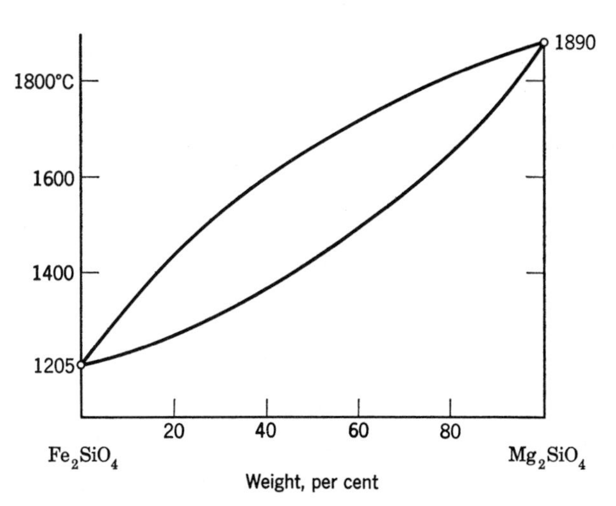

Fig. III-19. Melting relations of olivine (after Bowen and Schairer, 1935). The rock-making olivines form a continuous series of mixed crystals. The melting phenomena are analogous to those of the plagioclases (Fig. III-10).

The *rock-forming pyroxenes* are complex mixed crystals. The dominant molecules are $MgSiO_3$(en), $FeSiO_3$(fs), and $CaSiO_3$(wo). But the last molecule is not in itself a pyroxene. As shown by the diagrams of Figs. III-20 and III-21, the field of pyroxenes extends only to the join $CaMg(SiO_3)_2$—$CaFe(SiO_3)_2$, that is, the diopside-hedenbergite join, which represents pyroxenes of maximum content of lime.

A characteristic difference between pyroxene of the deep-seated rocks and pyroxene of lavas is that the lava pyroxenes are able to form extensive series of mixed crystals. The extreme cases are shown in Figs. III-20 and III-21. There are all transitions between the two cases. In many lavas the magnesium-rich members are only partially miscible, whereas complete miscibility exists among the iron-rich members; again some lavas carry orthorhombic pyroxene (for example, hypersthene-basalts). It should be remembered that the composition of the pyroxenes in these diagrams is so simplified that many of the natural pyroxenes cannot be adequately represented. But although many of the relations found in nature still await explanation, synthetic studies of simple pyroxenes represent important contributions to genetical petrology.

The following is a list of the molecules of the rock-making pyroxenes:

$Ca\ SiO_3$	Wollastonite, pseudowollastonite
$Mg\ SiO_3$	Enstatite, clinoenstatite
$Ca\ Mg(SiO_3)_2$	Diopside
$Fe\ SiO_3$	Ferrosilite, clinoferrosilite
$Ca\ Fe(SiO_3)_2$	Hedenbergite
$Mn\ SiO_3$	Rhodonite, bustamite
$Ca\ Mn(SiO_3)_2$	Johannsenite
$Na\ Al(SiO_3)_2$	Jadeite
$Na\ Fe(SiO_3)_2$	Acmite (aegirite)

Molecules containing

Al_2O_3	Augites
Fe_2O_3	Augites and babingtonite
TiO_2	Titaniferous augites

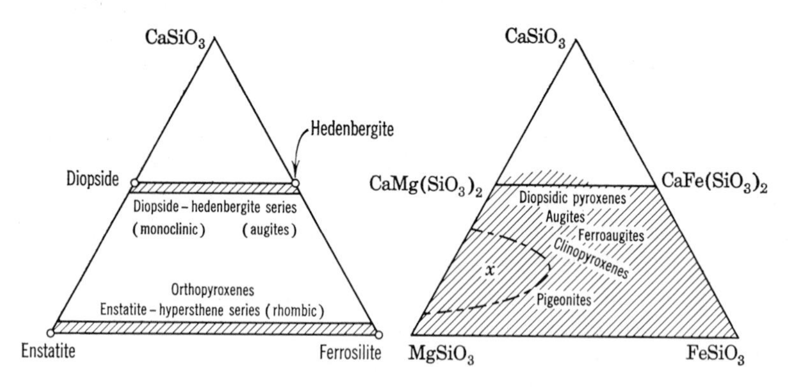

FIGS. III-20 and III-21. *Left:* Schematic presentation of the simplified pyroxene composition of subalkalic deep-seated rocks. The monoclinic pyroxenes lie close to the diopside-hedenbergite series. The orthorhombic pyroxenes are of the enstatite-hypersthene series. Between the two series, very limited solid solutions exist. *Right:* Schematic presentation of the simplified pyroxene composition of subalkaline effusive rocks. There are only monoclinic pyroxenes, representing ternary mixed crystals of the following molecules: $MgSiO_3$, $FeSiO_3$, $CaMg(SiO_3)_2$, $CaFe(SiO_3)_2$. With decreasing temperature, the field of mixed crystals is reduced; an exsolution field develops at x and expands with decreasing temperature. At the same time orthopyroxene develops in the $MgSiO_3$ corner.

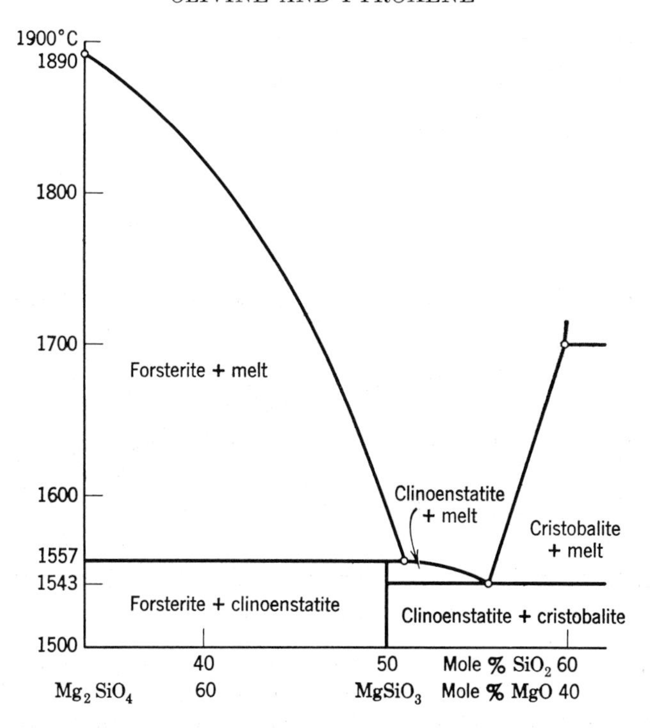

FIG. III-22. The melting relations of enstatite. (After Bowen and Andersen, *Am. J. Sci.*, 1914, and Greig, *Am. J. Sci.*, 1927.)

Enstatite. The melting relations of the pyroxenes are complicated by the fact that enstatite melts incongruently according to the scheme

$$2MgSiO_3 = Mg_2SiO_4 + SiO_2$$

which can be described only in terms of the binary system Mg_2SiO_4—SiO_2. (See Fig. III-22.)

The *join* $CaMg(SiO_3)_2$—$CaFe(SiO_3)_2$ corresponds to a complete series of mixed crystals, the melting relations of which are analogous to those of the olivine series (the iron compound showing the lowest melting point).

The *join* $MgSiO_3$—$FeSiO_3$ is no true binary system because of the incongruous melting of $MgSiO_3$, but the melting relations may still be demonstrated approximately by Fig. III-23.* Again the iron-rich

* The stable high-temperature form of pure $MgSiO_3$ is actually protoenstatite (another orthorhombic form of $MgSiO_3$). However, a small admixture of $CaSiO_3$ seems to stabilize the ordinary clinopyroxene.

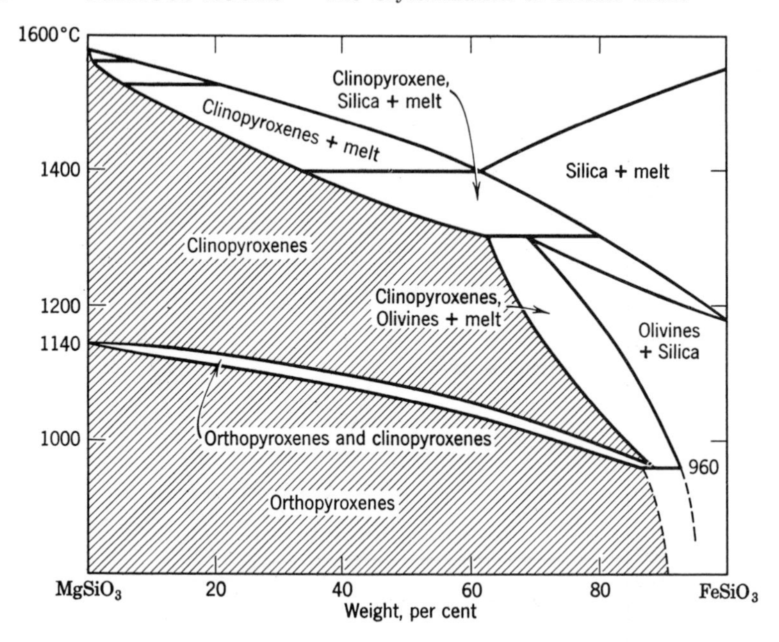

FIG. III-23. Diagram of the join $MgSiO_3$—$FeSiO_3$ in the system MgO—FeO—SiO_2. (After Bowen and Schairer, 1935.) Pure $FeSiO_3$ is not stable (at least not at elevated temperatures), but breaks up into olivine and silica. The field of mixed crystals, on cooling, invert into orthopyroxenes. The inversion curve runs from 1140°C in $MgSiO_3$ to 960°C in the most iron-rich member. Since this diagram neglects the existence of protoenstatite, it should best be regarded as representing the conditions under a small admixture of $CaSiO_3$.

compounds show the lower melting points. All the primary pyroxene phases are monoclinic; at lower temperature they invert into orthopyroxenes of the enstatite-hypersthene series.

The *clinopyroxenes* at elevated temperatures form with the diopside-hedenbergite series complete solid solutions extending over the whole field $MgSiO_3$—$FeSiO_3$—$CaMg(SiO_3)_2$—$CaFe(SiO_3)_2$. (See shaded area of Fig. III-21.) The orthopyroxenes, on the other hand, take very little lime in solid solution and are in the ternary diagram limited to a narrow field close to the $MgSiO_3$—$FeSiO_3$ join. (See Fig. III-20.) If we think of the ternary diagram with temperature axes erected perpendicular to the plane of the paper, we see that the clinopyroxene field extends like a domed roof over the diagram, whereas the orthopyroxene series forms the side wall. (See Fig. III-24.)

At temperature below the roof, two pyroxene phases coexist (as explained in the legend to Fig. III-24), the composition of which varies

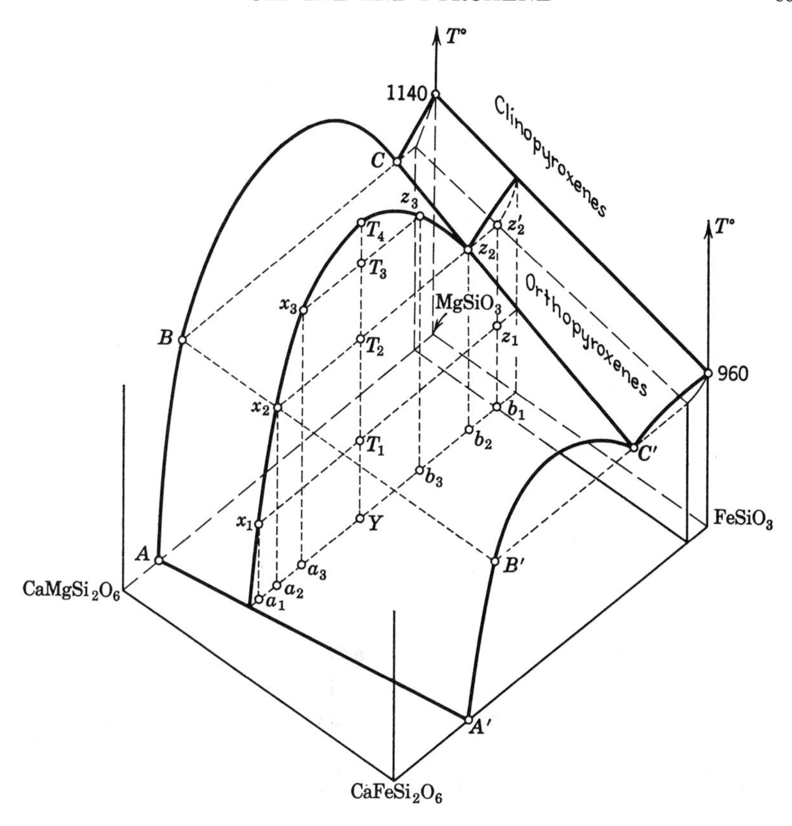

FIG. III-24. Schematic equilibrium diagram of the orthopyroxenes and clinopyroxenes at subsolidus temperature. The vertical plane $MgSiO_3$, $FeSiO_3$, 1140, 960 corresponds to the diagram of Fig. III-23, but it is here simplified, and the complications due to the breaking up of pure $FeSiO_3$ have been omitted. The curve A'—B'—C'—960 in the vertical plane rising from $FeSiO_3$—$CaFeSi_2O_6$ demonstrates the approximate inferred subsolidus equilibrium relations in the system of (nearly) pure ferrosilite-hedenbergite. The curve A—B—C—1140 in the plane rising from $MgSiO_3$—$CaMgSiO_2O_6$ gives the simplified inferred subsolidus equilibrium diagram of the system enstatite-diopside. Above the domed surface A—A'—B—B'—C—C' (the "roof") clinopyroxenes are stable. In the region below the roof and above the plane B—B'—C—C' (in the "attic") two clinopyroxenes coexist (in natural rocks they correspond to augite and pigeonite), and below the plane B—B'—C—C' clinopyroxene is in stable equilibrium with orthopyroxene. A homogeneous clinopyroxene, at temperature T_4 and composition given by Y, will, upon cooling to T_3, break up into two pyroxenes x_3 and z_3 (composition given by a_3 and b_3, respectively). On further cooling to T_2, the coexisting phases are x_2 and z_2 (composition a_2 and b_2). But this temperature is also the point of inversion of the Ca-poor clinopyroxene z_2 into the orthorhombic pyroxene z_2. Below this temperature, for example, at T_1, clinopyroxene x_1 (composition a_1) exists in equilibrium with orthopyroxene z_1 (composition b_1). (After Barth, 1949.)

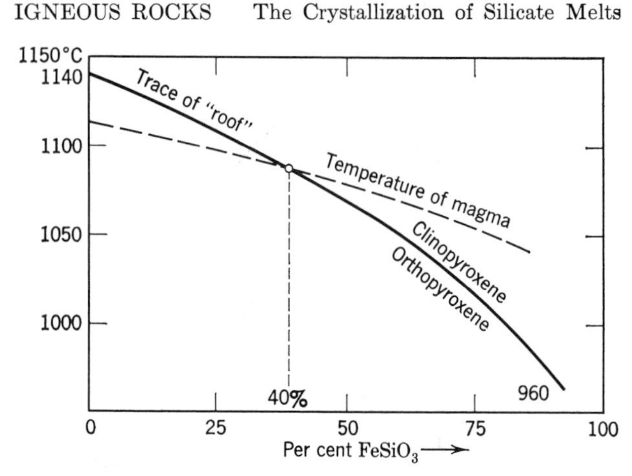

FIG. III-25. Diagram illustrates how the temperature of a magma at the beginning of the crystallization lies below—and in later stages of the crystallization lies above—the inversion curve for clinopyroxene-orthopyroxene. (From H. H. Hess, 1941.) Essentially the same diagram was published by S. Tsuboi, 1938.

with the temperature. It is evident that orthopyroxenes inverted from the monoclinic phase must have crystallized at temperatures above the plane *B—B′—C—C′*, whereas primary orthopyroxenes must have crystallized below this plane.

Finally we shall discuss the course of crystallization of the pyroxenes. If the temperature is high enough (above the "roof") only one pyroxene phase will separate. At lower temperatures two phases will appear. We have seen from Figs. III-19 and III-23 that fractional crystallization pushes the liquid toward the iron-rich parts of the diagram. In most magmas crystallization will start below the inversion plane *B—B′—C—C′*, with separation of magnesium-rich augite, which may be joined by magnesium-rich orthopyroxene. As the residual liquid becomes richer in iron, it will come closer to the inversion plane, for the inversion temperature decreases with increasing iron. Eventually the liquid may reach the plane and two clinopyroxene phases separate. What happens later is not clear. (See Fig. III-25.) Possibly the residual liquid may go beyond and above the "roof"; from then on only one (monoclinic) pyroxene phase will separate. According to this scheme, the normal course of crystallization is shown in Fig. III-26.

The *system diopside-forsterite-silica* serves to illustrate some of the relations between olivine and pyroxene. The equilibrium diagram (Fig. III-27) is divided into three fields by two boundary curves corre-

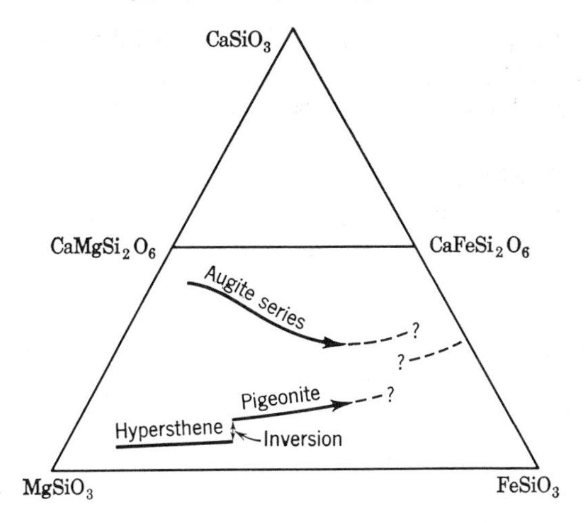

FIG. III-26. Normal crystallization course in pyroxenes. (After H. H. Hess, 1941.)

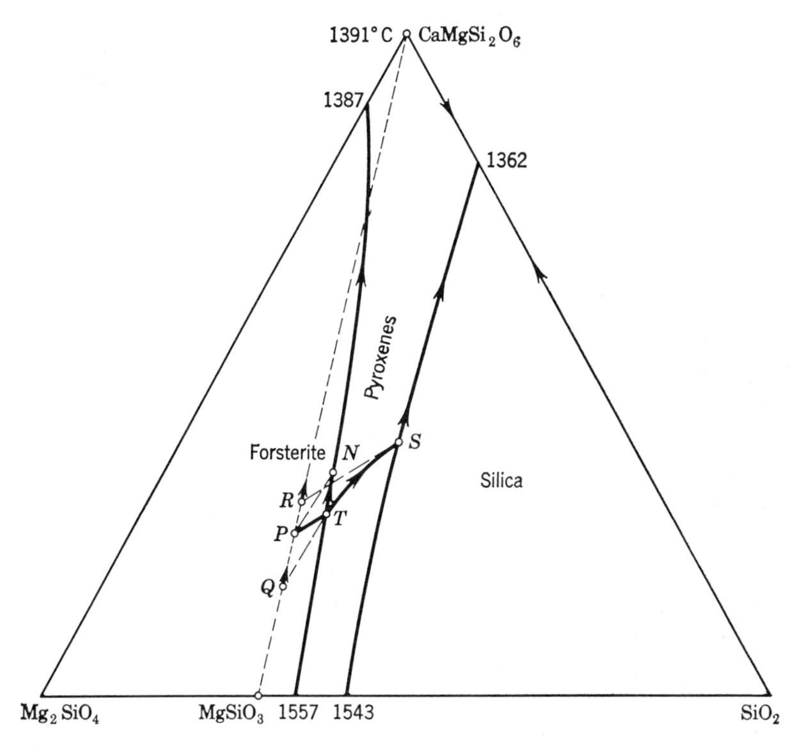

FIG. III-27. Schematic equilibrium diagram of the system diopside-forsterite-silica. (After N. L. Bowen, 1914.)

sponding to the primary crystallization of the following solid phases: (1) forsterite (Mg_2SiO_4), (2) pyroxene of varying composition from clinoenstatite ($MgSiO_3$) to diopside ($CaMgSi_2O_6$), and (3) silica (SiO_2).

Thus a melt whose projection point lies in the forsterite field will, on cooling, produce forsterite crystals; a melt in the silica field will give silica (either cristobalite or, below 1470°C, tridymite); and a melt in the field of pyroxene will give pyroxene, whose composition lies somewhere on the join $MgSiO_3$—$CaMgSi_2O_6$.

The distinctive feature of the diagram is that the compositions corresponding to pyroxene (along the join $MgSiO_3$—$CaMgSi_2O_6$) lie mostly outside their own field of crystallization and well inside the field of forsterite. The consequence is that most iron-free pyroxene must behave like clinoenstatite, that is, must dissociate on melting with formation of forsterite. These pyroxenes thus represent a series of binary solid solutions that melt incongruently.

A melt of the pyroxene composition P begins to crystallize with separation of forsterite, and the liquid changes along the straight line P—T. At T, pyroxene begins to crystallize, and Q represents the composition of the pyroxene. When the temperature is lowered further, pyroxene continues to crystallize, and forsterite begins to dissolve or react with the liquid to produce an increased quantity of pyroxene. The composition of the liquid now changes along the boundary curve, and the composition of the pyroxene in the act of crystallization, as well as that of the pyroxene which has already separated, changes toward P. When the temperature of the point N is reached, the last of the liquid and the last of the forsterite are used up simultaneously, and the whole consists simply of pyroxene of composition P.

Under equilibrium conditions no trace of this complicated crystallization history will be seen in the solid pyroxene. Fractional crystallization will leave its imprints, however, and extend the path of crystallization. If the early formed forsterite crystals are unable to react with the liquid, the liquid leaves the boundary curve at T and crosses the pyroxene field on the curve T—S. In the meantime the composition of the pyroxene separating has changed from Q to R, and there exist in the mixture, in addition to olivine, zoned pyroxene crystals of all compositions varying from Q to R. At S tridymite begins to crystallize, and the liquid changes along the boundary curve pyroxene-silica. Meantime the composition of the pyroxene separating changes from R towards pure diopside, and final crystallization takes place only when the temperature is that of the eutectic diopside-tridymite (1362°C) and the crystalline phases separating are tridymite and pure diopside.

Thus we see that the same original liquid, according to the physical conditions during cooling, may give in one instance a crystallization product of homogeneous pyroxene; in another a mixture containing olivine, pyroxene, and silica. This indicates what drastic effects fractional crystallization may produce in a rock magma.

Not always, however, is olivine resorbed. In melts of which x in Fig. III-28 is typical no conversion of olivine to pyroxene takes place. Olivine of composition o' starts to crystallize, and when the melt reaches the boundary line QL (at b) pyroxene of composition p' begins to crystallize. But thenceforth not only pyroxene but olivine also continues to separate (no resorption of olivine), the liquid following the boundary curve to B, where the pyroxene has the composition P, the olivine the composition O. In melts the composition of which lies within the triangle QRL, pyroxene begins to crystallize before olivine.

Many writers of petrological papers seem to assume that olivine always will crystallize before pyroxene in rocks containing these two minerals. It often does; in a great many oceanic olivine basalts we can see interstitial quartz as the last crystallization product obviously reflecting the failure of early olivine to react with the residual liquid to form pyroxene. (See page 119.)

But in rocks of other compositions no reaction need take place. Between olivine and diopside no reaction relation exists (see Fig. III-27).

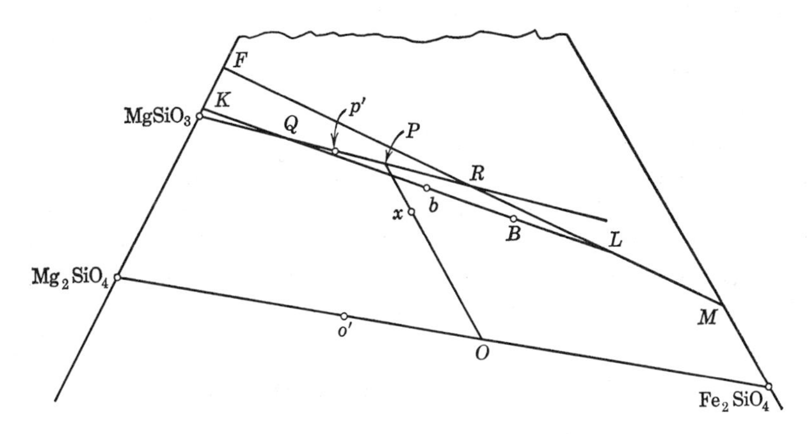

FIG. III-28. Diagram to illustrate crystallization in part of the system MgO—FeO—SiO_2. (After Bowen and Schairer, 1935.)

According to the present discussion it becomes apparent that neither will the mineral pair olivine-pigeonite (or olivine-clinohypersthene) always exhibit this relation. Certain natural rocks may, indeed, begin to crystallize with separation of pyroxene. Later in the course of crystallization, olivine appears as a stable product, and thenceforth continues to separate together with pyroxene *until complete solidification occurs.* A very important conclusion which is easily reached by study of the various equilibrium diagrams discussed in the present section is that, once started, no mineral can cease to crystallize in a cooling magma *unless it is replaced by its reaction product.*

In the case here under consideration olivine has no reaction product and must continue to crystallize to the very end. A good example of this process is afforded by the lavas of Gough Island (Barth, 1942). Olivine started to crystallize in the basaltic stage, no reaction product was formed, and so it continued to crystallize throughout the subsequent magmatic stages, forming olivine trachyandesite → olivine phonolite → olivine trachyte.

The physico-chemistry of the pyroxene minerals as here presented has been sketchy and incomplete. Only facts of particular importance to petrology have been discussed. The quaternary system CaO—MgO—FeO—SiO_2 forms the basis for a complete knowledge of the equilibrium relations of the lime, magnesium, and iron pyroxenes. This system is not known, but the limiting ternary systems CaO—MgO—SiO_2, CaO—FeO—SiO_2, and MgO—FeO—SiO_2 are known, and through them it is possible to get some insight into the very complicated crystallization, cooling, and inversion relations. They will not be discussed here, however. Nor will pyroxenes of different chemical composition be treated. It should be kept in mind, however, that for the evolution of certain rocks of "alkalic" affinities the alkali pyroxenes are of importance. In such rocks the molecular proportion of the alkalies is higher than that of alumina, which accordingly is used up in the making of feldspar molecules. The remaining soda then combines with Fe_2O_3, forming acmite ($NaFeSi_2O_8$), which melts incongruently and forms a very low-melting eutectic with quartz (760°C). (Bowen, Schairer, and Willems, 1930.)

6 · PLAGIOCLASE, OLIVINE, AND PYROXENE

The *system anorthite-forsterite-silica,* the equilibrium diagram of which is shown in Fig. III-29, is divided into five fields corresponding to the primary crystallization of the following solid phases: (1) silica

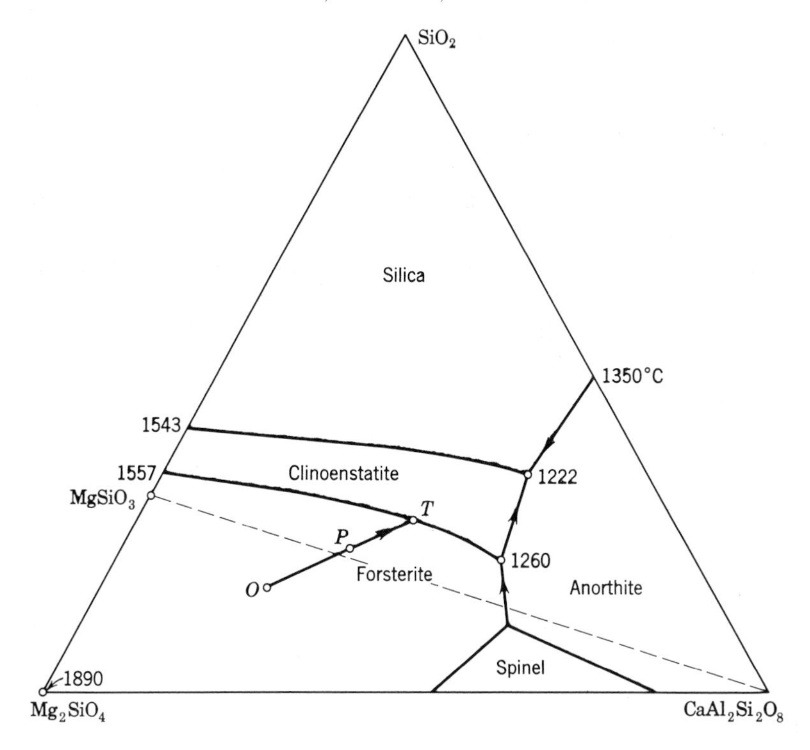

FIG. III-29. Equilibrium diagram of the system anorthite-forsterite-silica. (After Olaf Andersen, *Am. J. Sci.*, 1915.)

(SiO_2), (2) clinoenstatite ($MgSiO_3$), (3) forsterite (Mg_2SiO_4), (4) anorthite ($CaAl_2Si_2O_8$), and (5) spinel ($MgAl_2O_4$).*

It should be noted that the projection point of clinoenstatite ($MgSiO_3$) lies outside its own field and well within the field of forsterite. This is easily explained by the incongruent melting of clinoenstatite (see page 97).

A melt of composition O begins to crystallize at 1500°C, with separation of forsterite and the liquid changes along a straight line toward T. At T and 1375°C clinoenstatite begins to separate and forsterite to react with the liquid which converts it into clinoenstatite. While this action is going on, the composition of the liquid changes along the boundary curve T-1260°. When the temperature has fallen to 1260°, anorthite begins to separate. The temperature then remains

* Mixtures lying in the spinel field and certain other mixtures whose crystallization will carry them into this field cannot be treated as ternary, but for the present, these need not concern us.

constant at 1260°, anorthite and clinoenstatite increasing in amount, liquid and forsterite decreasing in amount, until all the liquid disappears. The completely crystalline mass now consists of forsterite, anorthite, and clinoenstatite.

If the original melt had the composition *P*, again forsterite would begin to separate. The same course would be followed, but at 1260° some liquid would remain after the forsterite was exhausted. This liquid would then continue its crystallization, with separation of clinoenstatite and anorthite, and would change in composition along 1260° to 1222°. At 1222°, tridymite (SiO_2) separates in addition to clinoenstatite and anorthite, and the temperature of the mass remains constant at 1222°, until all the liquid disappears. It should be noted that the early formed forsterite crystals during the process of further crystallization completely disappear without leaving any traces in the final crystallization product.

Throughout the foregoing discussions of crystallization perfect equilibrium is assumed. But it is easily seen that by fractional crystallization, if the early formed crystals of forsterite are prevented from reaction with the liquid either by being armored with a shell of clinoenstatite or by being removed from the liquid, the crystallization process will always proceed with continuous change in the liquid until the lowest point in the fusion surface, at 1222°, is reached.

In such cases, characterized by lack of equilibrium, forsterite and silica may occur together in the final crystallization product. But this can never happen under equilibrium conditions.

Diopside-anorthite-albite represents a very simple ternary system. The three fundamental binary systems are (1) anorthite-albite, (2) anorthite-diopside, and (3) albite-diopside.*

The ternary equilibrium diagram is shown in Fig. III-30. There is but one boundary curve, and it separates the field of the plagioclases from the field of diopside. Just as the temperature falls continuously from the melting point of anorthite to that of albite in the binary system, so there is a similar fall along the boundary curve from the anorthite-diopside eutectic (1270°) to the albite-diopside eutectic (1085°), which is very close to pure albite.

A melt, *O*, in the plagioclase field begins to crystallize at 1375°, with the separation of plagioclase *Q*. As the temperature falls, the plagioclase increases in amount and changes in composition until at 1216°

* Recent investigations by Schairer have shown that the system albite-diopside, which was supposed to be a simple eutectic system, is not strictly binary; but the deviations are so small as to be of no significance for this discussion.

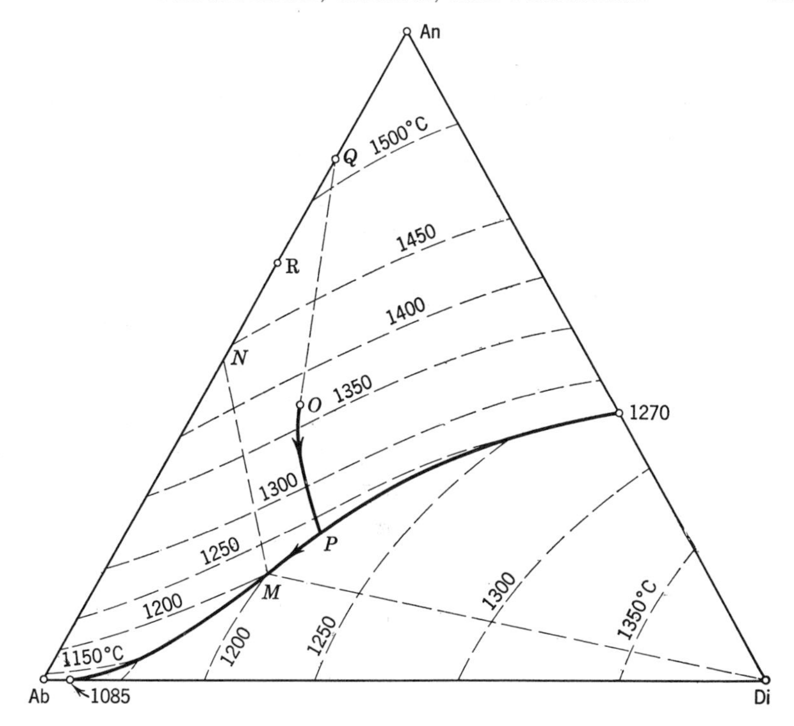

FIG. III-30. Equilibrium diagram of the system diopside ($CaMgSi_2O_6$)-anorthite ($CaAl_2Si_2O_8$)-albite ($NaAlSi_3O_8$). (After Bowen, 1915.)

the liquid has the composition *P* (on the boundary curve) and plagioclase the composition *R*. At this moment diopside begins to crystallize. With further lowering of temperature, the liquid follows the boundary curve, both diopside and plagioclase crystallizing, and at 1200° the liquid is all used up, the last minute quantity of liquid having the composition *M* and being in equilibrium with both diopside and plagioclase of the composition *N*.

If the early separated plagioclases were prevented from reacting with the liquid either through zoning or by being continuously removed from the liquid, fractional crystallization would produce a different course of the liquid. The residual liquid would reach the boundary curve at a point between *P* and *M*, and through continuous separation of diopside and plagioclase the liquid would finally reach the eutectic albite-diopside at 1085° and 97 per cent albite, which is the lowest point in the fusion surface.

The two ternary systems described in this chapter demonstrate that the course of crystallization of the colored minerals (olivine, pyroxene) and the course of crystallization of the feldspars proceed in their normal way without interferring with one another. This is approximately true also in polycomponent silicate systems similar to rock magmas. In the course of normal crystallization the feldspars will react with the complex liquid and be converted continuously into sodic feldspars. The iron-magnesium minerals will first separate as magnesium-rich olivines or pyroxenes and through reaction with the complex liquid, regardless of the presence of feldspathic silicates, will continuously become enriched in iron; and the olivines will be converted discontinuously into metasilicates (pyroxenes).

Vestiges of such reactions may often be observed with the microscope in natural rock. Basalts and diabases may show resorbed olivine, or olivine armored with a shell of pyroxene. Zoned feldspars with calcic core and sodic shells are normal constituents of almost all lavas. These natural phenomena appear to be identical with the resorption and reaction phenomena observed during the normal course of crystallization of artificial melts and find, as convincingly demonstrated by Bowen, their explanation in this way.

7 · HORNBLENDES AND MICAS

The hydrous igneous minerals are hornblende and mica. They are complicated in their crystal chemistry with wide diversification in composition and great polymorphic variability in their space lattices. See pages 243 ff. They contain (OH)-groups in their crystal structure and cannot, therefore, crystallize from dry melts; their stability is governed by vapor pressure just as much as by temperature, and, for the iron-bearing members, not the least by partial pressure of oxygen; indeed, only in specific ranges of P_{O_2} are (HO)-bearing iron silicates stable. For these reasons it has not yet been possible to investigate any multicomponent hydrous system which would shed light on the important reaction relations which are involved in the series pyroxene → hornblende → mica.

Figure III-31 gives the stability fields of various minerals. Muscovite can crystallize from a granite melt only at H_2O pressures higher than 2000 atm and below 700°–800°C. It cannot crystallize from a lava. Phlogopite, pargasite, and tremolite are the pure magnesian varieties. The corresponding minerals also have FeO and will break down at lower temperatures than indicated for the pure magnesian members. But still they would have a wider stability field than muscovite. This may explain the fact that biotite and hornblende are

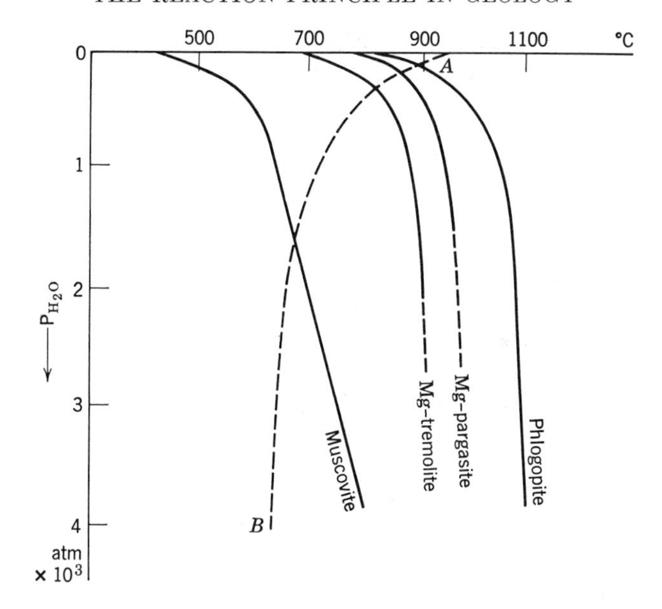

FIG. III-31. Equilibrium curves for the destruction of hornblendes and micas. Each mineral is stable on the left side of its curve. *AB* is the melting curve of granite. See page 128.

much more common in igneous rocks than is muscovite. So-called basaltic hornblende, common in certain lavas, is usually rather low in silica and contributes, therefore, to the concentration of silica in the residual melt.

8 · THE REACTION PRINCIPLE IN GEOLOGY

By piercing together the information obtained from artificial melts and from the study of natural rocks, it becomes evident that three reaction series are of prime importance.

1. The ferromagnesian minerals form a discontinuous reaction series, beginning with magnesium-rich olivine. The liquid reacts with olivine to produce pyroxenes, with pyroxenes to produce amphibole, and with amphiboles to produce biotites.

2. The plagioclases form a continuous reaction series from the calcic to the sodic members, the change in composition taking place by infinitesimal increments. These are the two famous reaction series of Bowen.

3. The alkali feldspars form another continuous reaction series from potassic to sodic members. In the lower end it merges with the plagioclase series.

These series are graphically indicated below:

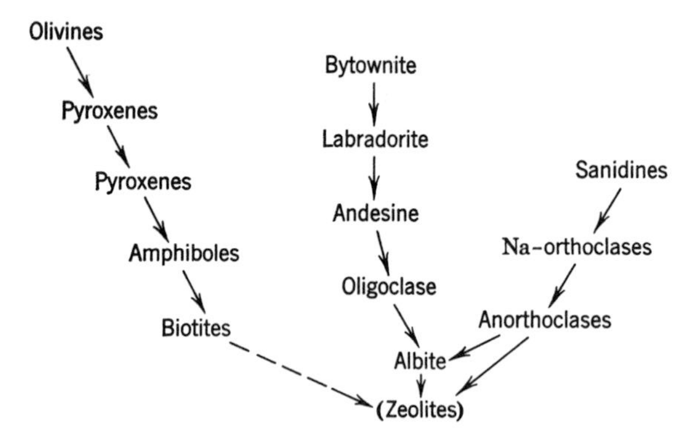

It is worthy of note that there is little in the nature of eutectic crystallization in these series. The system of the complex rock magma may be regarded as made up of three regions of polydimensional space, corresponding to the primary crystallization of members from each of the three series. Between the series there is some suggestion of the eutectic relation, in that a member of one series lowers the melting "point" of a member of the other series. But here the analogy with eutectic crystallization ends, for there is, within the series, no eutectic end point where final solidification must take place. The maximum depression in the fusion temperature between two series extends along a line (*boundary line*) representing the conditions of a three-phase equilibrium in which two solids and a liquid are in equilibrium (in the same manner the boundary "line" of three series determines a four-phase equilibrium). Crystallization along these lines is called *cotectic* crystallization.

The minerals have a reaction relation to the magma; each separated mineral tends always to change into a later member of the reaction series. This change of composition is effected by reaction with magma. According to the opportunity for reaction, the magma is entirely used up—in some cases sooner, in others later—and only then is solidification complete.

This discussion allows us to form a rough idea about the magmatic evolution of some common rock types through fractional crystallization. We start with a gabbroic magma. On cooling, olivine and calcic plagioclase (and small amounts of less significant minerals) will separate, thereby changing the composition of the still liquid magma phase into that of a dioritic magma. If the magma through orogenic movements or gravity is separated physically from the early formed minerals (the gabbroic minerals), it will act as a diorite magma, independent of the antecedent gabbro minerals. It will separate new mineral phases, and again change its composition and be converted into a granite magma. This kind of differentiation is shown graphically below:

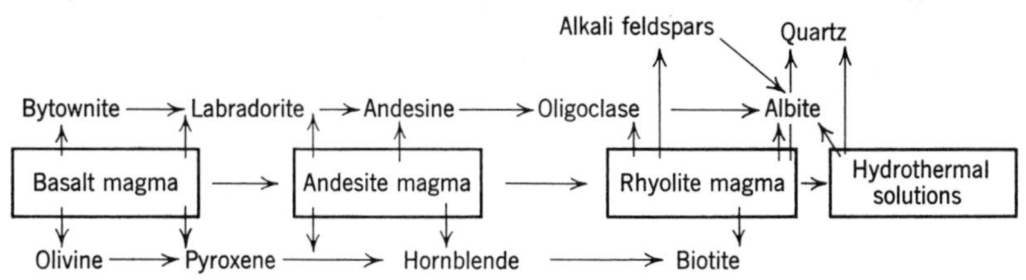

Diagram I. Basalt, andesite, and rhyolite correspond respectively to gabbro, diorite, and granite.

Assimilation and Differential Melting

In a general way this scheme also gives simple information about the behavior of *inclusions* in rock magmas. The various types of magma are placed in definite positions relative to the reaction series of the minerals. According to the reaction principle, any magma is effectively supersaturated in respect to all higher minerals in the reaction series. Inclusions made up of such antecedent minerals will not be dissolved, therefore, by the magma. But the magma may react with them in an effort to convert them into crystals that are in equilibrium with the magma. The heat consumed by the reaction is produced by the magma separating more crystals with which it is in equilibrium. Only inclusions made up of minerals belonging to lower stages in the reaction series can be dissolved directly by the magma. For example, a gabbro magma may dissolve granitic inclusion, but not vice versa.

By the same reasoning it becomes clear that when solid rocks are heated, differential melting will occur. The temperature need not be very high, for in most rocks a small amount of water is present, partly in the pores, partly in hydrous minerals. The melt thus formed would exhibit the composition of a granite (rhyolite) magma, for the composition last to crystallize is the composition first to melt. If more heat is introduced, more rock would melt and the composition of the melt would gradually change toward the composition of a diorite (andesite) magma as the minerals of the lower part of the reaction series are gradually dissolved. To visualize this, look at the diagram given above and turn all the arrows in the opposite direction. In this way we can see that much granite magma may form, but only rarely would the conditions be right for the formation of a gabbro magma. (See page 132.)

9 · THE CRYSTALLIZATION PROCESS OF BASALTIC MAGMA

According to what has already been said, it is possible to explain the diversity of igneous rocks by magmatic differentiation. In doing so we have to have a starting point, that is, we have to choose a primary magma. The geologic record reveals that basaltic magma occupies a unique position. It has, in all ages, broken through the crust of the earth in the form of dikes of great extension, has poured out on the surface in great floods, and has insinuated itself as sills and other concordant intrusives between the beds of layered rocks. According to the best estimates no less than 98 per cent of all effusive rocks are basaltic-andesitic in composition. On these grounds basaltic magma is accepted by many as a primary magma. See also pages 132, 221.

The Position of the Boundary Surface That Separates Basaltic Liquids Precipitating Feldspar from Those Liquids That Precipitate Pyroxene

Basalts are made up of plagioclase and pyroxene as dominant mineral phases. As a first approximation we shall define basalt magma as a silicate melt, which on cooling yields plagioclase and pyroxene as the chief crystalline products.

During the cooling of this melt two main reaction series will develop: (1) the series of the plagioclase feldspars from calcic to sodic (page 109) and (2) the series of the clinopyroxenes, which, at certain compositions of the magma, may start with an early formation of olivine that, on further cooling, is converted into pyroxene (page 109). Thenceforth pyroxenes continue to separate, the change in chemical composition being towards more iron-rich members.

A combination of these two series represents the main crystallization process of a basaltic magma. Graphically this can be expressed as in Fig. III-32. If the composition of the original basaltic liquid lies to the right or to the left of the boundary line (indicated by the line *O—P*) crystallization will begin with precipitation of only pyroxene or only plagioclase, respectively, until the boundary is reached. Thenceforth a simultaneous precipitation of both phases takes place.

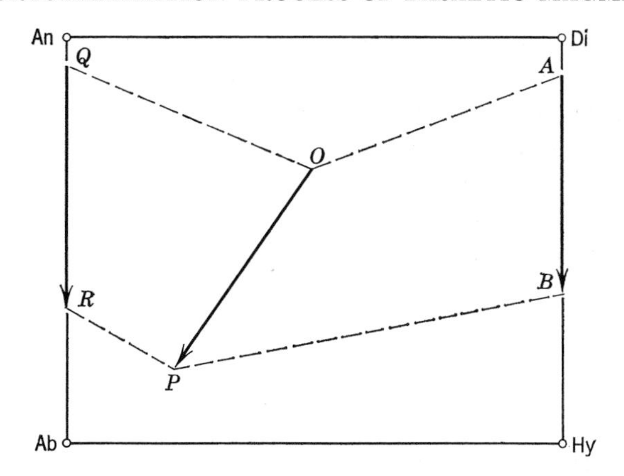

FIG. III-32. Schematic presentation of the crystallization process of basaltic lavas. A melt of composition *O* crystallizes by simultaneous precipitation of pyroxene of composition *A* and plagioclase of composition *Q*. As crystallization proceeds, the melt changes from *O* to *P*, and the two solid phases change from *A* to *B* and from *Q* to *R*.

With close approximation the composition of the pyroxenes can be said to vary from diopside to clinohypersthene (from *A* to *B*).

With equally close approximation the reaction series of the basaltic feldspars may be taken from bytownite to oligoclase (from *Q* to *R*).

Disregarding the remaining constituents (the sum of which rarely exceeds 10 per cent of the basalt), we can give the mineral composition of a typical basalt in terms of plagioclase and pyroxene; and for further discussions we shall be using the normative mineral constituents. The approximate composition of any typical basalt can thus be given in terms of normative plagioclase and normative pyroxene, that is, in terms of the symbols ab, an, di, hy.*

The four corners of Fig. III-32 may be regarded as representing these four constituents; but a quaternary system cannot be adequately presented by this type of diagram. We must resort, therefore, to some kind of tetrahedral projection in order to carry through a quantitative survey of the whole crystallization process.

* These are the ordinary normative symbols (ab = albite, an = anorthite, di = diopside, hy = hypersthene, see pages 65 ff.) which in a basalt usually add up to about 90 per cent. Since we want to express the total composition of the basalt in terms of these four symbols, they have to be recalculated to 100 per cent. Thus $ab' + an' + di' + hy' = 100$ are the four fundamental figures used in the further calculations.

Figure III-33 shows the concentration tetrahedron of the quaternary system ab, an, di, hy. The approximate composition of any ordinary basalt can thus be pictured as a plot inside the tetrahedron.

Thus point 3 of Fig. III-33 represents the average composition of the Deccan plateau basalt, which, disregarding TiO_2, H_2O, and other minor constituents totaling less than 5 per cent, corresponds to the following normative mineral composition:

$$\begin{aligned} ab' &= 27 \\ an' &= 29 \\ di' &= 22 \\ hy' &= 22 \end{aligned}$$

In Table III-13 the theoretical composition of this four-component system is compared with the actual chemical composition of the Deccan basalt. This example shows that the composition of typical basalts with fair approximation can be described in terms of said four components.

In the four-component system of Fig. III-33 the basal face of the tetrahedron represents the well-known ternary system albite-anorthite-

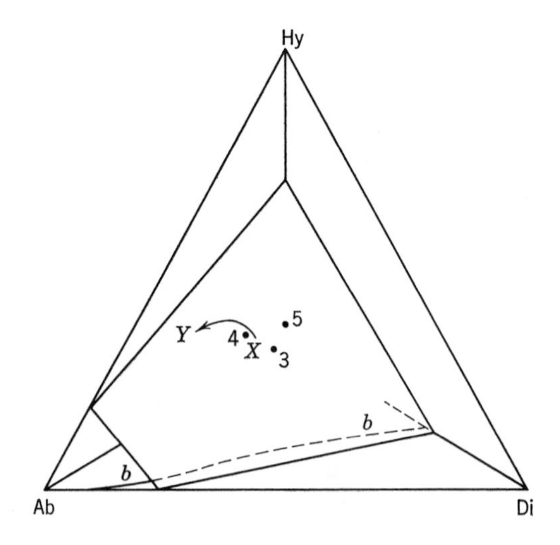

Fig. III-33. Tetrahedron illustrating the approximate composition of basaltic lavas. The plane represents the boundary surface which separates those lavas that precipitated plagioclase first from those that precipitated pyroxene first. *X-Y* indicates the change in the residual liquid with advancing crystallization. Points 3, 4, and 5 indicate the compositions of the Deccan trap rocks, the Oregon traps, and the Karroo dolerites, respectively. (After Barth, 1936.)

TABLE III-13

THE CHEMICAL COMPOSITION OF AN ARTIFICIAL MELT IN THE SYSTEM ab–an–di–hy, COMPARED WITH THE AVERAGE COMPOSITION OF DECCAN PLATEAU BASALT

	Artificial Melt	Deccan Basalt
SiO_2	53.4	50.6
Al_2O_3	15.9	13.6
FeO	12.0	12.8
MgO	4.1	5.5
CaO	11.5	9.5
Na_2O / K_2O	3.2	3.3

diopside (Fig. III-30) in which there is but one boundary curve separating the field of primary plagioclases from the field of the pyroxene.

If the projection point of the basalt had been in this plane the incipient crystallization would have yielded either only plagioclase or only pyroxene until the melt reached the boundary curve. In the tetrahedron a plane extending along this curve and upward divides the tetrahedron into two regions. In the front region we find the projection points of those magmas that precipitate pyroxene first; in the back regions those that precipitate plagioclase first. On the plane itself, corresponding to a two-phase boundary surface, are the projection points of such basalts that separate both plagioclase and pyroxene at the same time.

The boundary surface (if thought of as a plane) is mathematically defined by three coordinates. In this case it is convenient to take ab, di, and hy as the coordinates (an, which is the fourth component of the system, is thereby of course also defined, since the sum of all four components must equal 100). The equation of the boundary surface in terms of these coordinates is

$$ab' + 2di' + 2.3hy' = 123 \qquad \text{(III-3)}$$

The practical use of this equation is simple and obvious:

1. The norm is calculated from the chemical analysis in the usual way.

2. The normative constituents ab, an, di, hy are recalculated to 100 per cent, and the symbols thus formed are called ab′, an′, di′, hy′, respectively.

3. The sum ab′ + 2di′ + 2.3hy′ is formed.

If this sum equals 123 (approximately), the basalt lies on the boundary surface. If the sum is smaller or greater than 123, the basalt lies in the plagioclase field or in the pyroxene field, respectively.

This sum, which for short may be called *f* (norm), may be calculated for any basalt and will thus indicate the nature of the first crystallization products separating from the melt. When the melt reaches the boundary surface, both dark and light minerals will separate simultaneously, and the change in the melt will now be in the general direction X—Y within the boundary surface. But large changes in this melt cannot be read off the diagram, and it is absolutely not permissible to conclude that the residual melt of natural basalt is made up of sodic plagioclase and clinohypersthene, the chief reason being that natural magmas are much more complex than the model melt of the diagram. The presentation is obviously approximative in that only about nine-tenths of the actual basalt magma are included, that is, about 10 per cent of the components originally present are not considered. This is of no consequence for the main crystallization, however, and does not markedly influence the reaction series of plagioclase or pyroxene. It does become of great importance for the nature of the residual melts. The further crystallization process of a basaltic magma with the possible formation of a rhyolitic or a phonolitic residuum will now be given special mention.

The Formation of Quartz-Bearing or Nepheline-Bearing Residual Melts through Fractional Crystallization of Basaltic Magmas

In a quaternary system illustrated by the tetrahedron in Fig. III-34 we take the apex as representing the colored mineral phases, olivine and diopside; the three corners of the base represent anorthite, silica, and nepheline. Since silica and nepheline make albite, and the join albite-anorthite represents the plagioclases, the ternary system represented by the base is divided into two parts by this join, each part representing a smaller ternary system: (1) albite-anorthite-silica and (2) albite-anorthite-nepheline. The crystallization in each of these part-systems is known in principle. Each system contains a boundary curve separating the field of the plagioclases from the field of either silica or nepheline, as the case may be. From these two boundary curves on the base two boundary surfaces will rise into the tetrahedron.

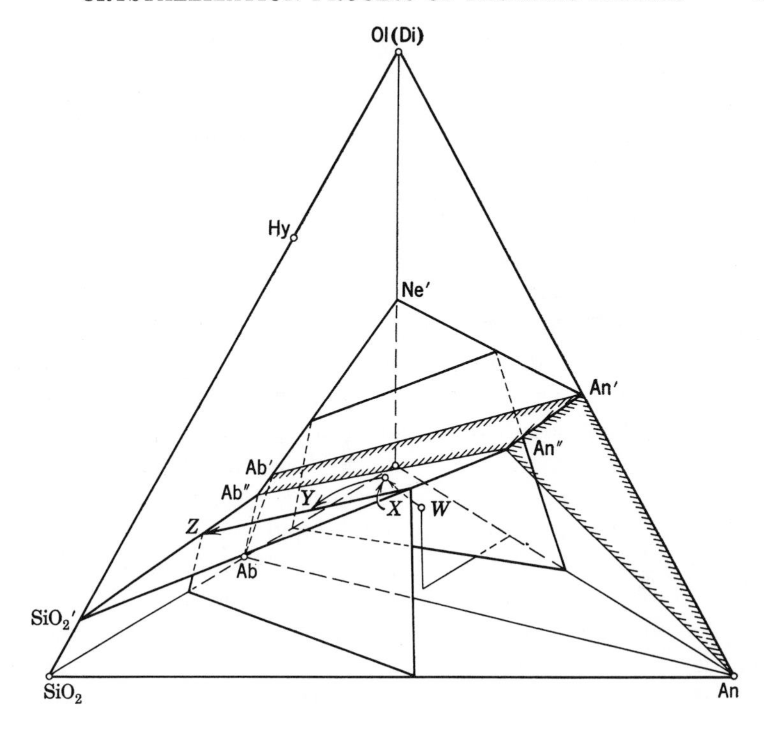

Fig. III-34. The principle of deriving alkaline, neutral, and quartzose residua by fractional crystallization of basalts. (After Barth, 1936.)

One separates those liquids that precipitate quartz from those that precipitate plagioclase; the other separates the plagioclase field from the nepheline field. Generally speaking, there must be a third boundary surface in the tetrahedron separating those liquids that precipitate the colored constituents from those liquids that precipitate quartz, plagioclase, or nepheline (it will be referred to as the horizontal boundary surface). Actually this is a composite boundary surface made up of three parts: one limiting the quartz field, another the plagioclase field, and the third the nepheline field. But since the apex has been taken to represent both olivine and diopside, no quantitative relation between the upper field of the colored constituents and the lower field of the colorless constituents can be shown. To simplify the figure, the boundary surfaces have been drawn schematically as planes, the positions of which have no direct quantitative relations to the actual boundary surfaces. Figure III-34 is, therefore, no true copy of the actual conditions in the quarternary system, but is a diagram from which the principles of fractional crystallization in basaltic lavas can be inferred.

Liquids of basaltic composition will be represented by plots not far from the horizontal boundary surface, and more or less straight above the albite-anorthite join. Such a liquid may be *W* of Fig. III-34. From this liquid, plagioclase will crystallize until the point *X* is reached by the liquid. At this point pyroxene will appear, or, if the temperature was high enough, olivine will crystallize first and later completely disappear through reaction with the residual liquid and formation of pyroxene. In any case the composition of the liquid will change toward the three-phase boundary *Y*—*Z*. At *Y*, quartz will appear, and will keep on crystallizing until the liquid phase has disappeared. Thus a quartzose residuum is formed.

No new assumptions have been made thus far. This is simply the old and well-recognized way of producing a siliceous residuum by fractional crystallization of a basaltic liquid.

But why should an original basaltic liquid be situated exactly at *W*? A position farther back should be just as reasonable. Many of the intra-Pacific basalts are approximately represented by such a position, and they presumably represent original basaltic liquids fully as well as mixtures closer to *W* which are more andesitic in composition. From such a liquid, also, plagioclase will precipitate first, but the crystallization path of the liquid will pierce through the horizontal boundary surface at a point that lies behind the join Ab′-An′. At this point olivine (and diopside) will appear, and the composition of the residual liquid is thereby thrown towards the three-phase boundary line plagioclase-olivine-nepheline. Thus an undersilicated residuum is formed by fractional crystallization of a basaltic liquid. These two examples show that a slight difference in composition of the original liquid may completely alter the nature of the residuum produced by fractional crystallization.

From Fig. III-34 we see that, if the original liquid is represented by a point within the body An-Ab-Ab′-Ab″-An′-An″, an initial precipitation of plagioclase will make the crystallization path of the liquid hit the horizontal boundary surface within the area Ab′-Ab″-An′-An″. If the original liquid is represented by a plot in front or in back of the said body, the horizontal boundary surface is met with in front or in back of the said area, respectively. The various cases thus arising can be discussed with the help of Fig. III-35. In Fig. III-35 the horizontal boundary surface is folded out to a plane.

Three main cases have to be considered. Because all of them can

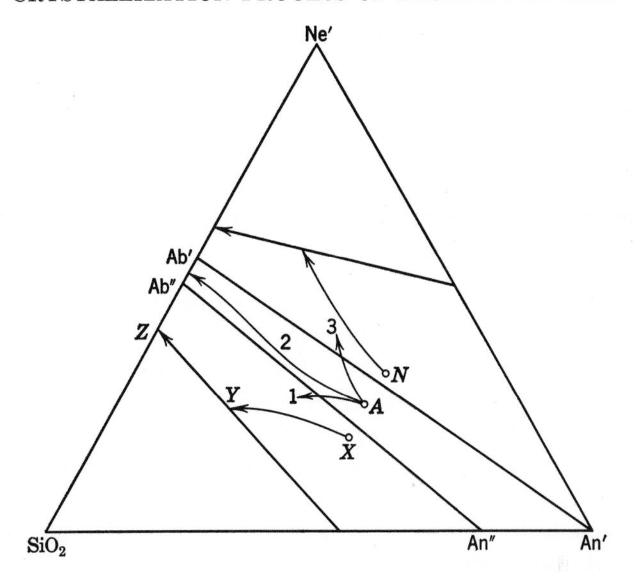

Fig. III-35. Graphic illustration of the different results of fractional crystallization of basaltic liquids. The letters correspond to those of Fig. III-34.

easily be read from Fig. III-35, a summary statement is all that is necessary.

1. Original mixture at *X*. Quartzose residuum always produced. Corresponds to the tholeiite-basalt series.
2. Original mixture at *N*. Undersilicated residuum always produced. Corresponds to the alkalic olivine-basalt series.
3. Original mixture at *A*. Three subcases must be considered:

A_1. Olivine separates in excess of its stoichiometric ratio and fails to react completely with the mother liquor. Quartzose (rhyolitic) residuum is produced. This case is very frequently observed in natural rocks.

A_2. Early olivine reacts completely with the mother liquor. Trachytic residuum (with neither quartz nor nepheline) is produced. This case is also frequently observed in natural rocks.

A_3. Pigeonitic pyroxene crystallizes in excess of its stoichiometric ratio and does not react with the residual liquid, which therefore becomes undersilicated. It seldom occurs, but may become of importance under high pressure stabilizing high-alumina enstatite-jadeite. According to E. S. Larsen, this reaction is important in the formation of submagmas in Highwood Mountains. (See page 204.)

Two results of this discussion are worth emphasizing:

1. The course of a liquid during fractional crystallization is capable of variation in accordance with the efficacy of fractionation, which itself varies in response to the controlling conditions.
2. Very slight variations in the composition of the initial liquid will, by fractional crystallization, give rise now to an oversilicated, now to an undersilicated residuum.

In a later chapter we shall see that the rocks of the Oceanic Islands are characterized by the following *alkalic series:* basalt-trachyte-phonolite (corresponding to N and A_2). The rocks of, for example, the circum-Pacific mountain chain represent a *calcic series:* basalt-andesite-dacite (corresponding to X and A_1). Finally the Thulean province exhibits: basalt-trachyte and basalt-granophyre in interdigitating associations.

Did these series develop in consequence of different external conditions, or do they reflect the composition of the several parental magmas? The magma of the reservoirs of the Oceanic Islands, being in contact with the presumably basic rock of the ocean floor, might itself be more basic than, for instance, the magma of the Thulean province, where patches of the sialic continental crust may be expected underneath the basalt plateaux. And more acid than any of these would be the magma of the circum-Pacific regions where sedimentary rocks of the Tertiary folding zone have been kneaded into the magma.

This evidence suggests one important conclusion: The geological environment is a decisive factor in the development and trend of differentiation of the several rock series. As a general rule it may be said that the uncontaminated primary basaltic magma produces rock series of alkalic affinities, whereas the more contaminated magma (incorporation of normal geosynclinal sediments) produces more calcic rock series. From this it follows that the alkali-lime index, as defined on page 172, in a general way may be regarded as a measure of the *degree of contamination* of the magma by the normal sediments of the geosyncline (clay and sand). The incorporation of special sediments, like limestone, will have a very different effect (see page 199).

A final critical remark: Natural rock series are always infinitely more complex than are the model melts, and additional factors which have not been considered in the present discussion may bring about notable modifications in the paths of crystallization. Important factors are, e.g., high water pressure or changes in the oxygen pressure during crystallization. Experiments by Osborn et al. (1959,

1960) suggest that during fractional crystallization under low oxygen pressure a basaltic liquid will change toward an andesitic composition (normal differentiation) but if total composition remains constant, so that oxygen pressure increases in the residual melts and more iron assumes the ferric state, liquids move in the direction of continual enrichment in iron oxides. The quantity and composition of the gas phase during fractional crystallization of basaltic magma is therefore of great significance in determining crystallization courses.

10 · PETROGENY'S "RESIDUA" SYSTEM

According to the previous discussion there can be no doubt that fractional crystallization of basaltic magma leads to residua rich in alkali feldspar ± quartz or nepheline.

Petrographic evidence indicates that the earliest minerals to crystallize are: forsterite and fayalite (olivine), diopsidic pyroxene, and a plagioclase rich in anorthite. By combining the four early crystallizing compounds with alkali-alumosilicates we get a series of eight simple systems which have been investigated. See Schairer and Bowen, 1938, 1947, 1955, 1956; Schairer, 1954, 1957; Roeder, 1951; Schairer and Yoder, 1960.

All these systems yield residual liquids from fractional crystallization which are very rich in alkali alumosilicates. We may conclude, therefore, that in the fractional crystallization of magmas the final residual liquids will exhibit a high concentration of alkali-alumina silicates.

For this reason the system having as its components $NaAlSiO_4$, $KAlSiO_4$, and SiO_2 was called by Bowen the "residua" system. The equilibrium diagram was shown in Fig. III-17. It is repeated here as Fig. III-36 in order to direct attention to the belt of low-melting temperatures, the lowest depression of the fusion surface, whose location is indicated by shading.

Bowen has pointed out that if liquids approaching in composition pure alkali-alumina silicate liquids do play the role of residual magmas, their composition will not have a random position but will be related to the belt of minimum melting liquids of Fig. III-36. This follows from the fact that these liquids, during fractional crystallization, must have approached the alkali-alumina silicate plane along troughs or valleys in polydimensional space that are directed towards the low trough in the alkali-alumina silicate plane. If, then, we take any rock which approaches such compositions reasonably close and

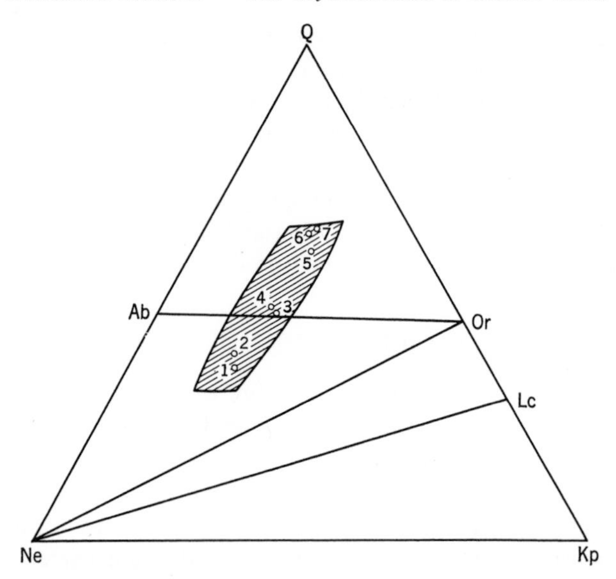

Fig. III-36. The diagram nepheline-kaliophilite-silica, with the low-temperature trough indicated by shading. The plots correspond to the following rocks (after Daly): (1) 15 tinguaites, (2) 25 phonolites, (3) alkaline syenites, (4) alkaline trachytes, (5) pantellerites, (6) 546 granites, (7) 102 rhyolites.

project its composition upon the triangle of Fig. III-36, the resulting projection point should lie in or close to the trough in this triangle, that is, within the shaded area. It is possible for some components to induce a marked offsetting of the position of a boundary curve or valley so that complete coincidence is not to be expected, yet the tendencies should be plain.

In Fig. III-36 the result of the test is graphically shown. Most phonolites, trachytes, syenites, granites, etc., lie in the trough area. The factor exerting the dominant control over the composition of these rocks was, therefore, crystal $\rightleftharpoons$ liquid equilibrium.

Instead of plotting all these rocks without discrimination, Benson (1946) has plotted only such rocks as belong to a field association indicative of comagmatic relationship between the basic and acid extremes of rock variation. Many of these rocks fall outside the low-temperature trough, and so do residual basaltic glasses analyzed by Vincent (1950). See Fig. III-37.

A normal basalt will, through fractional crystallization-differentiation, yield a residual trachytic or granitic magma representing less than a tenth of the volume of the original magma. The presence of the residual liquid within a framework of crystals of augite and plagioclase is, in a view expressed by Holmes (1936), fatal to the idea that it could collect to form a discrete body of, for instance, granite magma, except by intense squeezing.

The residual glassy mesostases of two tholeiites have been investigated chemically by Walker (1935) and Vincent (1950). They contain relatively large amounts of the anorthite molecule and do not fall in the trough area of Fig. III-37 (marked by crosses). It appears that these tholeiitic magmas have followed a course of differentiation which did not lead to a residual magma of granitic composition. However, a point to be considered is that certain aberrations must be expected in glass subject to alterations by weathering and/or hydrothermal action.

Objections may be raised to the view that acid magmas could be derived from basic magmas in the absence of any antecedent development of intermediate rocks. But again a point to be considered is that the shape of the melting curves for plagioclase is such that in the normal crystallization of plagioclase melts, calcic plagioclase will form in relatively large amount, followed by a small amount of intermediate plagioclase, and eventually ending with a large amount of sodic plagioclase. In the examples chosen by Benson for Fig. III-37 there is a general paucity of rocks of intermediate composition.

The marked concentration of rocks of rhyolitic, trachytic, and

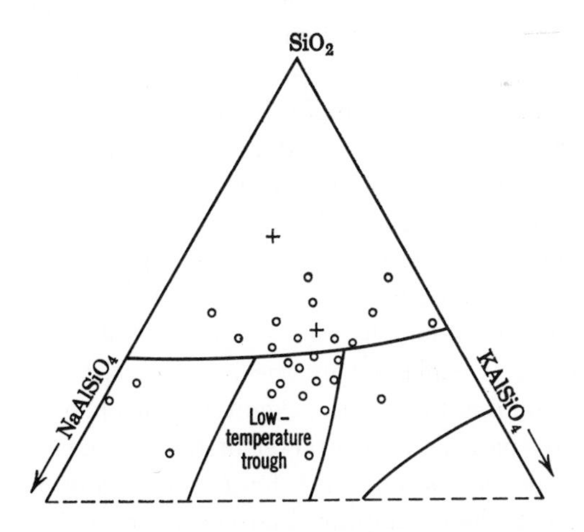

Fig. III-37. Upper part of Bowen's residual system diagram. Small circles are projection points of felsite granophyres, quartz-porphyries, and rhyolites which descend from basic magmas. Crosses are projection points of residual glasses in tholeiites.

phonolitic composition, whether intrusive or effusive, into the "low-temperature trough" is a statistical fact demanding an explanation. This does not mean, however, that no process other than fractional crystallization may have contributed. It will be remembered that assimilation takes place selectively. Crystalline phases with which the magma is saturated or supersaturated are precipitated, and substances toward which the magmas would change in the normal course of fractional crystallization are taken into solution and augment the later differentiates. Here also the laws of crystal $\rightleftharpoons$ liquid equilibria govern the process making it impossible, solely from a study of the qualitative composition of members of rock series, to determine the degree of contamination by refusion.

On the other hand, a series of fractional distillates or of non-consolute liquids would not yield such results. If such processes had been active, the projection point of the corresponding rock would have been thrown out of the low-temperature trough of Fig. III-36.

The fact that rocks whose field relations point to a close connection with basalts, often show a wide spread suggests that processes like fractional distillations, etc., played a contributing role. This is a reasonable conclusion, for vapor and other volatile constituents are by necessity concentrated in the residua systems. On the other hand, granites of igneous-anatectic relations, the great bulk of which makes it difficult to accept them as derivatives of any basalt magma, often lie in the trough area and may, therefore, best be explained in terms of remelting of siliceous crustal material.

Pegmatites are rocks of unusually coarse grain. They may have any composition, but are usually light colored and high in alkali feldspar, corresponding to the residual system of petrology.

Magmatic pegmatites may develop in the last stages of the normal evolution of the igneous rocks from residual solutions rich in alkali-alumina silicate and volatile constituents. The volatiles (H_2O and other gases, as described on page 126) effectively lower the temperature of crystallization and the viscosity of the solutions. The coarse-grained structure is explained by the low viscosity; the composition, usually corresponding to granitic, syenitic, or nepheline-syenitic rocks, is explained by the residual character of the magma; and the content of rare minerals, a distinguished feature of many pegmatites, has been explained on the assumption that certain rare chemical elements, whose ionic radii are different from those of the rock-making elements, concentrated in the residual solutions.

However, many of the minerals in pegmatite bodies are secondary, that is formed by replacement. Generally speaking, processes of replacement must have played a large part in the formation of most granite pegmatites; beryl, albite, all lithia minerals, all manganese minerals, and all phosphates seem to have been introduced into the pegmatite by replacement.

Famous occurrences are the lithium-rich pegmatites in the Black Hills of South Dakota and in several places in California. The replacing minerals are lepidolite, spodumene, petalite, amblygonite, and others. Worthy of special note are the Li-Rb-Cs-As-Sb minerals in the pegmatite at Varuträsk in northern Sweden.

It is worth mentioning that many pegmatites, which used to be regarded as typical residua systems, likewise fall well outside the trough of Fig. III-36. (See Fig. III-85.) It is believed, therefore, that such rocks do not result from strictly magmatic processes but actually represent products of metasomatism, metamorphic differentiation (pages 297–298), and/or granitization (pages 355–356).

11 · SILICATE WATER SYSTEMS

Most magmas contain dissolved water which has a profound effect on the constitution of silicate melts (see page 143), and is important in determining the course of crystallization. Another conspicuous effect of water is the great lowering of the liquidus temperature. See Fig. III-38. At water pressures corresponding to a depth of 20 km, the lowering of the liquidus may be 100°C (diopside), or as great as 400°C (albite). See Fig. III-39.

The solubility of water in melts of granite and obsidian has been determined experimentally and follows very nearly the curve shown in Fig. III-40. In melts of 25 per cent quartz, 75 per cent orthoclase, and in melts of mixtures of albite and orthoclase (corresponding to syenitic magmas) the solubility of water is from 1 to 1.5 per cent lower.

Figure III-40 demonstrates that rock magmas are able to dissolve some water (up to about 8 per cent, depending upon the composition of the magma) with the help of moderate pressures, but that extremely high pressures are necessary to keep more than about 10 per cent of the water in solution. This situation is also demonstrated in Fig. III-41. It should be observed that granite does not melt at a point, but like other polycomponent systems it melts (and solidifies) over a temperature interval limited by the liquidus temperature (T_L) and the solidus temperature (T_S). Above the liquidus everything is molten. In granite, however, no true solidus temperature is observed,

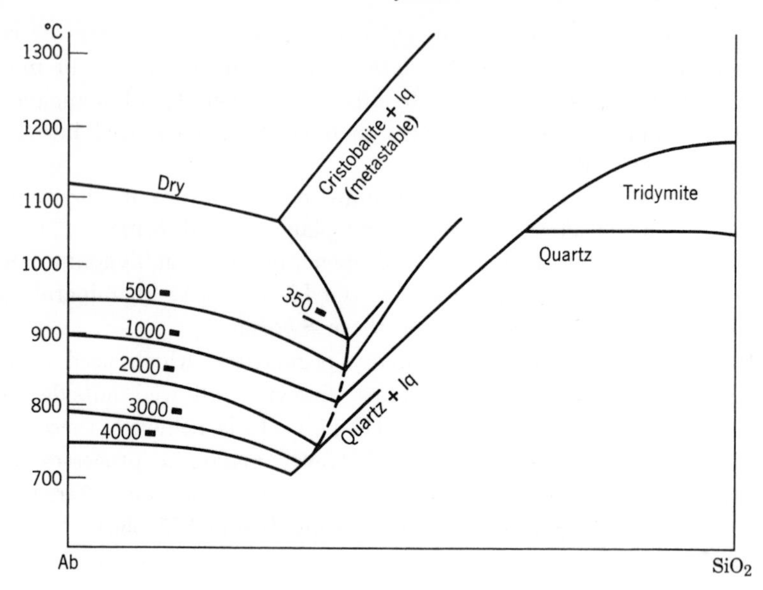

FIG. III-38. Isobaric equilibrium relations in the system Albite-SiO_2 under different water pressures. Pressures are indicated thus: 500 ▬ = 500 atm. Increasing pressure brings increasing amounts of H_2O into the silicate melt, which again causes increasing depressions of the melting temperatures. But the eutectic relation between albite and silica remains. (After Tuttle and Bowen, 1958.)

and T_S represents the point at which the residual liquid becomes separated from the crystalline phases.

With reference to Fig. III-41, we shall analyze the crystallization history of a completely liquid granite magma of the composition of the ternary eutectic at about 850° and containing 3 per cent of water in solution (point *a*). Let us assume that the magma is located at a depth of 4 km; this corresponds to an incumbent load (external pressure on the magma) of about 1000 atm, but the internal vapor pressure in the magma is much lower. On cooling the magma will begin to crystallize at about 780°C (point *b*), and continue quietly until a temperature of about 720° is reached (point *c*). At this temperature about half of the system would be crystalline, the residual melt containing about 6 per cent water in solution. The pressure necessary to hold this amount of water in solution at 720°C just corresponds to the external pressure at this depth, and thus further crystallization would be accompanied by an ebullition of water (and the formation of a magmatic vapor phase). This is the so-called second boiling point of a magma, induced by increase in the vapor tension of the residual liquid through separation of anhydrous crystals, as distinct

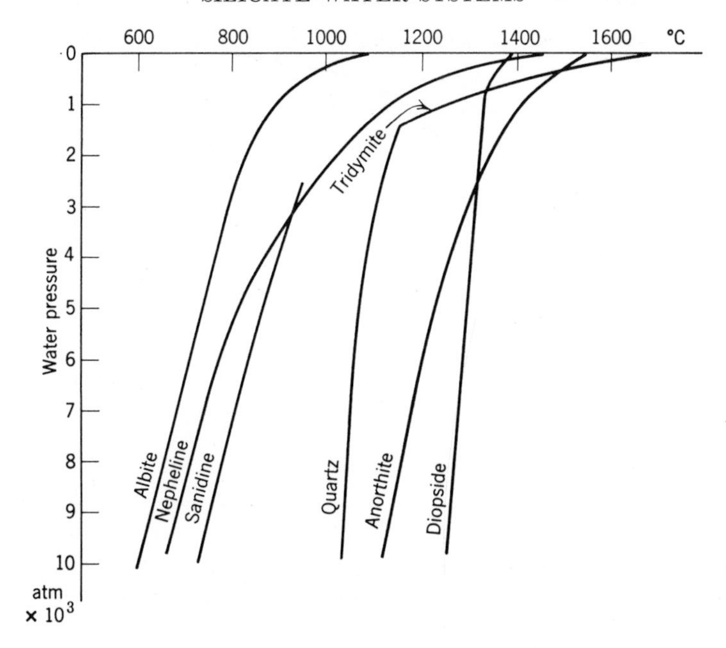

FIG. III-39. Liquidus temperatures in systems of H_2O in combination with $NaAlSi_3O_8$, $NaAlSiO_4$, $KAlSiO_8$, SiO_2, $CaAl_2Si_2O_8$, $CaMgSi_2O_6$. (Yoder, 1958.)

from ordinary boiling, which is caused by increase in the vapor tension through addition of heat. The boiling will continue until the temperature-pressure curve drops below the external load pressure.

At this point it should be kept in mind that the melting temperature decreases with increasing pressure of water vapor owing to the

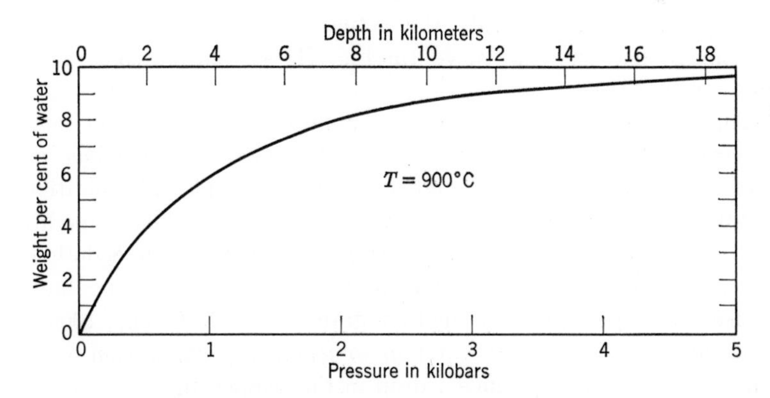

FIG. III-40. The solubility of water in molten granite as a function of pressure at the 900°C isotherm. (After R. W. Goranson, *Am. J. Sci.,* 1931.)

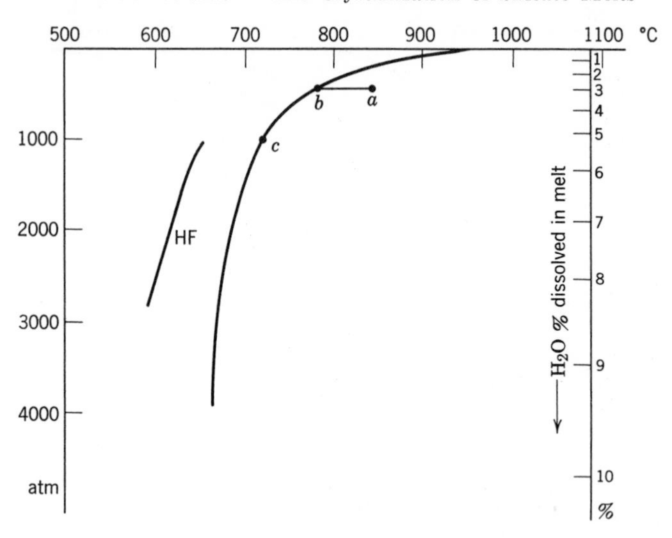

Fig. III-41. Beginning of melting of natural granite under H_2O pressure. The curve HF illustrates the effect of adding 8 per cent hydrofluoric acid to the water. (Tuttle and Bowen, 1958.)

increased amount of water dissolved in the liquid phase at the higher pressure. But high pressure has another effect: According to the Clapeyron equation the melting point must increase with increasing pressure when the solid phases (the minerals) are heavier than the liquid (magma). And whereas the effect of the dissolved water on the melting point is at first very marked, it diminishes rapidly (i.e., it is counteracted by the pure physical effect of high pressure), and the difference between the value for 3000 atm and that for 4000 atm is only about 10°. Plainly the actual effect is soon reduced to a vanishing quantity—the curve (in Fig. III-41) flattens out and will soon start to rise.

This enables us to draw a very important conclusion of general applicability: The lowest temperature of most silicate magmas in the earth is about 660°C. It should be noticed, though, that addition of certain gases (for example HF) will further lower this temperature.

The degassing of water vapor from a magma is an endothermic process. The heat absorbed increases with the temperature but decreases with pressure. In molten granite at 900°C and 1000 atm containing 5.7 per cent of water in solution, the evaporation of 1 per cent of water would produce a drop in the temperature of about 3°C.

However, if the evaporation of water entails crystallization, as in lavas which are not superheated, the heat drop will be more than offset by the heat of crystallization generated. In systems in which a volatile substance is held in solution by pressure a release of pressure with consequent loss of volatiles will promote crystallization whether or not heat is lost from the system (crystallization proceeds adiabatically). At first thought it would appear absurd to suggest that a molten silicate liquid could crystallize while the temperature is being raised, but this happens if the heat generated by crystallization cannot escape. In an average case the heat of vaporization will take up only a small fraction (1–3 per cent) of the heat of crystallization, and the remaining heat will go into the melt (lava) whose temperature will rise on the order of 100° to 300° during adiabatic crystallization. See pages 139–140.

Volatiles other than water also affect the melting relations. In the presence of 8 weight per cent HF solution granite begins to melt at 595°, which is 75° lower than in the presence of H_2O alone (Wyllie and Tuttle, 1961).

The data so far discussed have formed the basis for various conclusions concerning gas content of rock magmas and explosive energies associated with eruptions. However, additional complications arise from the fact that natural rock magmas do not represent truly closed systems; the gaseous and "volatile" molecules or atoms often find a chance to diffuse away and percolate toward the surface. This general rule in geology explains the degassing of the earth as a geological process that has been going on since the beginning of time.

Figure III-42 shows graphically the amount of water that can be dissolved by albite melts (or granite melts) under differential hydrostatic pressure, that is, when the melt is subjected to a pressure higher than that resting on the vapor phase. This occurs in nature if the country rock surrounding a magma is pervious to water but impervious to the silicate melt, the magma therefore being under a confining pressure equal to (or greater than) that due to the weight of overlying rock, whereas the water would be under smaller confining pressure.

In such situations the water will be "squeezed" out of the magma before the saturation point is reached. To give an example: at a depth of 3.7 km, corresponding to a pressure of about 1000 atm (*a* in Fig. III-42), a magma is able to dissolve about 5.5 per cent of water if the walls surrounding the magma are impervious to water. But if the walls are semi-permeable to water, the effective pressure on the water phase decreases and some of the water diffuses out of the magma, the exact amount being determined by the degree of permeability (position of point *b* in Fig. III-42). The same principle will cause

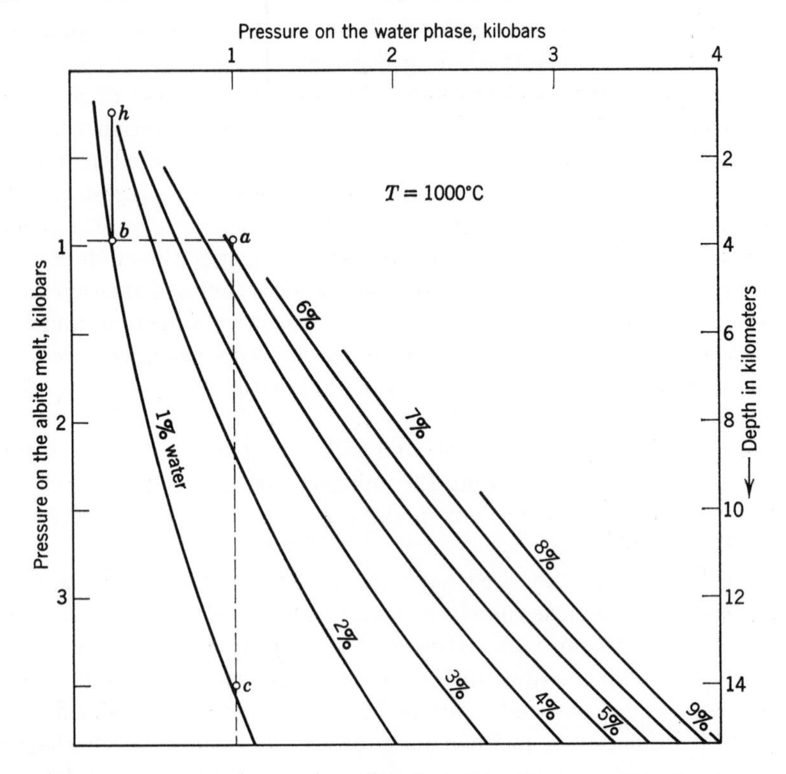

FIG. III-42. Solubility of water in albite melts at 1000°C under differential confining pressure. The curves, 1% water, 2%, 3%, . . . , 9%, indicate the conditions at which the albite melt is able to dissolve 1, 2, 3, . . . , 9 weight per cent of water. (Drawn from thermodynamic data of Goranson, 1937.)

water and other volatiles to concentrate at the top of magma reservoirs.

Incidentally, intermittent volcanic activity may be explained by a mechanism opening up, and again partly closing, the avenues of diffusion.

Observations in nature indicate that most lavas and magmas contain less water than the amount corresponding to saturation. Although all observers are impressed by the apparent amount of gases visibly liberated during volcanic eruptions, volumetric estimates indicate that less than 1 per cent by weight is discharged from the lava (a content of 40 cc per gram rock corresponds to 0.7 per cent by weight, the average density of volcanic gases at 1200°C and 760 mm Hg being about 0.18 kg per m^3). The obvious explanation is that the system magma-gas is subjected to differential pressures corresponding to the conditions pictured in Fig. III-42. Contrary to common belief, the solubility of the gases therefore *decreases* with depth (corresponding to the line *a–c* in the diagram). On this theory Verhoogen (1948) has interpreted the eruptions of Nyamuragira in Central Africa, including the formation of numerous small lava lakes of Hawaiian type.

The thermal activity in Iceland or in the Yellowstone Park is commonly regarded as a manifestation of boiling of the subjacent magmas. Since the waters and other gases in these areas find egress through the overlying rocks, the subjacent magma cannot be "saturated" with water; but water vapor is squeezed out of it (intermittently) according to the principle of differential confining pressure.

C. MAGMAS AND LAVAS

1 · ORIGIN OF MAGMAS

Here in the light of the fiery lava which has risen 1800 miles through the crust of the earth, which makes earthquakes for our seismographs twenty miles underground, which pours out under the sea or spouts up colossally on top of Mauna Loa, ought we not to be reminded of earth power?

T. A. Jaggar, *Volcanoes Declare War,* 1945.

Magma is the stuff out of which igneous rocks are born. According to a definition of Rittmann (1958), it is a completely or partially molten mass of silicates with dissolved gases, occurring within or beneath the crystalline crust of the earth and being capable of intruding as such into fissures and erupting at the earth's surface, splitting itself into lava and volcanic gases.

What are the sources of magmas? In the absence of any definite knowledge of the early development of the planet, no satisfactory basis can be found in cosmogony. Present indication of the quantity of radioactive material in the rocks, and present knowledge of the geothermal gradient, suggest that the earth is manufacturing heat (see page 7). Thus there seems to be more than enough radioactive energy for effecting a periodic refusion of parts of the crust, aided perhaps by periodic release of pressure (see page 134), or by blanketing due to a local thickening of the overlying crust.

The calculations on page 304 demonstrate that a layer of 25 km having the radioactivity of an ordinary granite would produce heat at a rate corresponding to that of the heat flow of the earth. Many sedimentary layers in geosynclinal areas have a high radioactivity, indeed. By their own radiogenic energy they will materially contribute to the heating of the deep parts of the geosynclines, and within reasonable time a temperature at and above that of the melting of the rock material may be attained.

Thus it seems reasonable to assume that magmas, mostly of granitic-granodioritic composition, are generated in the deeper parts of the geosynclines by *anatexis* or differential melting of rocks.

However, other kinds of magma must be postulated. The dominant role of basaltic lava has repeatedly been stressed. There have been copious effusions of basaltic lava at all times; indeed, it seems that whenever a deep rift develops in the crust, there is basaltic lava ready made to ascend and flood the surface. For these reasons the hypothesis of an earthwide substratum of basaltic magma was early developed. It was for a long time advocated by Daly (1914–1933), and became generally accepted by geologists as a reasonable explanation of observed facts.

The physical properties of the substratum was supposed to be that of a melt at high pressure. However, the fact that the transversal seismic waves were able to penetrate this layer made geologists (including Daly) abandon the idea, as well as the idea of a glassy (molten) peridotite shell.

But again new knowledge may reverse the conclusion. Experiments also demonstrate that transversal waves are transmitted through liquids if the periods of the waves are shorter than the time of relaxation. According to geological evidence, the time of relaxation in the substratum must be of the order of thousands of years (v.i.) while the periods of seismic waves are measured in seconds; so, the substratum, though liquid in the physico-chemical sense (molten), behaves seismically like a solid body.*

In this way we may tentatively revert to the hypothesis of a "glassy" basaltic sublayer. Thus there is, in addition to the orogenic magmas produced by local remelting of the crustal rocks, a (worldwide?) primary basaltic magma. Its locale is below the crystalline

* Early experiments by Birch and Bancroft (1942) seemed to disprove that rock "glass" under these conditions could transmit the *S* waves, and they found the Maxwell relaxation theory to be inadequate to account for the relationship between rigidity, viscosity, and internal friction of glasses. However, Kuhn and Vielhauer (1953) have performed new experiments, found the propagation of *S* waves possible, and have concluded that the theory of Maxwell holds well in a new form given by Kuhn. The fact that wherever the crust is deeply rifted, be it in continents, oceans, or geosynclines, basaltic magma is available and capable of invasion is a proof of the existence of a subcrustal basaltic magma stratum. The properties and behavior of this submagma are elucidated by Rittmann (1960), as explained in Fig. III-43.

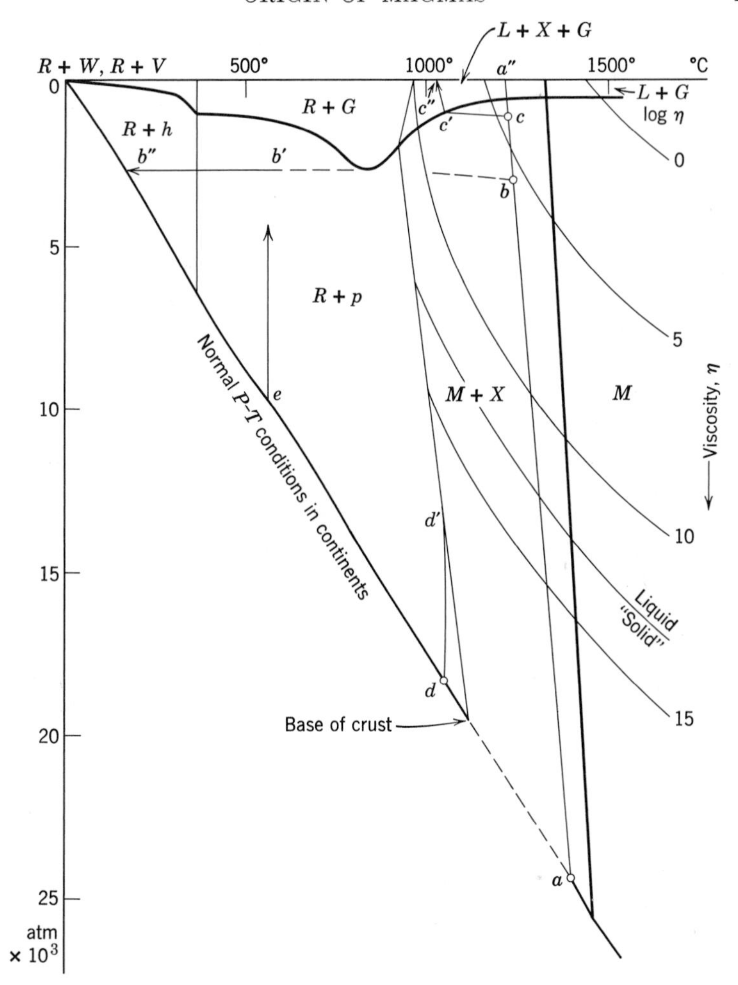

FIG. III-43. Schematic *P-T* diagram of olivine-basaltic matter. (After Rittmann, 1960.) M = Molten magma; L = Lava; R = Rock (solid); V = Vapor; W = Water; X = Phenocrystals; G = Gas; p = pneumatolytic solution; h = hydrothermal solution.

1. The opening of an abyssal fissure to the magma stratum causes a sudden release of pressure from a to b, thereby the viscosity drops from 10^{20} to 10^5 poises, so that at b the magma is seismically liquid; however, no change of state has taken place; under all conditions during the ascent a-b-c, the melt with suspended crystals remains a magma.

a. By further lowering of pressure, or temperature (to c'), the magma arrives in the field $(L + X + G)$ and becomes a foamy magma transitional into a lava (at a'' or c'').

b. If the magma does not reach the surface, but is arrested at b, it will cool slowly and by fractional crystallization the melt will change from basaltic

(*Continued p. 134.*)

crust at about 60–70 km depth in continental areas, shallower under the oceans. See Fig. III-44.

The change of the melting point of basalt with pressure is important. It may be computed from the well-known equation of Clausius-Clapeyron $dT/dP = T\,\Delta V/\Delta H$. The numerical values are: T (beginning of melting at 1 atm) = 1273°K (1000°C), $\Delta H = 100$ cal/g, $\Delta V = 0.033$ cm^3/g. Consequently $dT/dP = 10°$ per 1000 atm. Suppose a slab of basalt lies at the depth of 60 km; here the temperature may attain 1190°C, and the pressure is 19,000 atm (point d on Fig. III-43). The point of initial melting of the basalt is raised from 1000° to 1190°, so the basalt is still crystalline. On release of pressure the slab will start to melt. If the release is taken to be 5000 atm, the slab will be 50° above its initial melting point at the new pressure. The heat capacity is 0.2 cal/g, so the heat available to melt a portion of the slab is $0.2 \times 50 = 10$ cal/g. The heat of melting is, however, about ten times larger, or about 100 cal/g. Therefore only 10 per cent of the slab will melt. It will be a differential melting (anatexis) of the rock and the melt thus produced *will not have the composition of a basalt,* but that of a trachyte or a phonolite.

It is necessary, therefore, that the basaltic substratum is in a molten (non-crystalline) state at its subterranean locale. Otherwise it would not reach the surface with a homogeneous basaltic composition as we actually observe. (See legend to Fig. III-43.)

An estimate of the depth can be made at the Hawaiian volcanoes if one assumes that the molten magma column of height H and density d' will balance the column of the surrounding solid rocks of height h and density d. Equilibrium requires that

$$d \cdot h = d' \cdot H$$

(*Fig. III-43, continued.*)

to andesitic-trachytic, while its relative gas content continuously increases. At b' in the field $(R + p)$ the magma has congealed to a plutonic rock (olivine gabbro with an intergranular pneumatolytic solution, which on further cooling (b'') passes into a hydrothermal solution $(R + h)$.

2. If a fissure opened only down to the *crystalline* basaltic crust, at d, initial melting would start (d'). But the small amount of interstitial liquid which could form would have a trachytic or phonolitic composition—never would it be basaltic. The point is that only a *small* fraction (the order of magnitude is 10 per cent) of the rock can melt, for the heat of melting must be taken from the rock itself.

3. Under P-T conditions of the lower sialic crust (point e), the formation of magma from a pre-existing rock of basaltic composition upon release of pressure is absolutely impossible. The rock remains as it is. Only at very low pressures may the intergranular pneumatolytic solutions eventually find egress and emerge at the surface as fumaroles or hot springs $(R + G)$.

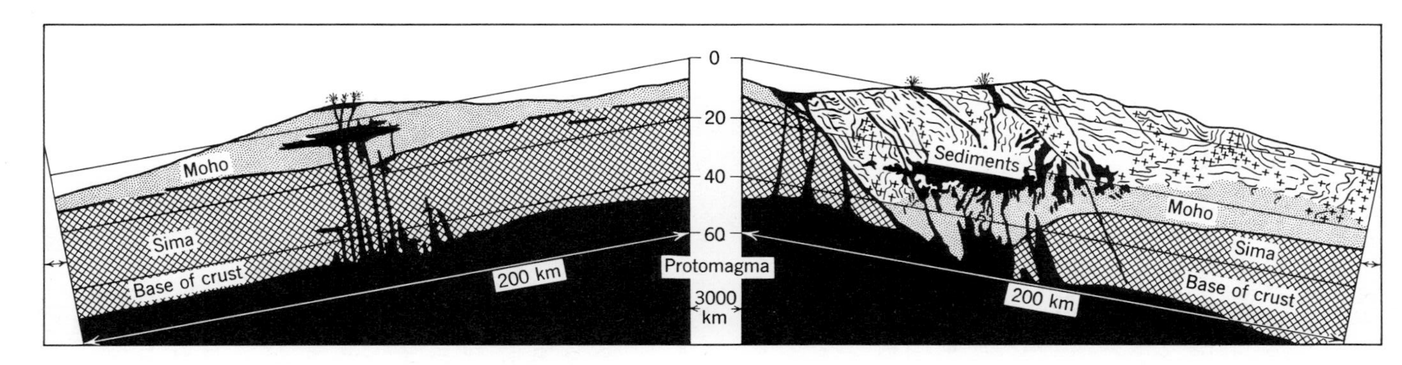

FIG. III-44. Schematic sections through the oceanic crust (*left*), and through the continental crust (*right*). The volume of extruded basalt is enormous; it covers the floors of the oceans, even plateau basalts in the continents are enormous; the basalt effusions are homogeneous in chemical composition in space and time. All basalts derive from the subjacent protomagma. Along the mid-ocean ridges it ascends through sima without suffering contamination. Along rift valleys and other major fissure systems in the continents, it may occasionally penetrate to the surface, but in most situations the thermal energy and other "emanations" of the protomagma contribute to the heating of the superincumbent rocks—for example, of the geosynclinal sediments—and produce metamorphites and anatectites with the concomitant formation of diapiric plutons, and of sialic magmas and attendant volcanism.

If in Hawaii $H - h$ is about 10 km, $d = 3.3$ and $d' = 2.8$, then $h = 56$ km. Earthquake evidence in Hawaii indicates that the basalt lavas rise from a depth of 40–60 km.

A schematic picture is shown in Fig. III-44, giving the author's present ideas of the relations between the world-wide, homogeneous, deep-seated basaltic submagma and its rocky products on the one hand, and the anatectic sialic magmas of the orogenic areas with their derivatives of greatly diversified rock complexes on the other.

2 · VISCOSITY AND TEMPERATURE OF MAGMAS AND LAVAS

The *viscosity* of natural magmas and of silicate melts depends on temperature, pressure, and chemical composition. Basic magmas and magmas high in volatiles show a lower viscosity than acid magmas ("basic" and "acid" refer to the content of silica). See page 142.

Thus basaltic lavas of Hawaii are relatively fluid. Down a gentle grade of 2° a velocity of flow of 400 m per hour was measured. In ideal Newtonian liquids the velocity of flow, v, and the viscosity, η, have the following relation:

$$v = \frac{dh^2 \sin \alpha}{3\eta} \cdot g$$

where α is the angle of inclination, g the gravity constant, d the density, and h the depth. But in a lava flow no such relations hold, for the steep temperature gradients, the solid skin constantly forming at exposed surfaces, and other crystallization phenomena greatly complicate the mechanism of propagation. In Hawaii, velocities of about 1 km per hour are usual, and the measured maximum is probably about 60 km per hour. The temperature of the fresh Hawaiian lavas is about 1000° to 1100°C.

Another example of the fluidity of basaltic lava is afforded by an extensive lava flow along a water grade in the western part of the Grand Canyon, involving a descent of 210 m in 135 km.

A much higher viscosity is usually exhibited by acid lavas as, for example, those encountered in many of the Italian volcanoes, as in Sakurashima, Japan, or at Mt. Pelée in the West Indies. Such lavas either flow very slowly (Sakurashima) or they do not flow at all, as observed at Mt. Pelée. Through several orifices of this volcano the lava rose slowly, formed large monoliths, which because of their high viscosity were unable to flow but grew into the air as weird pinnacles, broke, and so rolled down the slope as solid lava blocks. (Lacroix, 1904.)

Various viscosity data are entered in Table III-14. Unfortunately our data reveal very little as to the viscosity relations of true magmas. Magmas are inaccessible, and thus far natural lavas likewise have confounded most attempts at direct viscosity measurements. During the recent eruption of Hekla in Iceland, Einarsson (1949), by using a blunt stick as a "lava-penetrometer," found viscosities from 10^5 to 10^7 poises. Field measurements in Hawaii indicate a viscosity of about 4.5×10^4 poises, and Minakami's (1951) results on the basaltic lava of Oo-sima give about 2×10^4 poises at 1100°. Unfortunately the petrologist in most cases is limited to the study of the rock. And it is a long extrapolation from the dead rock to the living magma. No reheating can return to the sample the volatile constituents and the high vapor tension it enjoyed in its magmatic life.

The velocity of diffusion and of crystallization is a function of the viscosity. With increasing viscosity the lava shows an increasing tendency to congeal as glass. Acid rock glasses (obsidian, pitchstone) are much more common than basaltic glasses. Viscous magma does not easily react with unstable phases. Zoning in crystals is therefore common. On page 83 an explanation is given of how zoned crystals of plagioclase (basic core, acid shell) develop in a cooling magma. In certain lavas intricately zoned feldspars with sodic center, or with alternating zones as shown in Fig. III-45, are observed. In order to

TABLE III-14

VISCOSITIES MEASURED IN THE LABORATORY

	Viscosity, g/sec/cm	Temperature, °C
Water	1×10^{-2}	20
Pyroxene	5×10^{0}	1450
Glycerin	1×10^{1}	20
Basalt	2×10^{2}	1400
Olivine basalt	3×10^{3}	1200
Albite	4×10^{4}	1400
Albite	$n \times 10^{8}$	1150
SiO_2, glass	1×10^{10}	1440
SiO_2, glass	1×10^{12}	1300
Obsidian	1×10^{12}	800

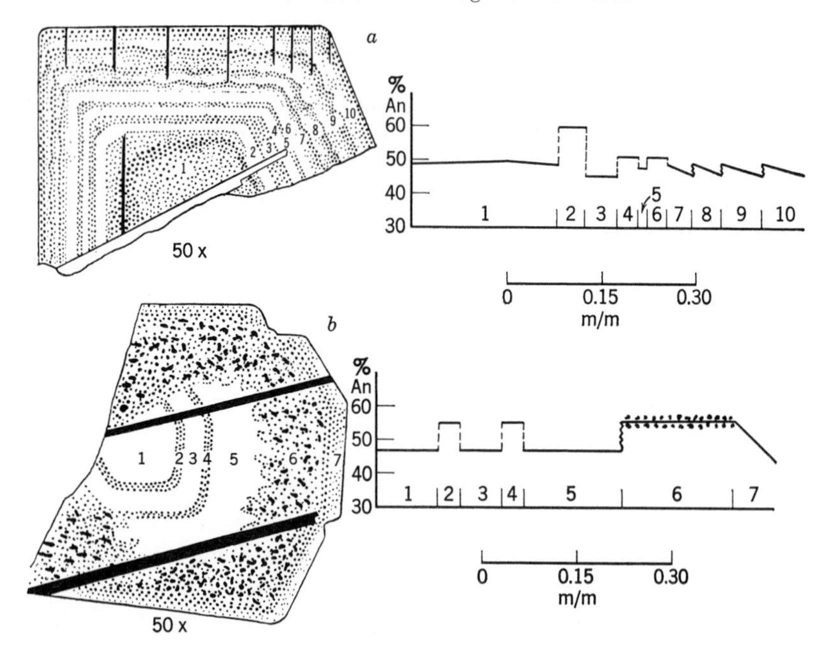

Fig. III-45. Examples of zoned plagioclase crystals (after G. Paliuc, 1932). The curves represent the anorthite contents of the consecutive zones. Ordinate = anorthite content. Abscissa = width of the zones.

explain this it has been postulated that lava pulsations are brought about by discharge of mineralizers, or by incomplete decantations in zones of magma with different temperatures and viscosities, or else that the feldspars did not crystallize from a magma of the composition of the rock in which they are found. Again our lack of exact knowledge of the properties of magmas is frustrating.

Movements of crystals in the magma are of great importance. The early crystals usually exhibit a density different from that of the magma and will, accordingly, float or sink under the influence of gravity.* The following relationship is known as Stokes' law:

$$v = \frac{r^2(d - d')}{\frac{9}{2}\eta} \cdot g$$

where v is the constant velocity of sinking (or upward propagation) of a sphere with radius r and density d, while the melt has the density

* Most solid silicates are heavier than the magma from which they crystallize. The only exception is probably leucite and sanidine which are known to rise in the magma under Vesuvius (Rittmann, 1960).

d' and the viscosity η. Studies of thin sections of lavas with the microscope have shown that even rather small crystals (some millimeters across) in a partly glassy ground mass can move either up or down relative to their surroundings with a force sufficient to cause a front current and a wake to develop. Images of these currents are now stamped in the congealed glass.

There are many examples supposed to demonstrate large-scale gravitative separation of olivine and other heavy minerals to form strongly mafic phases of floored injections. Among them are the Palisades sheets along the Hudson River, New Jersey, the Mt. Wellington sill, Tasmania, many of the Karroo dolerite sheets of South Africa, and many of the Tertiary sills in West Scotland.

In deep-seated rocks similar textures are unknown. However, concentrations of heavy crystals in the lower part of deep-seated massive rock bodies have been observed occasionally, and interpreted as the result of a sinking of the heavy crystals in a magma chamber.

The viscosity will increase with pressure, but decrease with temperature. If temperature and pressure are raised simultaneously so as to keep the volume constant, the viscosity decreases, as a rule. At depths of about 60 km, corresponding to 1200°C and 17,000 bars the magmatic viscosity is probably around 10^{13} poises. According to the rate of deformation of the Scandinavian Peninsula and according to other data of isostasy the viscosity of the ultradeep "magmas" seems to be about 2×10^{21} poises (see page 133).

It is, therefore, a very unsatisfactory account that we can give today of the deeply buried stuff of which the igneous rocks are born. The stuff may ascend, in the deeper parts very slowly, in the higher parts much more rapidly, at the same time transforming itself into more ordinary rock melts. But the deep convection currents effecting the growth of mountains have their place, not in rock melts but in the viscous disordered substratum (state of *oligophase*, page 14).

The *temperature* in the zone of dynamomorphism follows the geothermal gradient and increases with depth. At higher levels, in the zones of plutonism and volcanism, more ordinary magmas individualize and may assume temperatures slightly higher than their walls, but we do not know whether or not this superheat is appreciable; the temperature of crystallization is usually in the range 1100°–700°.

Lavas are rarely superheated. The volatile constituents originally present boil away, converting the lava into nearly dry melts similar to those studied in the laboratories. During such adiabatic crystallization of a hydrous lava the temperature may rise 100°–300° due to the heat of crystallization. It is worthy of note that this behavior is very different from what has been discussed so far in the melting diagrams. In hydrous lavas, crystallization proceeds with *increase* of temperature and decrease of the volatile constituents. Lavas of the 1947 eruption of Hekla on Iceland kept a temperature of about 1000°C for many months; spongy lava which was squeezed out of the semi-solid front of the flow, thereby losing its gases, crystallized rapidly in heating itself to about 1150°. In places where gases were released at a great rate the temperature was increased to 1250° (Einarsson 1949). In the crater of Kilauea, Hawaii, there long existed a lava lake, known as Halemaumau. Flames were observed rising from the general surface of the lake, indicating combustion of the emanating gases with the air. In such places the temperature was particularly high. Temperatures of the lava lake itself and of the burning gases over the lake are graphically shown in Fig. III-46.

Lavas will retain some heat at long distances away from their source areas, and flows have been reported which were still fluid enough to maintain their motion 10 years after the cessation of emission (Scrope, 1825).

The *time required for crystallization* of a large magma chamber is indicated by computations of E. S. Larsen (1948). Assuming that the batholith of Southern California (exposed over a distance of about 350 miles) was magmatic, and that it reached the surface, it would

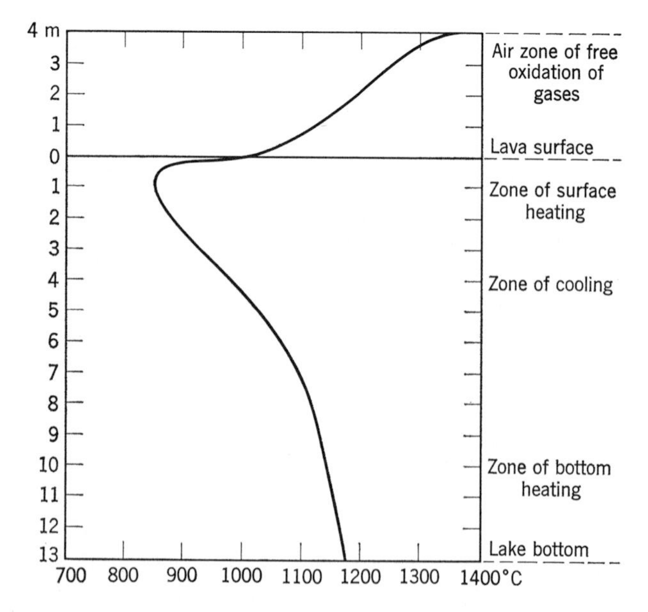

FIG. III-46. Thermal gradient of the Kilauea lava lake. (After Jaggar, 1917.)

have crystallized to a depth of 6 km in about one million years, if the loss of heat was by conduction and if no source of heat, such as radioactive heat, was present. An estimate of the effect of the heat carried by the batholith on the present temperature gradient at the surface indicates that the temperature gradient would be raised about 10°C per kilometer, a small amount compared with the range in temperature gradients found in different places.

Actually the picture is much more complicated: The batholith must have had a cover, radioactive heat is present, and the crystallization takes place through loss of gases as well as heat.

Such complications may explain the different value found by Bederke (1947) on a deductive basis from studies of the regional metamorphic mineral facies surrounding an inferred deep-seated granite intrusion in Altvatergebirge, Silesia. He thinks that the metamorphic recrystallization was effected by post-tectonic granitic intrusion, that the depth to the magma was 2000 to 3000 m, and the time of heating about 50,000 years, and that the geothermal gradient increased ten times (attaining 1°C per 3 meters). The time required for the crystallization of the whole granitic body is of the order of magnitude of one million years.

Wegmann has criticized these conclusions and calculations and thinks that the regional metamorphism is unrelated to deep granitic intrusions, but controlled by other events producing the necessary amounts of heat, for example, intrusions of swarms of basic dikes often accompanied by deep-seated masses of rock, remelting, hybridization, syntexis, ion mobilization and migration.

3 · CONSTITUTION OF SILICATE MELTS

The entropy of fusion of silicates is not high, so that no major changes in structure are to be expected on melting. On general grounds one would, therefore, expect the liquid structure to resemble the solid. Experiments have confirmed this.

Molten silicates possess an ionic constitution in analogy with solid silicates. The extent of ionic dissociation is comparable to that of aqueous solutions of strong electrolytes. Typical cations are Na^+, K^+, Ca^{2+}, Mg^{2+}, Fe^{2+}, which occupy no fixed position in the melt but exhibit a high degree of mobility. They are called network modifiers in contradistinction to the anions, which are network builders.

The anions are predominantly silicon-oxygen tetrahedra linked together by oxygen bridges to one-, two-, or three-dimensional networks. The network is similar to that found in crystalline silicates but more irregular. See Fig. III-47. In basic melts (low in silica)

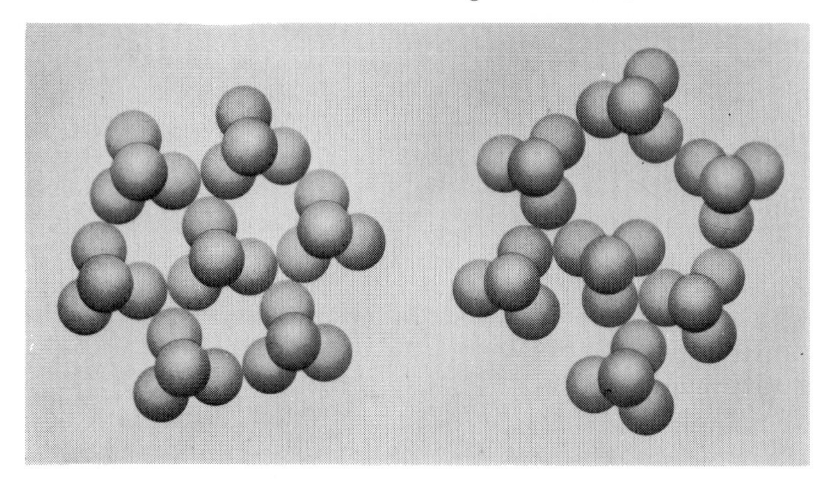

Fig. III-47. *Left:* the regular pattern of the $(SiO_4)^{4-}$ tetrahedra in crystals (cristobalite structure). *Right:* the irregular pattern of the $(SiO_4)^{4-}$ tetrahedra in glass (silica glass structure). (Photo Carl Zeiss.)

individual tetrahedra of $[SiO_4]^{4-}$ are present, but with increasing silica content polymerization (formation of network) increases. The higher the polymerization, the higher the viscosity of the melt. See Table III-15. The viscosity is found also to increase upon adding Al_2O_3, proving that Al can replace Si in the polymerized anion, as it does in some crystalline silicates, in feldspars, for example.

The degree of polymerization is markedly influenced by the nature of the cations present. In simple silicate melts the viscosity increases in the sequence Fe-Mg-Ca-Sr-Li-Na-K (cf. Fig. III-49); that is, the

TABLE III-15

Relation between Viscosity and Degree of Linkage of the SiO_4–Tetrahedron

Composition	Si:O	Viscosity in poise
SiO_2	1:2	10^{10}
Na_2SiO_5	1:2½	28
Na_2SiO_3	1:3	1.5
Na_4SiO_4	1:4	0.2

smaller the polarizing power of the cation, the higher the polymerization of the silicon-oxygen anion.

Addition of H_2O radically reduces the viscosity, for $(OH)^-$ will break the oxygen bridges and destroy the network. Upon release of pressure, water will exsolve from a silicate melt. Thus, when a deep-seated magma loaded with water and of low viscosity erupts, the water will boil out, and the magma will immediately stiffen and change into a viscous lava.

4 · LIQUID IMMISCIBILITY

It is a well-known fact that many substances, which are capable of mixing as liquids in all proportions at a certain temperature, may separate into two liquids upon cooling or heating. Water-nicotine is a familiar example. Fig. III-48 explains the general phenomenon.

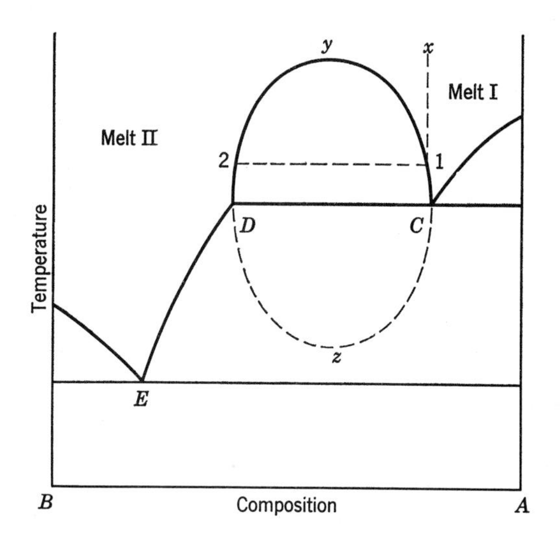

Fig. III-48. A binary system with two immiscible liquids. A melt, x, is cooled. At 1, small drops of liquid 2 separate in equilibrium with liquid 1. At the temperature CD a liquid of composition C is in equilibrium with a liquid of composition D. At this temperature crystals of A begin to separate from both liquids, until liquid C is used up. The mixture is now composed of crystals A and liquid D. Through further cooling, the liquid will follow the course D-E, continuously separating crystals A until E, the eutectic point between A and B, is reached. The immiscibility gap has, in principle, the shape of a ring (in ternary systems the shape of a spheroid). It may have an upper and a lower critical solution temperature (y and z, respectively); but, in the figure, z is not realized because these parts of the exsolution ring are cut by the stability fields of crystalline phases.

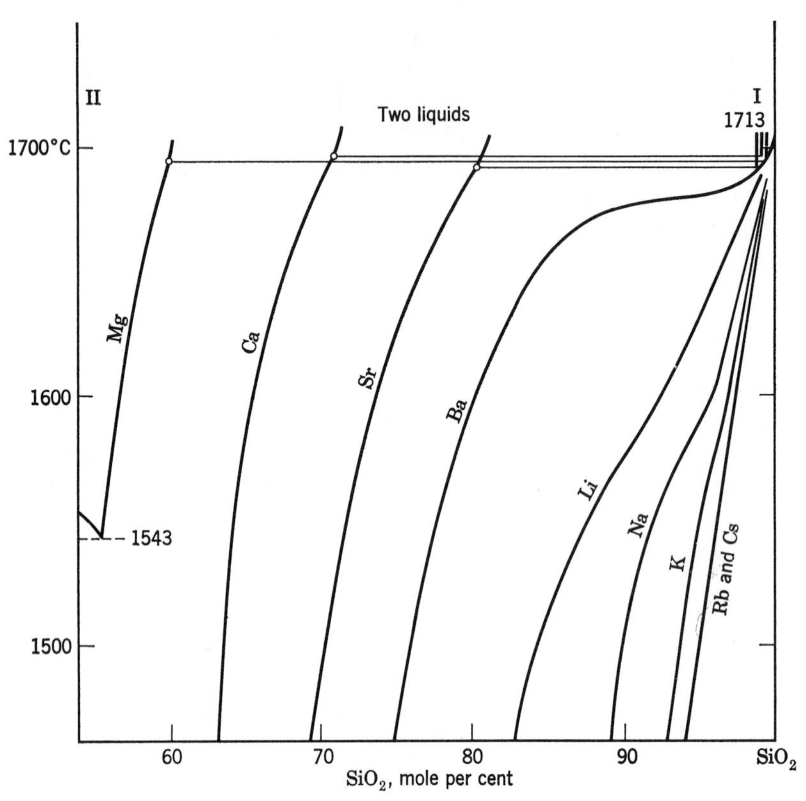

FIG. III-49. Melting curves for the systems (1) $MgO—SiO_2$, (2) $CaO—SiO_2$, (3) $SrO—SiO_2$, (4) $BaO—SiO_2$, (5) $Li_2O—SiO_2$, (6) $Na_2O—SiO_2$, (7) $K_2O—SiO_2$, (8) $Rb_2O—SiO_2$, (9) $Cs_2O—SiO_2$. (After F. C. Kracek, *J. Am. Chem. Soc.*, 1930.) The diagram shows that the system $MgO—SiO_2$ exhibits a large gap of immiscibility, that the gap becomes successively smaller by replacing Mg with more alkalic cations (Ca and Sr), that $BaO—SiO_2$ exhibits a highly anomalous melting curve but no gap, that still more alkalic cations produce more normal melting curves until, at Rb and Cs, the normal shape is restored. The curve for Ra would be found between Ba and Li, if Ra-mixtures could be studied.

Liquid immiscibility in silicates, discovered by Greig in 1927, has aroused much discussion. He demonstrated that silica shows only partial miscibility with melts of CaO, MgO, FeO, Fe_2O_3; at high contents of SiO_2 an area of immiscibility exists (see Fig. III-49).

There is an interesting difference in the physical properties of the two conjugate melts. Melt II of Fig. III-49 exhibits a high fluidity caused by individual SiO_4-groups that unmix from melt I, which is viscous and consists of highly polymerized silicon-oxygen networks.

These two melts have been referred to as ortho-melts and para-melts, respectively.

Small amounts of water affect the immiscibility relations; for the OH-ions, in destroying the network, stimulate the formation of ortho-melts. It has been shown by Friedman (1950, 1951) that the system Na_2O-SiO_2, which is miscible in all proportions, by addition of H_2O splits into two immiscible liquids, one very rich in water (corresponding to an ortho-melt[?]) one rather rich in SiO_2 (corresponding to a para-melt). This also happens if small amounts of Al_2O_3 are present.

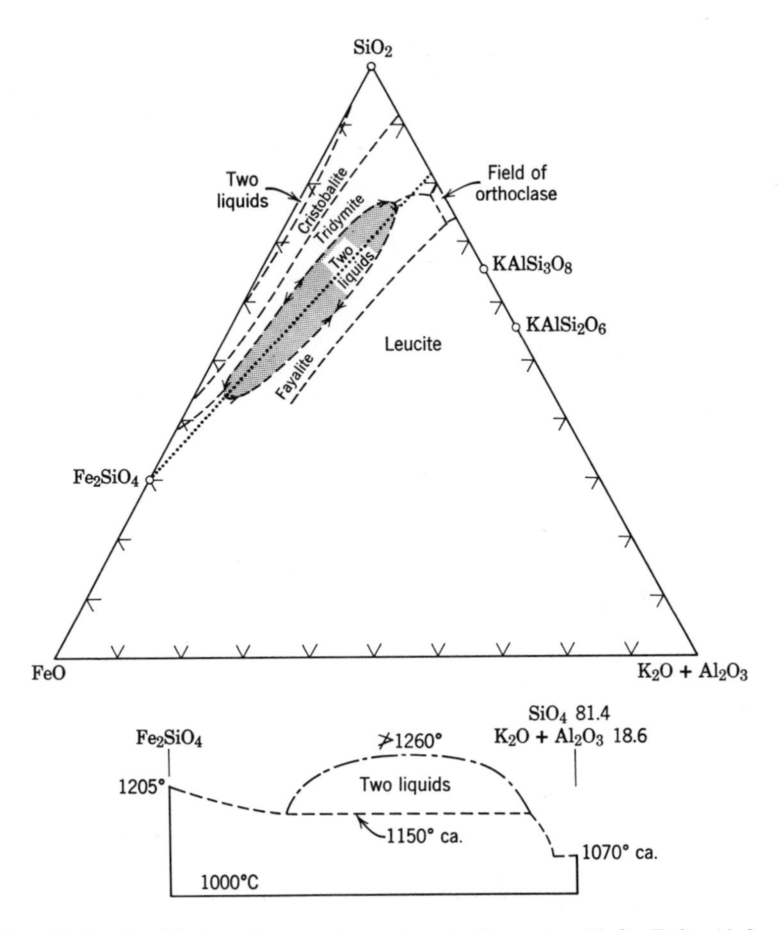

FIG. III-50. Equilibrium diagram of one plane in the system K_2O—FeO—Al_2O_3—SiO_2. K_2O and Al_2O_3 are present in the ratio 1:1 (after Roedder, 1951). *Below:* the profile Fe_2SiO_4 ..... shows the temperature relations of the fusion surface.

The immiscibility region has the shape of a spheroid. The upper extent is not known, for parts of it are cut by the stability fields of $Na_2Si_2O_5$ and quartz. The lower portion is not cut off by any stability fields of crystalline phases, but shows a lower critical solution temperature at approximately 225°C, below which complete miscibility occurs.

In the early history of geology, immiscibility in igneous magmas was assumed in order to explain the diversity of rocks. A homogeneous mother magma, on cooling, was supposed to split into two liquids which remained mutually immiscible, one corresponding to a basaltic magma, the other to a rhyolitic magma. But experiments have shown that this cannot be true. It is to be remembered that unmixing is a manifestation of phase equilibrium and that two liquids which constitute an immiscible pair are in equilibrium with each other, *as well as with any additional phase that may be formed.*

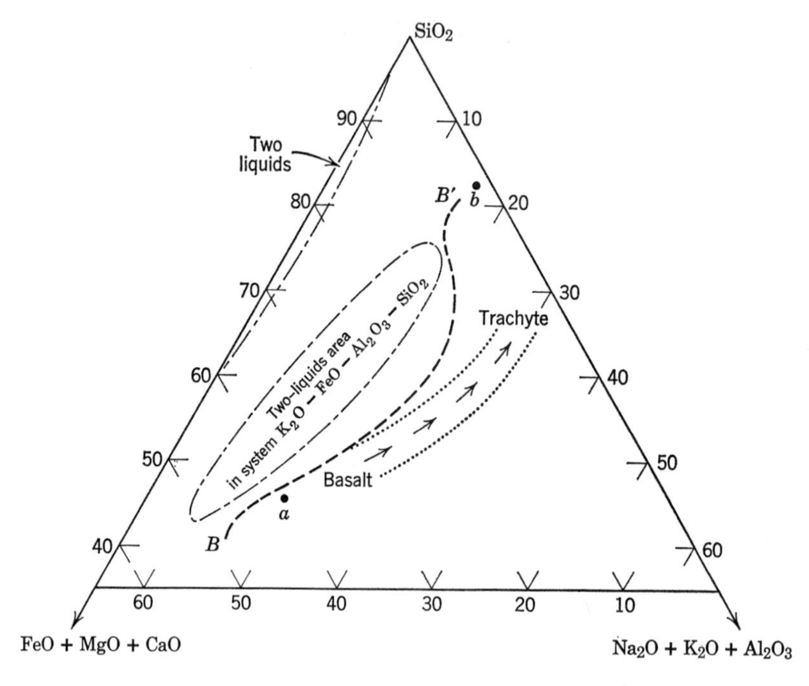

FIG. III-51. Diagram (after Holgate, 1954) to show the relation of the composition of mixtures which show immiscibility to the composition of igneous rocks. The line *B–B′* limits the composition of extrusive rocks. The path indicated by arrows represents the normal liquid line of descent from basalt to trachyte. *a* and *b* correspond, respectively, to basalt and glass investment on a transfused quartzite xenolith immersed in basalt. See text.

Thus, in the association basalt-rhyolite, if it is assumed that their liquids constitute an immiscible pair, there will be a stage when the basalt liquid is in equilibrium with a plagioclase of labradoritic composition, and the associated rhyolite liquid should also be in equilibrium with the very same labradorite. But from what has been explained in section III-B we know that there is no stage at which rhyolite liquid is at equilibrium with so calcic a plagioclase.

Experiments have shown that in dry silicate melts approaching the composition of common types of igneous magmas, immiscibility phenomena occur. See Figs. III-50 and III-51. If water is added, the areas of immiscibility may increase considerably; this is indicated by the fact that the compositions of effusive rocks always lie to the right of the line B–B' in Fig. III-51. No lava can have a composition lying within any area of immiscibility.

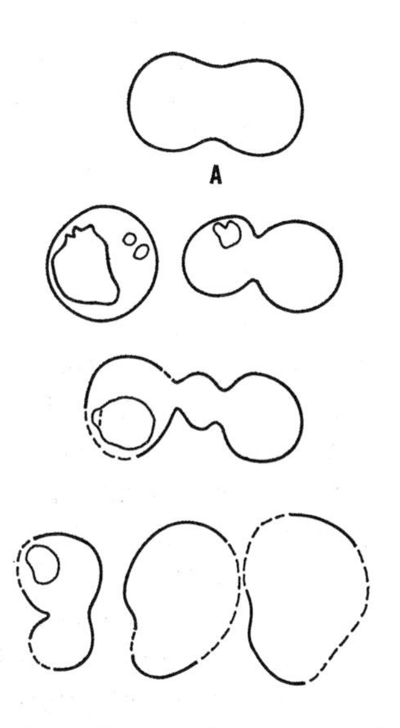

Fig. III-52. The shapes of large globules (×5) as observed by Drever (1960) illustrating coalescence arrested at various stages. **A** is the theoretical shape proposed by Greig (1928).

Thus a basalt of composition *a* (Fig. III-51) will not be able to dissolve a siliceous xenolith, but the basalt will react with it and form a glassy or quartzo-feldspathic rim around it, the composition of this rim is shown by point *b*. This is one of the many examples of the relations of siliceous xenoliths and their basic host rocks given by Holgate (1954).

Another feature observed in rocks and taken as a criterion of liquid immiscibility is the existence of globular structures. A melt splits up into two immiscible parts, so that one part exsolves by forming countless small nuclei dispersed in the other melt. Eventually the nuclei flow together, forming drops and globules such as observed for example by Drever (1960) in a calcic basalt (eucrite) of West Greenland, and taken as evidence for two immiscible magmatic liquids. See Fig. III-52.

5 · THE VAPOR PHASE AND HYDROTHERMAL DIFFERENTIATION

The outstanding factor in determining the character of modern volcanism is the gas content of the crystallizing magma. If this be mainly of steam released in a closed chamber, then only steam explosions are to be expected as the surface manifestation of the crystallization of the magma below; if to the steam are added such chemically active gases as chlorine, sulphur, hydrogen and the hydrocarbons, then chemical reaction between these will be sufficient cause of higher temperatures in lava flows of the character well known at Vesuvius, Stromboli or Kilauea.

A. L. Day, *J. Franklin Inst.*, 1925.

Let us assume a magma containing a certain quantity of water in excess of that finally taken up by the crystalline igneous rock; this original amount would probably be less than 1 per cent. As crystallization proceeds, the remaining molten magma becomes relatively enriched in water until either water is squeezed out, owing to differential pressure, or else the true saturation point is reached. Thereafter further crystallization will result in the exudation of water vapor as a separate phase (the second boiling point), and this process will continue until crystallization is complete. Thus three phases are present in the magma chamber: (1) solid phase, the rock; (2) molten phase, the magma; and (3) vapor phase, which may develop gradually into hydrothermal solutions.

In a natural magma chamber not only does the hydrothermal solution contain water, but also every compound in the magma, in accordance with the distribution law, will distribute itself between the three phases in proportion to the solubilities in each phase.* Furthermore, the composition of the vapor phase will change as the crystallization of the magma proceeds. For this phenomenon of a magma giving off gases (and solutions) of different chemical compositions at different times, H. Neumann (1948) has used the term *endomagmatic hydrothermal differentiation.*

Let us study the concentration of a certain compound (n) in the vapor phase as a function of (1) the distribution coefficient, k = solubility in molten phase/solubility in vapor phase, and (2) x = the fraction of magma changed into solid phase + vapor phase.

* The water, when above its critical point and hence a gas, is highly compressed. At 400°C and 1000 atm pressure the density of water is 0.71, not greatly different from that of water under ordinary conditions. It has a significant solvent power on non-volatile mineral components, for example, on sodium and silica.

Two different cases must be considered: (*a*) The vapor phase continuously disappears from the magma chamber immediately on its development. (*b*) The vapor phase remains in the magma chamber in equilibrium with other phases until the magma is completely frozen.

We shall denote with t the fraction of water dissolved in the magma (in Fig. III-53, $t = 1/100$, corresponding to 1 per cent of water dissolved in the magma). It can be shown that a compound (n) with a coefficient of distribution of k_n, will be enriched in the youngest or in oldest hydrothermal solutions according to whether $k_n > t$, or $k_n < t$, respectively. In the diagram (Fig. III-53) a curve for $k = 1/50$ illustrates the first case, the curve for $k = 1/500$ illustrates the second case, and, if $k = 1/100$, the compound (n) will exist in the same

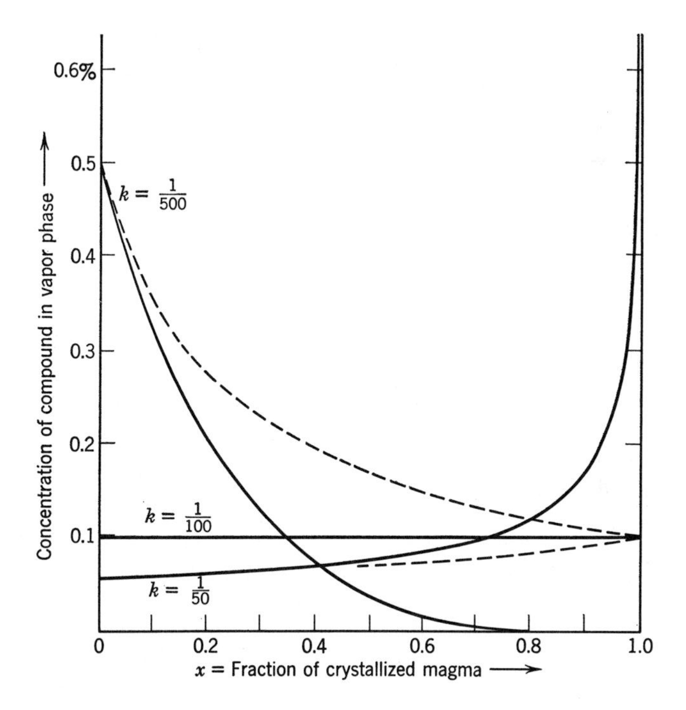

Fig. III-53. Curves showing the variation of concentration in the vapor phase of volatile compounds with different solubilities. k = distribution coefficient (solubility in magma/solubility in vapor phase). The diagram is computed under the assumption that the quantity of the several volatile compounds in the magma at stage 0 (at the moment of incipient separation of the vapor phase) is 0.001 per cent, and that 1 per cent of water is dissolved in the magma. Solid lines are for hydrothermal solutions leaving the magma immediately on their appearance. Broken lines are for hydrothermal solutions remaining in magma chamber. (After H. Neumann, 1948.)

concentration in the oldest and in the youngest hydrothermal solutions.

Thus we see that in the general case the composition of the vapor phase expelled by a magma must vary with time. In the early stages the compounds of high relative solubility in the vapor phase, that is, rather volatile compounds, will dominate; in later stages the relative amount of less soluble compounds will increase.

Chemical reactions in the gases and between the gases and their environment make the situation more complicated, but they do not detract from the principal value of the discussion.

It has been contended that various heavy metals are transported by vapors giving rise to ore deposition. The metals in any event will be only minor constituents of the vapor, and the form in which they occur will be partly determined by the nature of the major constituents. An idea of the concentration of the major constituents can be obtained from analyses of gases that presumably have some resemblance to magmatic gases at an intrusive contact. See Table III-16.

Intrusive rocks differ among themselves in composition, and so do the associated vapors, but the commonly occurring assemblage of minerals is sufficiently uniform to permit calculating the composition

TABLE III-16

ANALYSES OF GASES

All Figures Are Given in Atmospheres Relative to 1000 Atmospheres of Water Vapor (After Krauskopf, 1957)

	Volcanic Gases from Kilauea and Mauna Loa (1)	Gases from Heated Diabase and Basalt (2)	Gases from Obsidian, Andesite, and Granite (3)	Gases from Fumaroles and Geysers (4)	Volatiles from the Earth's Interior during Geologic Time (5)
C as CO_2	167	99	27	0.2	22
Cl as HCl	1	12	11	0.6	5
F as HF	...	91	50	0.6	72
S as H_2S	124	26	4	0.2	0.8
N as N_2	64	24	13	0.3	2

TABLE III-17

POSSIBLE EQUILIBRIUM COMPOSITIONS OF A "STANDARD" MAGMATIC GAS PHASE AT 600°C *

	Oxidizing Conditions		Reducing Conditions	
O_2	10^{-15}	10^{-17}	10^{-19}	10^{-21}
H_2	0.04	0.4	4	40
HCl	10	10	10	10
Cl_2	$10^{-8.5}$	$10^{-9.5}$	$10^{-10.5}$	$10^{-11.5}$
HF	50	50	50	50
CO_2	50	50	50	10
CO	$10^{-2.1}$	$10^{-2.1}$	$10^{-1.1}$	0.2
CH_4	$10^{-9.6}$	$10^{-5.6}$	$10^{-1.6}$	40
N_2	10	10	10	9
NH_3	$10^{-4.5}$	$10^{-3.0}$	$10^{-1.5}$	1
H_2S	1	29	30	30
S_2	$10^{-1.4}$	10^{-2}	10^{-4}	10^{-6}
SO_2	29	0.8	$10^{-3.1}$	$10^{-6.1}$

* Figures are pressures in atmospheres. Each vertical column shows the composition corresponding to the oxygen pressure at the head of the column. Assumed total pressures: H_2O 1000 atm, HCl 10 atm, HF 50 atm, N_2 10 atm, CO_2 50 atm.

of a sort of "standard" equilibrium vapor from which minor deviations are to be expected.

This has been done by Krauskopf (1957), see Table III-17. For details the student should consult Krauskopf's paper, which is also useful as an illustration of the method of collecting thermochemical data, and an exercise in handling thermodynamic relations. Table III-17 gives the major gases in the "standard" vapor. The concentration of a volatile metal compound in this vapor is determined not only by the volatility of the compound, but by its possible reactions with various solids and with the gases in the vapor. Zinc chloride, for example, is present in the vapor, and its amount is limited, not by equilibrium with solid or liquid zinc chloride, but by equilibria with zinc sulfide and with HCl and H_2S in the vapor. Finding the metal content of the vapor, therefore, involves two steps: deciding which solid compounds are stable at 600°, and then calculating equilibrium

TABLE III-18

Possible Volatilities of Metals in "Standard" Gas Phase at 600° C. (in Atmospheres)

(After Krauskopf, 1957)

Metal	Total Vapor Pressure in Equilibrium with Sulfides, Except as Noted
Hg	High
Sb	0.7
As	0.1
Bi	$10^{-2.2}$
Pb	$10^{-2.3}$
Sn	$10^{-2.9}$ *
Mn	$10^{-3.4}$ †
Fe	$10^{-3.9}$
Cd	$10^{-4.6}$
Zn	$10^{-4.7}$
Co	$10^{-5.0}$
Mo	$10^{-5.1}$ ‡
Cu	$10^{-5.7}$
Ni	$10^{-6.2}$
Ag	$10^{-7.3}$
Au	10^{-16} §

* $SnCl_4$ in equilibrium with SnO_2.

† $MnCl_2$ in equilibrium with $MnSiO_3$.

‡ $MoO_3(g)$ in equilibrium with $MoO_3(s)$. Vapor pressure of chloride in equilibrium with MoS_2 probably higher.

§ Au(g) in equilibrium with Au(s).

constants for reactions with the principal gases present. The results are shown in Table III-18.

These calculations indicate that at an intrusive contact, at 600° and at a depth of a few kilometers below the surface the ore-forming metals would be present in both the solid and fluid phases; in the solid phase usually as a sulfide, in the vapor phase at a concentration indicated by Table III-18, the most volatile compounds should eventually be carried farthest from the igneous source. Observations show that mercury, arsenic, and antimony are found in low-temperature deposits far from an igneous source; this is in complete agreement with Table III-18. Again lead (as galena) is often found farther from the igneous source than are ores of zinc and copper. Iron, zinc, and cadmium have fairly similar volatilities and are often found together in ore deposits. Cobalt and nickel are seldom significant in low-temperature deposits and have low volatilities.

On the other hand, the observed behavior of copper, silver, and gold show no relation to their volatilities; furthermore, tin is most abundant in high-temperature deposits, but has volatility relations similar to zinc and iron; and manganese having concentrated farther from an igneous source than lead shows a considerably lower volatility.

The major magmatic gases having left their magmatic source and partly condensed to hydrothermal solutions, will change in chemical

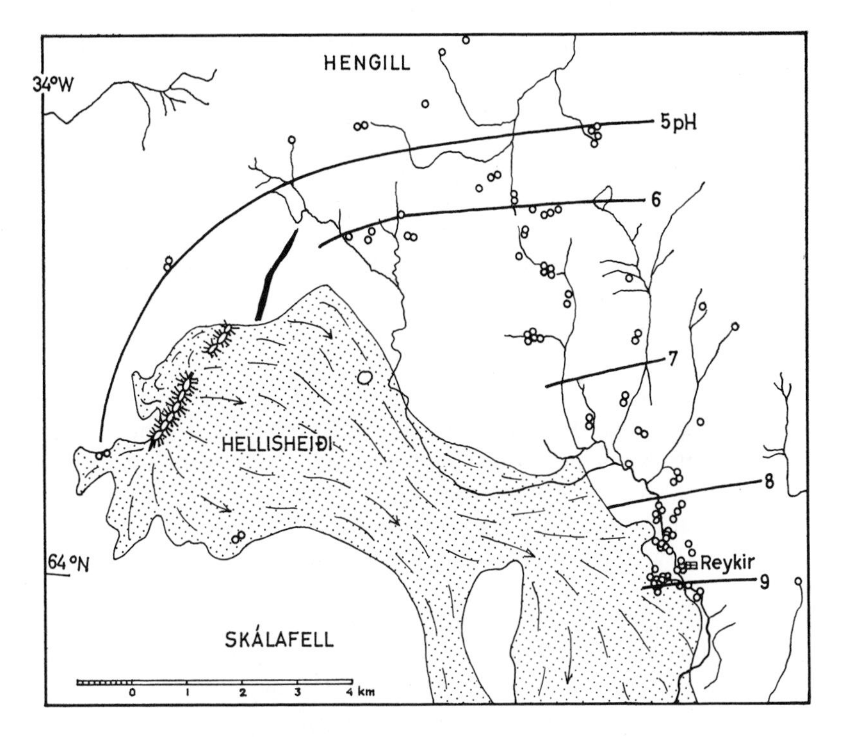

Fig. III-54. Map of the volcanic area of Hellisheidi in S.W. Iceland demonstrating the regular increase in the pH value of hot springs with increasing distance from the loci of volcanic activity. The recent lava flow issuing from the craters at Hellisheidi is shown by shading, the hot springs are shown by circles, and the regular increase in alkalinity of the hot-spring waters is shown by consecutive lines of increasing pH values. (After Barth, 1950.)

composition during their progress along fissures in the surrounding rocks, partly owing to fractional precipitation, partly because of reactions between the solutions and the wall rocks (leading, for example, to the ore geologists' phenomenon of "telescoping"). This kind of exomagmatic hydrothermal differentiation explains why the composition of the volcanic emanations varies rather rapidly from place to place and from time to time.

This last stage in the exomagmatic differentiation has also been studied, and can be shown to follow a regular pattern. The chemically active constituents in the gas phase are strong acids (HCl, HF, H_2S, SO_2 etc.); they react with the adjacent rocks, silicates and, of course, most effectively with limestone (with the formation of "pneumatolytic contacted minerals," often in perfect crystals in great profusion). As these reactions go on, the acids in the original gas phase will be used up and only chemically inert compounds will be left. This evolution has been observed and described, e.g., from Iceland, see Fig. III-54.

6 · PYROCLASTIC ROCKS

Widespread deposits of blankets of fine-grained to very coarse-grained volcanic fragments: dust < 0.05 mm $<$ ash < 4 mm $<$ lapilli < 32 mm $<$ agglomerates, bombs, and irregular blocks all testify to explosive volcanism. Ejecta, including bombs which have suffered a systematic modification of their original form and structure as a result of their flight through the air, are, together with gases and vapors, ejected from volcanic vents and thus distributed over large areas.

An idea of the quantities involved in the formation of pyroclastic rocks is furnished, for example, by the explosion of the volcano Bezymianny (Kamchatka) on October 22, 1955 (Data from Gorshkov, 1959): Total energy 2.2×10^{25} ergs, weight of ejected material 24×10^9 tons, volume of agglomerate flow 1.8×10^9 m^3 (4.3×10^9 tons), initial velocity of explosion 500–600 m per sec, initial pressure 3000 atm.

The expansion of gas in a magma is slow at great depths (H_2O over 500°C and 1100 atm is almost incompressible), but increases rapidly with decreasing pressure at low pressure. Consequently, the juvenile glossy-solid fragments in tuffs are intratelluric.

Ignimbrites or *welded tuffs* are compact rhyolitic rocks, forming sills or flows extending over large areas (Tertiary ignimbrites in Australia, New Zealand, Japan, Sumatra, in the Cordilleran region of U.S.A., etc.). They exhibit peculiar parallel structures which have given rise to names like eutaxite, ataxite, piperno a.o. Air transport, sedimentation, and subsequent welding of the pyroclastic material have been suggested. This is probably not true in all cases, some of them may have formed intratellurically, possibly from a "fluidized" magma; they do represent a stage between lava and tuff.

Tuffs are compacted pyroclastic fragments, being indurated equivalents of volcanic ash and dust. The activity of tuff vents has been thoroughly discussed by H. Cloos (1941).

The mineralogical composition of tuffs indicates the nature of the original rock (trachyte tuffs, basalt tuffs, etc.). Palagonite tuff is known primarily in Iceland. Palagonite is formed from sideromelane (basaltic glass) by hydration. The agent causing the hydration is here the melted ice formed by subglacial volcanic activity.

Nuées ardentes were first described in 1902 in the destruction of St. Pierre, Martinique. (See A. Lacroix, 1904.) They are generated in volcanoes producing viscous (acid) lava that easily clogs the conduit. In this situation the volcanic gases cannot escape but, instead, accumulate at the top and will, with eventual rupture of whatever restraint may have become imposed, go into paroxysmal explosion. The liquid viscous lava in the conduit, highly charged with gases that through the release of pressure suddenly come out of solution, now becomes autoexplosive as a whole and thus capable of lifting itself out of the conduit in a stupendous "en masse" expansion. The liquid lava becomes an infinitely subdivided mass, that is, a cloud of vapor and solidifying particles, a nuée ardente, the original temperature of which (more than 1000°C) tends to be maintained by progressive crystallization and by the formation of vapor films between all solid particles, preventing all solid contacts, leaving the moving mass quite frictionless and capable of flowing upon the slightest of inclines. Its great weight will precipitate this upon the volcano's flank in a downrushing avalanche of "block and ash" amid rising clouds of steam and dust.

Mud flows consist of water-saturated pyroclastic material. The *jökulhlaup* in Iceland, the result of subglacial volcanic eruptions, consists of colossal amounts of melt water mixed with icebergs, volcanic products, and rocks sweeping down the mountain side, burying everything in its way, and flooding the lowlands.

7 · MAGMA TECTONICS

By congealing, the lava (or magma) is transferred into an igneous rock whose properties can be studied in the field and in the laboratory. Observations on both volcanic rocks and deep-seated igneous rocks enable the geologist to determine the origin, source, and direction of movement of formerly hot liquid, viscous magmas.

This field of research has been called magma tectonics; it was introduced and developed by Hans Cloos and his school in a long series of papers from about 1920 onward. Special reference is given to his textbook, *Einführung in die Geologie* (Berlin, 1936), and to Robert Balk's *Structural Behavior of Igneous Rocks* (*Geol. Soc. Am. Memoir* 5, 1937). The method is based on a detailed study of the visible flow features which igneous rocks produce before final consolidation. All visible structures which date from the time of fluidity are carefully measured in the field, and plotted on maps. Whole mountain ranges are so surveyed, and it is possible to reconstruct reliably the mechanics and directions of flow from the field observations. The "flow picture" of deep-seated intrusions, thousands of cubic miles in volume, is not essentially different from the dynamic pictures observed in wind tunnels, or in experiments with viscous liquids in laboratories.

Primary Flow Structures in Igneous Rocks

Prior to consolidation, almost all igneous rocks pass through a stage where they carry in suspension a certain proportion of solid bodies. These may be fragments of the surrounding solid rocks; or the lava may carry crystal grains which, for some reason or another, had formed before the bulk of the lava froze and had attained appreciable size. In other cases, the lava itself has been inhomogeneous, some portions being rich in gas bubbles or mixed up with foreign rock fragments; or two differently colored varieties of lava occur side by side, in which case their viscosities may have differed slightly.

Long after volcanic lavas have consolidated into hard and brittle rocks it is possible to reconstruct the directions and mode of flow, thanks to two extremely important visible features which result from this mixture of lava with solids or gas: (1) flow lines or linear parallelism; (2) flow layers, or platy parallelism, also called foliation.

Flow lines are shown in Fig. III-55. The scattered mineral grains, especially if they have a prismatic shape (hornblende, augite, also feldspar), point in the same direction.

Foliation is shown in Fig. III-56. There are not many igneous masses in which the minerals are not, at least locally, aligned in subparallel planes. But not in all igneous rocks is the foliation readily noticed. There are degrees of perfection and intensity of foliation, and the same applies to the flow lines.

Combination of flow lines and foliation frequently occurs. Almost always the flow lines lie in the foliation planes, although for a few

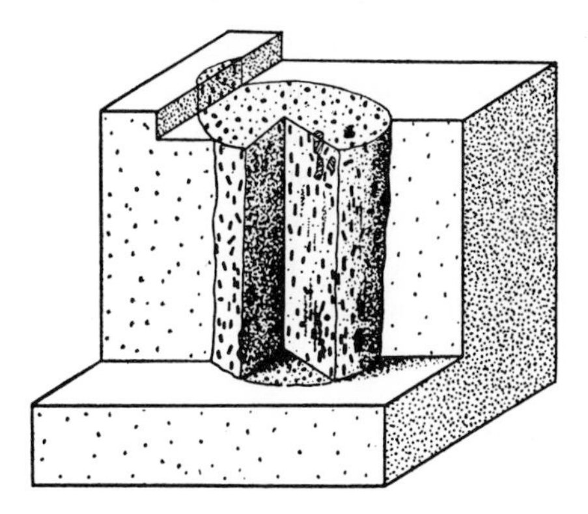

FIG. III-55. Diagram showing the linear parallelism in a vertical neck or chimney of an igneous rock. The principal flow has been vertically upward, and admixtures (prismatic crystals, gas bubbles, or fragments of the wall rock) orient their longest axes in this direction. (After R. Balk, 1937.)

localities discrepancies have been recorded. For instance, in the high Sierra Nevada the schlieren planes vary in strike and dip, whereas the linear parallelism of hornblende and biotite crystals in the same area remains constant throughout.

As a general rule we may assume that a spherical element of the original lava was stretched most in the direction of the flow lines,

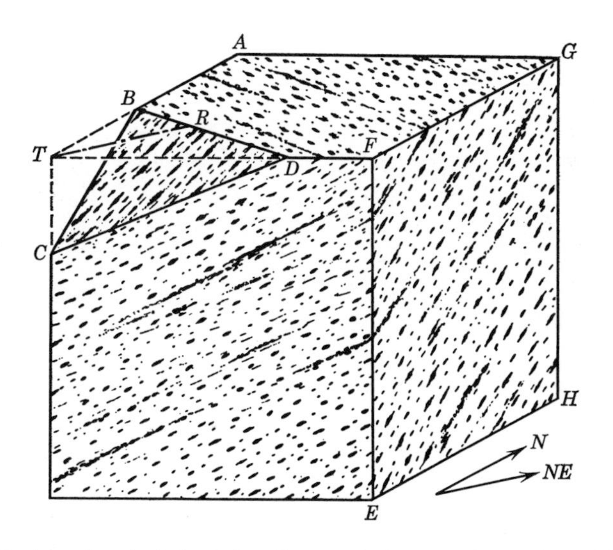

FIG. III-56. Trend and pitch of flow lines, and their relation to foliation. (After Balk.)

somewhat less in a direction normal to it within the plane of foliation, and least perpendicular to the plane of foliation. Thus the two larger axes of the ellipsoid of deformation are within the plane of foliation, the smallest axis normal to it.

The arrangement of primary flow structures in dikes is most readily understood, because all elements can be referred to the border walls. By far the most common pattern is so arranged that, on approaching a wall, all elements become parallel to that wall (*conformable patterns*).

In most larger rock masses flow structures may be directly related to a direction of principal flow in some portions, but may bear an indirect relation to it in other portions of the same mass. Many intrusions are characterized by foliated or even gneissic borders, caused by mechanical friction engendered by the upward flowing mass.

A correct interpretation of *disconformable* flow patterns in dikes and massifs is difficult and involves complex hypotheses of various processes working in succession.

1. Dome and arch structures are due to an upward current of magma.

2. Fans of cross joints and tension joints are perpendicular to flow lines and prove that intrusions continued to expand the frozen shell in the same direction. The magma of such masses pushed persistently upward long after the upper portion consolidated.

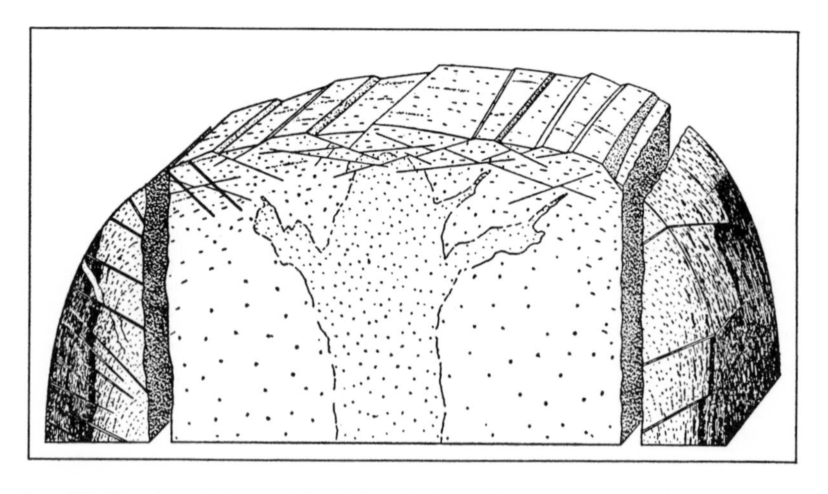

Fig. III-57. Association of flat-lying faults with marginal fissures and thrusts in granite massifs. (After Balk.)

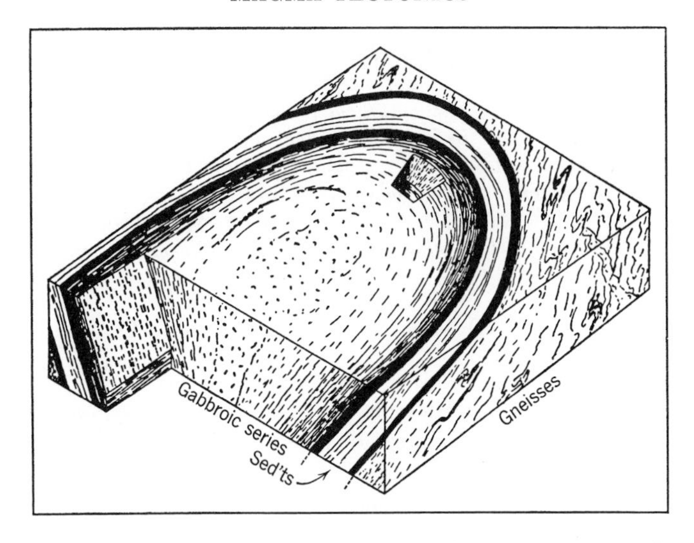

FIG. III-58. Significant structural features of many gabbro massifs. (After Balk.)

3. Zones of marginal fissures and upthrusts, which accompany steep-wall intrusions, require an intrusive force (see Fig. III-57) which was active while the core was still liquid enough to emit granite dikes, aplites, and pegmatites.

4. Gneissic borders are evidence for intense mechanical forces.

5. Flat-lying normal faults are so arranged that they effect a more or less horizontal expansion of intrusive bodies. They antedate or are equal in age to aplitic and pegmatitic dikes.

6. A structureless rock, devoid of flowage features, may, nevertheless, have domed its cover. Virtually all discordant intrusions, formerly interpreted as partly or entirely emplaced by stoping, when examined structurally, show evidence of an organized upward flow of magma.

7. The emplacement of each mass should be judged on its own evidence, and we should make use of the great body of structural criteria now available.

Significant differences in structural type of acid and basic massifs can be correlated with differences in the degree of viscosity of the several magmas. Only magmas as viscous as the granitic ones are able to create and sustain broad arches and thus force their way upward. And only less viscous magmas like the gabbroic ones are able to flow into the narrower passages and openings in the crust and assume funnel-shaped structures. See Fig. III-58.

D. ROCKS AS PRODUCTS OF DEFINED PROCESSES

. . . das Studium der natürlichen Gesteinsassoziationen und ihrer geologischen Bedingtheit eröffnet uns einen Ausblick auf die Möglichkeit, dereinst zu einem natürlichen System der Eruptivgesteine zu gelangen. . . . *

V. M. Goldschmidt,
Stammestypen der Eruptivgesteine, 1922.

1 · SURVEY OF MAGMATIC DIFFERENTIATION

Bunsen explained the varieties of igneous rock revealed through his analyses by assuming the independent existence of two magmas—the "normal pyroxenic" and "normal trachytic"—and by supposing a process of intermixture to account for the intermediate varieties. Von Waltershausen thought that igneous magmas were arranged in a series of concentric shells, according to specific gravity. Durocher, in his celebrated essay on comparative petrology, maintained "that all igneous rocks, modern and ancient, were derived from two magmas which coexist below the solid crust of the globe, and occupy there each a definite position."

Charles Darwin (1844), in his important work on volcanic islands, pointed to fractional crystallization as a means of rock differentiation: (1) the movement of crystals in a magma under the influence of gravity and (2) the squeezing out of the more fusible constituents from a partially consolidated mass. G. F. Becker further developed the idea and compared a laccolith in which the marginal parts are different from the center to a barrel of cider which has been frozen from the outside.

No variety of igneous rocks simply happened by an act of creation. Each of them represents the end product of long and complicated petrogenetic processes operating in the crust of the earth and constantly producing rock differentiation. In each case the last stage of this process was the congealing of a magma (or lava) into a solid rock.

Petrologists agree that the differentiation cannot be attributed to any one mechanism, and for most of the igneous bodies actually studied no complete or satisfactory explanation has been proposed. Crystallization differentiation, assimilation, mingling of magmas, etc.,

* The study of natural rock associations and of their geological causality exposes to us a vista of the possibilities of some day arriving at a natural system of igneous rocks.

may each have played its part, and there may be still other important processes of which we know nothing today. The student should recognize this situation and be on the outlook for new evidence and for new principles in the interpretation of petrogenetic problems.

1. *Fractional crystallization,* the simple principle of which is demonstrated in most of the mineral series discussed in Part III, Section B, obviously greatly influences the crystalline end products. In the cooling history of basaltic magma, as described on page 119, the effect of fractional crystallization is strongly in evidence. Many petrologists today regard it as the main factor controlling igneous rock differentiation. Various examples of rocks, the evolution of which is believed to depend on this process, will be adduced in later chapters. First we shall briefly mention some other processes competing with crystal fractionation in producing differentiation. Says Fenner: "The insistence by prominent geologists that all igneous theory must conform to the tenets of crystal fractionation has been stultifying to independence of thought and has delayed for many years progress toward an understanding of magmas and their processes" (*Am. J. Sci.,* **248,** 1950).

2. *Gaseous transfer* from lower to higher levels of a large magma body and farther upward to the surface of the earth is active wherever the internal gas pressure is sufficient to create a gas phase (see page 126). The efficacy of magmatic gases as agents of differentiation lies both in their ability to carry away material entirely when they escape, and in their ability to make selective transfer of material from lower to upper levels. A gas bubble leaving the depths and finding its way to the surface will become contaminated by the volatiles resident in the various regions through which it passes; they will distill into the bubble almost as if it were a vacuum. Thus the magma, rather effectively, will get rid of all compounds of high vapor tension.

3. *Liquid immiscibility* is supposed to be an effective means of separating sulfide melts from silicate melts. At an early stage of the magmatic evolution a heavy sulfide melt may separate and tend to sink to the bottom. It is believed that "magmatic" sulfide ore deposits are formed in this way. It is also possible that liquid immiscibility plays a part in the development of silicate magmas (see page 147).

4. *Thermodiffusion.* At one time the *Soret effect* was considered to be the principal cause of differentiation (Logario, 1887; Teall, 1888; Brögger, 1890), that is, a diffusion of certain dissolved molecules towards the chilled borders of the magma. But later Bowen (1915, 1921, 1928) declared it to be of no practical importance because the diffusion velocities were too small. However, the experimental in-

vestigations of the Soret effect is difficult, and it now seems possible that, under certain conditions, it can bring about considerable inhomogeneity in magma. A related mechanism, thermal diffusion combined with convection, has been shown to be more powerful (W. Wahl, 1946). This combined process, which may be called simply thermodiffusion, was probably active in the so-called complex dikes and composite lava flows whose mode of formation so far has received no satisfactory explanation.

5. *Effect of gravity.* The structure of silicate melts is similar to that of crystalline silicates, but more irregular (see page 142); but the irregularities are only on a molecular scale. In order for equilibrium to prevail, the magma must be homogeneous, so that compositional differences will, in time, be eliminated by diffusion.

However, in a magma chamber of great vertical extension, conditions become more complicated. Equation 25 in the Appendix shows that the partial free energy of a chemical species changes with the position in the field of gravity.

Therefore, a homogeneous magma of great vertical extension is

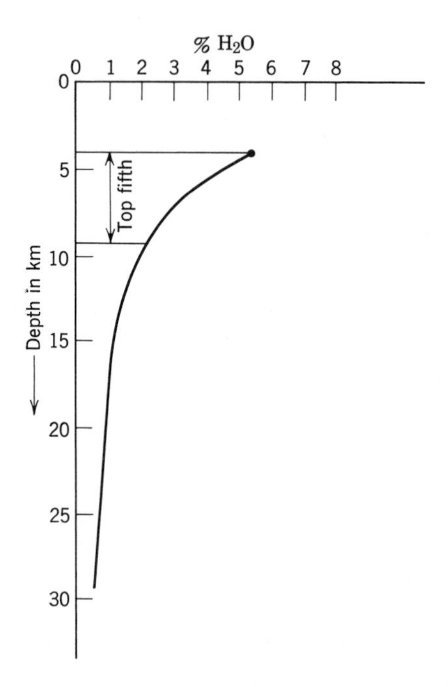

Fig. III-59. The distribution of H_2O in albite melts at 1000°C, extending down to 30 km depth, terminating upward 4 km below the earth's surface, and covered by roofs impermeable to the vapor.

not stable thermodynamically. The "tendency of escape" of atoms and molecules of large fictive volumes increases rapidly with pressure; consequently they are squeezed out of the lower levels where the pressure is high, and move upward. Ideally, elements and compounds with large fictive volumes (e.g., H_2O, see Fig. III-59) will be gradually concentrated at higher levels; elements with smaller fictive volumes will diffuse downward. However, interionic bonding forces (chemical affinities) are strong enough to disturb this gravitative arrangement, in that heavy elements with strong affinity to light phases will, if the light phases are more abundant, concentrate in these light phases at the top, and vice versa for less abundant light elements with strong attraction to heavy phases. The affinity relations of the various phases in a magma are not known, quantitatively. But qualitatively the volatiles together with alkalies and silica will concentrate in the top parts, while iron, for example, will sink (Brewer, 1951). In addition to the volatiles, one should frequently expect a concentration at the top of strong bases: Na_2O, K_2O, BaO.

6. *Assimilation,* which means the incorporation of rock masses by magmas, is opposed to true differentiation. It is instructive to regard differentiation merely as a decrease of entropy, and assimilation an increase in this quantity. Physical chemistry teaches that entropy tends to increase in the course of any spontaneous process unless special forces or restraints are imposed; assimilation, therefore, is the natural and universal tendency.

As explained on page 111, the laws of assimilation are governed by the laws of fractional crystallization. If a rock mass subject to assimilation by a magma exhibits a chemical composition comparable to that of any of the usual igneous rocks, the process of assimilation will not be able to radically change the composition of the magma or its products of fractional crystallization. But if rocks of special chemical character are subjected to assimilation, the composition of the magma may be radically changed, and the ensuing rock may distinguish itself as a product of assimilation. Thus, as first proposed by Daly in 1910, assimilation of limestone or dolomite will cause desilication of the magma and the formation of alkaline (nepheline-bearing) rocks. Assimilation of water by a dry magma likewise alters the normal crystal fractionation and generates special rock types (see page 222); so does the assimilation of carbon dioxide (page 203). Incorporation of salt deposits must needs affect the magma, but thus far no field evidence is known to indicate this type of assimilation.

It is correct, therefore, to distinguish between all differentiation processes listed under points 1 to 5, on the one hand, and assimilation processes on the other. Rocks produced or obviously affected by assimilation are often regarded as *anomalous*. In the Niggli scheme of classification the *normal magma types* are defined as derived by pure differentiation of a crystallizing magma complex.

2 · VARIATION DIAGRAMS

If we want to compare groups of igneous rocks, the chemical analyses are of great value. They may be recalculated in terms of Niggli values, normative minerals, or in any other way serving the purpose.

By directly using the analytical values in constructing an ordinary variation diagram many characters of a group of rocks are fully displayed. In the diagrams the values for the oxides are plotted as ordinates against the silica values as abscissas. (See Fig. III-60.)

In petrographic provinces, where the rocks range from basalt or gabbro to rhyolite or granite, the silica increases regularly. In most batholiths the earliest intrusions were low in silica, and later intrusions were progressively more siliceous; thus the arrangement of the members along the silica axis corresponds, generally speaking, to the

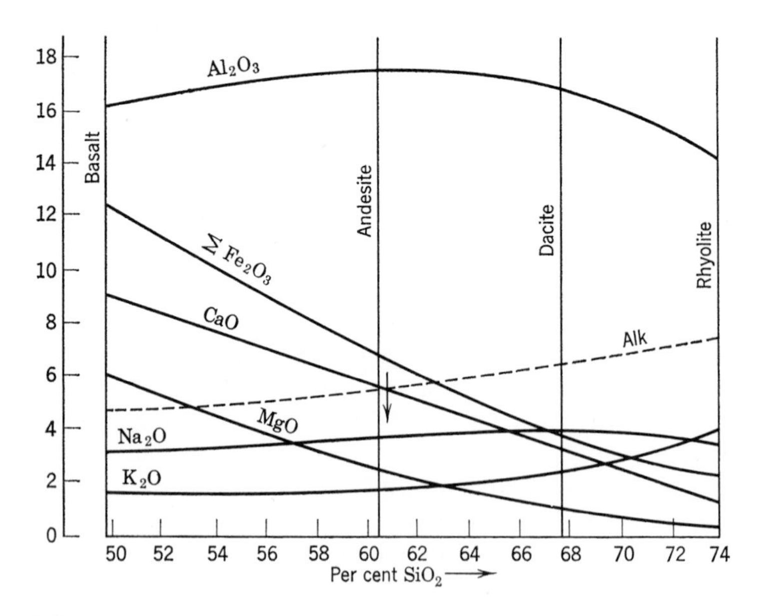

FIG. III-60. Variation diagram for Daly's average basalt-andesite-dacite-rhyolite. The different oxides are plotted as ordinates against the silica values.

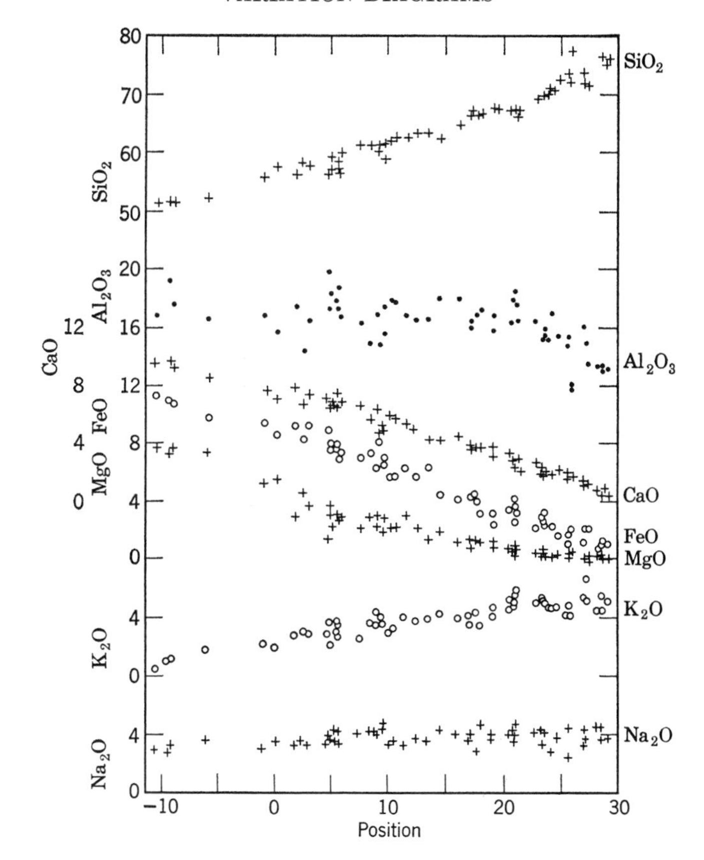

FIG. III-61. Variation diagram for the lavas of the San Juan Mountains, Colorado. Abscissa $\frac{1}{3}SiO_2 + K_2O$—FeO—MgO—CaO. (After E. S. Larsen, *J. Geol.,* 1938.)

age sequence. Instead of using weight percentages, petrologists of the Niggli school would plot the Niggli symbols, *al*, *fm*, *c*, *alk* as ordinates against the *si*-values.

In this type of diagram potash increases, and lime, total iron, and magnesia decrease about as regularly as silica changes. When this fact is taken into consideration, the position of the rock, according to the Larsen method, is determined, not by silica alone but by the sum of one-third silica plus potash minus the sum of the magnesia, lime, and total iron calculated as FeO, that is, $\frac{1}{3}SiO_2 + K_2O - MgO - CaO - \Sigma FeO$. (See Fig. III-61.)

A very simple way is to use the color index as defined on page 53 as abscissa. Thorntone and Tuttle's method is to use what they call the differentiation index, which is quite analogous to the color index but more complicated. By this method the normative content of Or + Ab + Q determines the position of the rocks.

Finally various ratios may be used: FeO/FeO + MgO, or

$$\frac{K_2O + Na_2O}{K_2O + Na_2O + CaO}$$

A useful diagram in studying *basaltic* rock series is obtained by Murata's method. He plots per cent MgO against the weight ratio of Al_2O_3/SiO_2. See Fig. III-62.

The variation diagrams, whatever their basis of construction, by depicting the regular change from rock to rock in the sequence of their age at the same time reflect the gradual change of the parental magma in the process of differentiation. The curves themselves are suggestive, therefore, of the mechanism of differentiation, whether by reaction crystal ⇆ melt, by gaseous transfer, or by liquid immiscibility, etc.

Let us consider the artificial system albite-anorthite-diopside. It

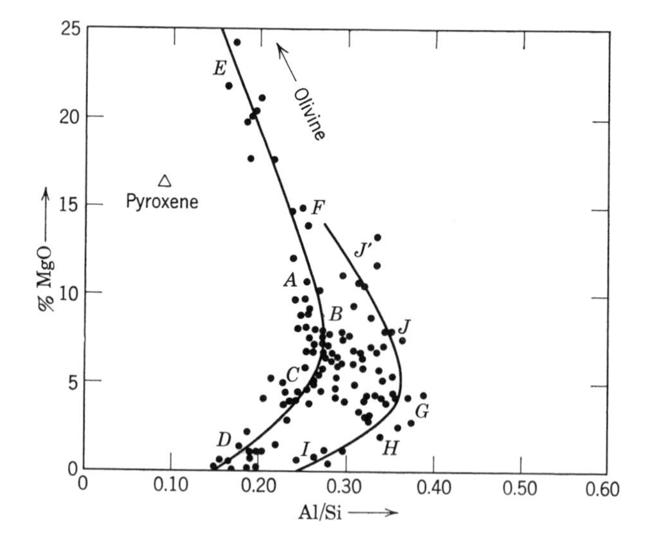

FIG. III-62. Variation diagram of basaltic rocks according to Murata (*Am. J. Sci.*, 1960).

Tholeiite series: A, Tholeiite olivine basalt; *B,* Tholeiite basalt; *C,* Quartz basalt; *D,* Granophyre; *E,* Oceanite.

Alkalic basalt series: F, Ankaramite; *J′–J,* Alkalic (olivine) basalt; *G, Hawaiite* (andesine andesite); *H,* Mugearite (oligoclase andesite); *I,* Trachyte.

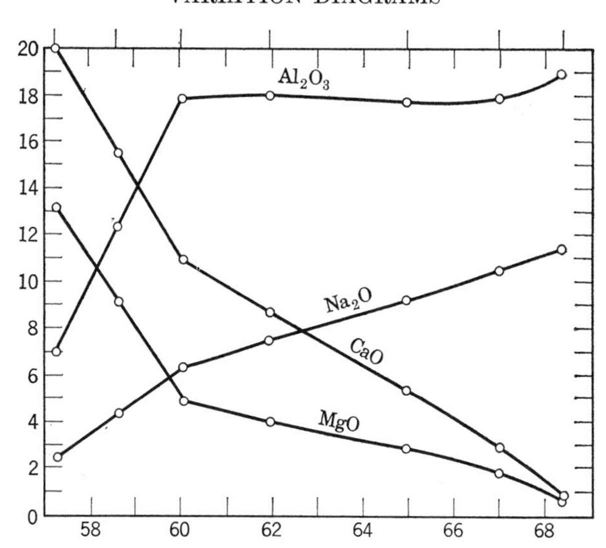

FIG. III-63. Variation diagram illustrating change of composition of liquid during crystallization in the system diopside-anorthite-albite. (After Fenner, *J. Geol.*, 1926.)

has been called a haplobasaltic mixture because its composition is sufficiently close to basaltic magma to throw considerable light on the crystallization of this magma. (See pages 107 and 114.) Beginning with a certain mixture of this system, we plot in Fig. III-63 the changing composition of the liquid as it crystallizes. We note that there is a break in the curves corresponding to the point at which the boundary line of the equilibrium diagram is encountered. Now if we were to plot the changing composition of a natural magma which is not composed of only five, but at least of ten essential oxides, we should expect to find many breaks in the curves, each corresponding to the separation of a new mineral phase. The separation of mixed crystals of gradually changing composition will be reflected in a curvature of the lines rather than breaks.

In this way it can be shown that, with increasing silica, the curves for soda and potash rise and are, in most natural rock series, convex upward. The lime curve, on the other hand, falls and is usually concave upward. The magnesia curve likewise falls and is typically concave upward. The curves for both ferrous and ferric iron fall, but they, as well as the curve for alumina, are more irregular and depend for their shape on local conditions of the special magma.

If, on the other hand, the chemical variation in such rock series were due to, say, a comingling of magmas, the variation curves would be straight lines.

In order to analyze accurately the processes of magmatic differentiation Rittmann (1960) has introduced a new serial index, s.

$$s = \frac{(Na_2O + K_2O)^2}{SiO_2 - 43}$$

By plotting s against the SiO_2-values (weight percentage is used throughout) he can distinguish between the following cases:

Process	s	SiO_2
Crystal settling	constant	increasing
Gas transfer, top (+ alkalies)	increasing	oscillating
Gas transfer, bottom (− alkalies)	decreasing	increasing
Assimilation of shale, etc.	increasing to 2.3	increasing
Assimilation of lime	increasing	decreasing

Triangular variation diagrams have also been constructed. Larsen recommends a method of superposing two triangles, thus surveying at the same time six variables. In the same triangle he makes one

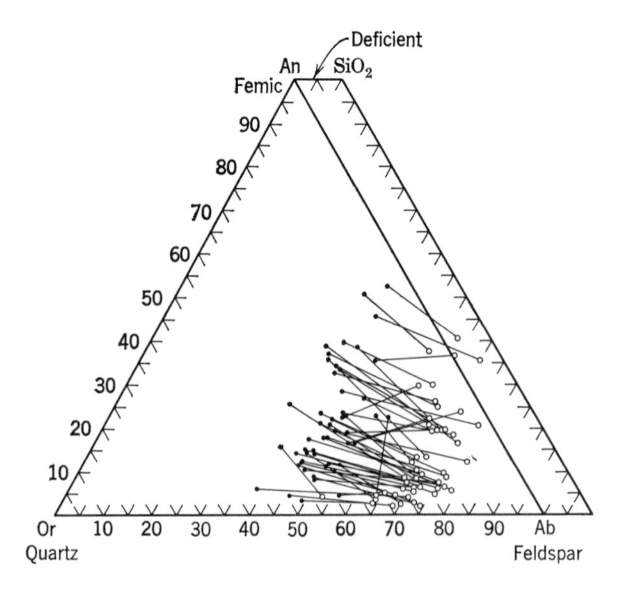

Fig. III-64. Triangular diagram for the lavas of the San Juan Mountains, Colorado. Dots are or-ab-an; circles are quartz (or deficient SiO_2)-feldspar-femic.

plot for the normative feldspars, Or, Ab, An, and another plot for the normative quartz, total feldspar, and total ferric minerals. (In silica-deficient rocks feldspathoids are calculated to feldspar, and the degree of deficiency of SiO_2 is plotted outside the triangle as a negative quartz value. See Fig. III-64.)

3 · ROCK SERIES

In 1886 Judd introduced the concept of a petrographic province. The name is self-explanatory. The rock members making up a province are believed to be comagmatic or consanguineous, implying that they are all derived from a hypothetical common magma, often called a parent magma. Within a rock province a chemical or a mineralogical character may be *absolute* or it may be *serial*. "Superposed on the absolute characters . . . are . . . the serial . . . which consists in a definite variation in the amounts of one or more constituents concomitant with a variation in one or more others, which variations may be either in the same or opposite direction" (H. S. Washington, 1906). Some igneous provinces are characterized, for example, by a high content of soda in all members, or by an anomaly in one of the minor chemical constituents, as high selenium or low phosphorus. In these properties the rock suites may exhibit both lateral and temporal dispersion. It is important to note that a petrographic province must be defined in respect to both time and space, and that it may be made up of a group of subprovinces.

Such relations naturally led to the hypothesis of a common origin of all members. Additional investigations revealed that the same characteristic serial features often were present in a great number of provinces, which, although geographically widely separated, were shown to represent analogous geological milieus. For instance, we find the charnockite series of rock, commonly associated with anorthosite, always surrounded by deep-seated pre-Cambrian granitic gneisses in the Adirondacks of northern New York State, in the province of Volhynia of southern Russia, in the Madras area of peninsular India, in the Egersund district of southwestern Norway, and in other gneissic areas.

Generally speaking, rock suites associated with vertical faulting of non-orogenic areas are *alkalic*, whereas rocks intruded into fold mountains of orogenic regions are *calcic*. We must assume, therefore, that analogous types of rock series have developed in consequence of the same physico-chemical and geological processes.

Before an adequate discussion of a general genetic theory can be presented, it is necessary to survey and classify the various rock suites.

Igneous rocks have been divided into two great groups. It was noticed by Harker (1897), Becke (1903), and others that the characteristic rock suites of the orogenic and non-orogenic areas are strongly contrasted in their chemical and mineralogical characters. For example, the folded mountains of the circum-Pacific orogeny were made up of calcic intrusive and extrusive rocks, whereas in the coastal districts and islands of the non-orogenic Atlantic basin alkalic rock series were dominant, the petrographical differences thus being matched by differences in the tectonic geology. The igneous rocks of the Atlantic coast were erupted along faults and fissures or through explosion vents, often associated with block sinking and great crustal instability. The circum-Pacific rocks, on the other hand, were associated with the folded mountain chains fringing the ocean. For this reason the two groups of rock suites were called Atlantic and Pacific. This was unfortunate, for it soon became evident that the Atlantic rock types showed no intrinsic connection with the Atlantic Ocean. They did show, however, a close connecting to faulting and movements of subsidence, and were gradually encountered all over the globe in connection with this kind of tectonics. The Hawaiian lavas, for instance, which are associated with fissure eruptions and faulting, are "Atlantic" in their relations. The intrusive rocks of the Caledonian mountain chain of the Atlantic seaboard of Norway and of Scotland are "Pacific." The geographical adjectives are now gradually becoming obsolete, and are replaced by *alkalic, subalkalic, calcic,* etc.

Table III-19 is a survey of the present usage.

TABLE III-19

DIFFERENT CONNOTATIONS

Harker Becke	Different Authors	Tyrell	Peacock	Niggli
Pacific	subalkalic, calc-alkalic	calcic	calcic	Kalk-Alkalireihe = Pazifisch
			calc-alkalic	
Atlantic	alkalic	alkalic	alkali-calcic	Natronreihe = Atlantisch
			alkalic	Kalireihe = Mediterran

Recent investigations have served to confirm the fundamental truth of this generalization, and tectonic environment is now recognized as an active factor in the evolution of petrographic provinces.

It should be borne in mind that the twofold division is arbitrary in so far as there are here, as well as in all other fields of petrography, gradual transitions. Many rock suites do not show clear calcic or alkalic affinities. Rather they are transitional, and it is quite arbitrary whether we want to divide the whole range into two, or three, as Niggli did, or even four, as proposed by Peacock.

The substance of Peacock's proposal goes beyond a fourfold division, however. According to his scheme it is possible to determine quantitatively the degree of alkalinity. To this end he has used the

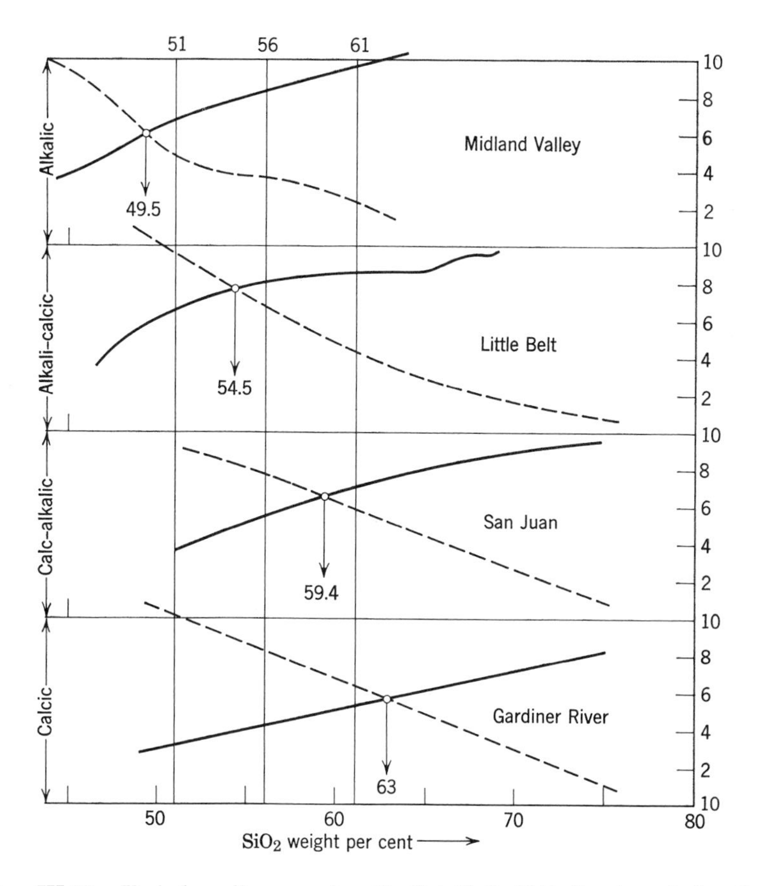

Fig. III-65. Variation diagrams for $Na_2O + K_2O$ (full lines) and for CaO (dashed lines) from four different rock series, which, according to the alkali-lime indices, are divided into four groups. Abscissa = weight percentage of SiO_2. Ordinate = weight percentage of CaO and of $(Na_2O + K_2O)$.

so-called *alkali-lime index,* which is a very good index by which to characterize igneous rock suites. It can be used as a quantitative parameter of a rock suite, and eliminates in the first place the application of any names.

In an ordinary variation diagram the curves for CaO and for ($Na_2O + K_2O$) always intersect. In general, the soda + potash curve starts low and increases with increasing silica; lime, on the other hand, starts high and varies antipathetically to the alkalies, and decreases with increasing silica. This is a direct consequence of the mechanism of crystal fractionation (see page 167). Thus the curve for total alkalies intersects the lime curve in some point of the diagram. The value of the abscissa of the point of intersection expressed in terms of percentage of SiO_2 is the alkali-lime index. In most rock series the value is in the range 45 to 65. The smaller the alkali-lime index, the more alkalic (altantic) the rock suite, and the higher the index the more calcic (pacific) the rock suite. The boundaries for the fourfold classification should be drawn at the following SiO_2-values (as shown in Fig. III-65).

$$\text{alkalic} < 51 < \text{alkali-calcic} < 56 < \text{calc-alkalic} < 61 < \text{calcic}$$

Why is it that the non-orogenic rocks are alkalic and the rocks of the fold mountains are calcic? There is much evidence suggesting

TABLE III-20

RITTMANN CLASSIFICATION OF ROCK SERIES

Proportion of Alkalies	s	Serial Character	
1. $Na_2O \sim K_2O$	1	extreme	Pacific or Calcic
	2	strong	
	3	medium	
	4–5	weak	
2. $Na_2O > K_2O$	4	transitional	Atlantic or Sodic
	8	medium	
	17–∞	strong	
	–6–0	extreme	
3. $Na_2O < K_2O$	4	weak	Mediterranean or Potassic
	8	medium	
	14–∞	strong	
	negative	extreme	

that the rocks of the non-orogenic regions are derived through simple crystal fractionation of a deep-seated primitive magma, whereas the rocks of the fold mountains have become contaminated by sedimentary materials (mostly SiO_2 and Al_2O_3 from sandstone and shales, respectively) which alter the path of crystallization and decrease the alkalinity of the rock suite. See pages 120 and 222. On the basis of the Rittmann serial index (see page 168), we obtain the classification in Table III-20.

4 · EXAMPLES: NON-OROGENIC ROCK SERIES DERIVED FROM THE BASALTIC PRIMARY MAGMA

In the following pages a number of igneous rock provinces and rock series will be described, illustrating how far the diversification of rock types may be explained by petrogenic processes whose mode of operation is known from laboratory experiments or from general physico-chemical laws.

The experiments have clearly shown that a great number of rock types derive from a homogeneous melt of basaltic composition through fractional crystallization. There probably exists a world-wide homogeneous basaltic melt at a depth of some 60–70 km (see Fig. III-44); rocks derived from this melt will be considered first. Later, the rocks of the anatectic magmas of the orogenic regions will be treated.

The whole earth is divided into two fundamentally different rock provinces: (1) the deep-sea ocean basins and (2) the continents and their shelves. Continental lands composed of sial material float on the heavier substratum or sima. They have been compared to ice floes floating in the sea. About one-half of the earth's surface is sialic. The other half, outside the continental slopes, seems to lack sial, the ocean floors beneath the oozes and red clay being basaltic of composition. The basalts of the oceans, therefore, would seem to be the most unadulterated representatives of the underlying primary magma.

Observations fit this assumption. But it does not mean that the primary submagma never intrudes into the continents, only that it then is apt to become more contaminated and in places will be so modified by assimilation that it completely loses its identity (Barth, 1936).

Chemical and mineralogical studies indicate that basalts may be divided into three "series": tholeiitic basalts, high-alumina basalts, and alkalic olivine basalts. On a view expressed by Q. W. Kennedy

(1938) tholeiite and alkalic olivine basalt originate in different crustal levels, and are therefore rather unrelated.

However, the distinction between them now seems difficult to maintain. As explained on page 120 (see also page 182) there is much evidence to support the conclusion that the alkalic olivine-basalt series is pristine in composition, whereas the tholeiitic series (as well as the high-alumina series) are modified. The two series can be best separated by the use of a Murata variation diagram. See Fig. III-62. While the diagram nicely shows the difference between the tholeiitic and the alkalic olivine-basalt series, there is nothing to indicate that the two series derive from different mother magmas; rather it seems that the settling of olivine and clinopyroxene explains the difference between them, and that by intermediate series—for example, as reported from Easter Island, Ardnamurchan, and other places—a continuous passage exists between typical alkalic olivine-basalt and tholeiite.

In continental areas there is complete gradation from tholeiite to calc-alkali series, as indeed there should be if the calc-alkali series is derived by assimilation of sial in tholeiite.

Oceanic Provinces

The central Pacific province is characterized by a great number of volcanic islands situated along ridges rising from the depths of the ocean basin. Olivine basalt, which makes up by far the largest part of these islands, is regarded as the parent magma from which all the other central Pacific rocks have been derived. The great similarity of the rocks of this enormous province, occupying about one-sixth of the surface of the earth, is really remarkable. There are minor differences; rocks of Tahiti are distinctly more alkalic than those of the other islands, and they are probably also richer in titania. Notwithstanding such differences, the striking feature is the similarity. General conditions must have been essentially uniform throughout the area, and the parent magma must have been practically the same in composition.

In general, a Pacific lava contains olivine, pyroxene, feldspar, and ore as the only constituents (hornblende and biotite are very rare, quartz is as good as absent, nepheline occurs in some rocks as a late crystallate), the mineral composition of the rocks thus becoming very simple.

The colored constituents form a reaction series:

diopside ↘
olivine ⟶ pigeonite → (hornblende?, mica?)

and the colorless constituents form another series:

bytownite → labradorite → andesine →

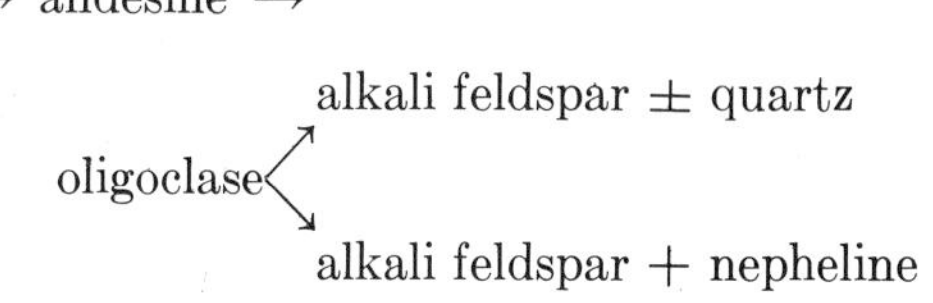

The series of the colored minerals is the same in all rocks, whether over- or undersaturated. The feldspar series is always the same as far as the plagioclases are concerned. However, at the very end the alkali feldspar will associate with either quartz or nephelite, the series thus being split into two, the former characteristic of the (over)-saturated rocks, the latter characteristic of the undersaturated rocks.

These reaction series are in evidence in two different ways. On a small scale they are manifested in each individual rock, because the sequence of crystallization parallels the sequence of the reaction series. On a large scale they become apparent when the different rocks are compared, for all lava types can be arranged in sequences that correspond to the scheme of these series.

This shows that the sequence of crystallization is parallel to the sequence of differentiation, or that the theory of crystal settling, as especially advanced by Bowen, is adequate to explain the differentiation of the Pacific lavas.

Gravitative settling is always possible; a hypothetical diagram showing the supposed result is given in Fig. III-66.

The study of the thin sections makes it obvious that if the early accumulative crystal fraction, which is particularly rich in olivine (*D* in Fig. III-66), is able to separate, an independent, olivine-rich rock comes into existence. This rock has received the name oceanite (may be described as a picrite basalt) and has been found as a characteristic component in all oceanic provinces. (See Table III-21.)

Furthermore, if the last mother liquor to crystallize interstitially in an olivine basalt (for instance, the last 5 per cent of the original magma) had been squeezed out, it would have crystallized as a trachytic rock, closely related to the actual Pacific trachytes. In the great majority of cases the trachytes seem to represent crystallization differentiates of a common magma of a composition corresponding to that of an olivine basalt.

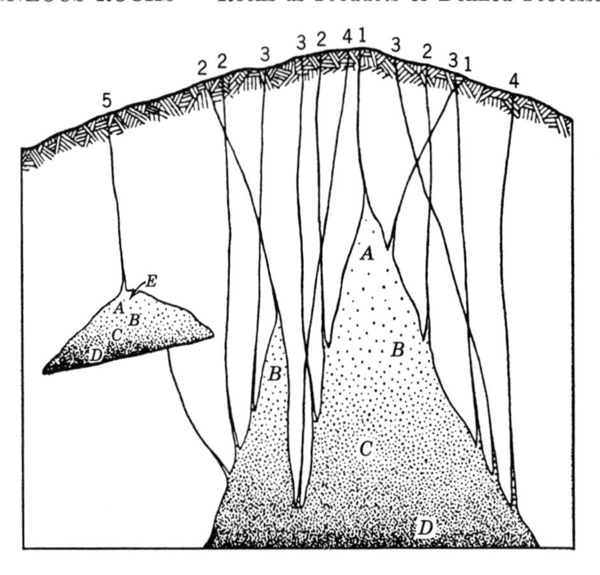

Fig. III-66. Diagram illustrating a possible manner of nearly simultaneous eruption of lavas of differing composition from the same magma chamber. *A*, andesite; *B*, basalt; *C*, olivine basalt; *D*, picrite-basalt; *E*, trachyte. (After G. A. Macdonald, 1949.)

In order to make a survey of all rock types, the crystallization series of the colored minerals may be disregarded for a moment, and the two series of the colorless minerals may be projected graphically on the triangular diagram that was used in Fig. III-35. The plagioclases are represented by the join An-Ab, along which the reaction series of the colorless minerals runs. In the lower end it splits up, one branch yielding only alkali feldspar (trachyte), the other becoming undersaturated in silica and yielding some nepheline (phonolite). (They correspond, respectively, to case A_2 and case N in Fig. III-35.) Since the colored minerals are essentially the same in all lavas, the rocks themselves may be projected on this same diagram (Fig. III-67), and in a qualitative way the diagram thus shows the genetic relationship of the various Pacific lavas. Obviously the parental olivine-basalt magma is represented by a point very close to the join Ab′-An′. Small local variations would then throw it from one side of the join to the other, and the crystallization fractionation would, accordingly, result in now a trachytic, now a phonolitic differentiate (corresponding to slight variations in the position of point W of Fig. III-34, page 117, where the details of this fractionation are discussed).

TABLE III-21

OLIVINE BASALTS AND DERIVATIVES

(Chemical Composition)

	(1)	(2)	(3)
SiO_2	45.6	46.50	48.35
TiO_2	1.7	1.70	2.77
Al_2O_3	8.3	9.37	13.18
Fe_2O_3	2.3	2.47	2.35
FeO	10.2	10.79	9.08
MnO	0.1	0.11	0.14
MgO	21.7	21.00	9.72
CaO	7.5	6.25	10.34
Na_2O	1.3	1.52	2.42
K_2O	0.4	0.22	0.58
H_2O	0.6	0.17	....
P_2O_5	0.3	0.10	0.34

Norm		(Mineralogical Composition)		
	(1)	(2)	(3)	
	....	2	trace	glass
q	....	1	trace	quartz
or	2.2	5	10	alkali feldspar
ab	11.0	24	37	plagioclase
an	15.6			
di	17.5	30	39	pyroxene
hy	12.1			
ol	34.1	33	9	olivine
mt	3.2	5	4	ore
il	3.2			
ap	0.8	0.3	0.6	apatite

(1) World average of oceanite (Tyrrell, *The Principles of Petrology*, 1926).

(2) Oceanite from Kilauea, Hawaii (Washington, Petrology of the Hawaiian Islands III, *Am. J. Sci.*, Vol. 6, 1923).

(3) Average of olivine basalt, Hawaiian Archipelago (G. A. Macdonald, 1949).

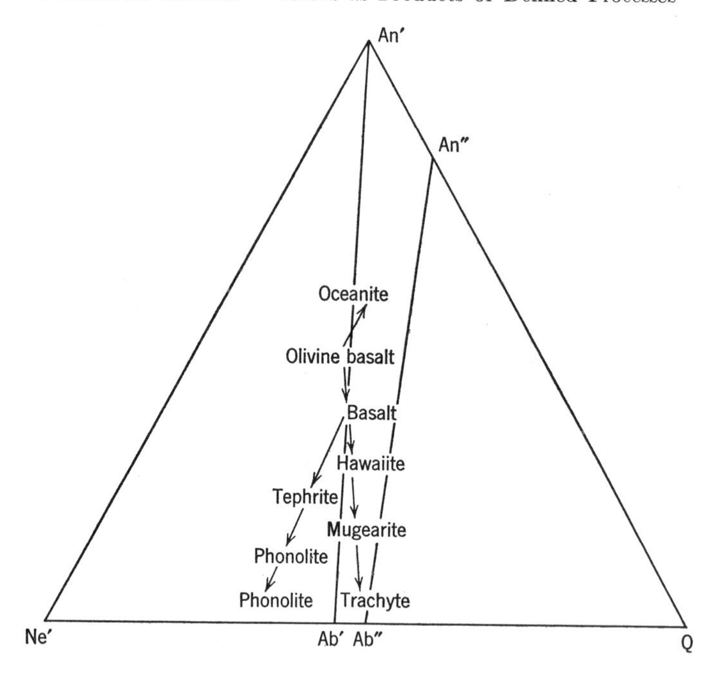

Fig. III-67. The genetic relationship of the various lava types in the central Pacific province. The triangle corresponds to that of Fig. III-35. The basalt-trachyte series and the basalt-phonolite series correspond to case A_2 and N, respectively.

The Atlantic Ocean. The Mid-Atlantic Ridge is a well-known representative of a most remarkable feature of the ocean floors. It is a long and narrow submarine ridge which extends from about 70° N (Jan Mayen) to latitude 50° S, where it bends eastward, parallel to the South African coast, into the Indian Ocean. In its N-S direction it parallels the American, European, and African coasts. At the equator this parallelism is strikingly illustrated by a pronounced E-W flexure.

The high parts of the ridge shoal to less than 1000 fathoms in places, and the highest peaks form groups of islands such as the Azores, St. Paul's Rock, Ascension Island, Saint Helena, Tristan da Cunha, and Gough and Bouvet islands. Olivine basalt and common basalt dominate. In addition there are small quantities of trachyte and phonolite, obviously representing "residua" systems produced by fractional crystallization of the olivine-basaltic parent magma.* The similarity to the central Pacific rocks is obvious. The only general difference seems

* One exception is St. Paul's Rock, which consists of peridotite.

to be that the south Atlantic rocks are somewhat higher in alkalies. A possible explanation is that the magmatic substrata of the Ridge exhibit a large vertical extension, thus inducing conditions stimulating diffusion differentiation. The volatile constituents and the alkalies would gradually diffuse through the substratum and, being lighter than the surroundings, concentrate in the upper parts of the Mid-Atlantic Ridge.

Rocks of very similar petrography occur on the continental shelf west of Africa. The rock province of the volcanic islands, Madeira, Canary, and Cape Verde, is dominated by basaltic rocks associated with derived alkalic lavas (for example, soda-rich phonolite).

The Indian Ocean. The deep-sea islands, Kerguelen and others, display in their general lithology the same characteristic features as those of the other oceanic islands. The submarine rise, of which they form the peaks, is a direct continuation of the Mid-Atlantic Ridge.

The rock series of all ocean basins are thus remarkably similar. Small differences occur, but the similarity of all these rocks occupying an area of about one half of the total surface of the earth is a conspicuous fact of great petrogenetic significance. At all times and over all these areas the parental magma was of the composition of an olivine basalt.

Iceland and Thulean Basalts

In more than one way Iceland is an island of transition. Geographically it is located midway between Europe and America and served as a stepping stone for the early vikings who tried settling America. Petrographically it is transitional into a continental province, although placed in the Mid-Ocean. It is overwhelmingly basaltic but not alkalic, the residual differentiation products being liparitic; nepheline-bearing lavas do not exist.

In early Tertiary times colossal volcanic eruptions took place in the North Atlantic basin. An island much larger than the present Iceland almost filled the basin, and the volcanic activity extended into Jan Mayen and Greenland on the north, Iceland and Scotland on the south, all in all more than one million square kilometers, and swarms of basaltic dikes were injected as far east as Scania in Southern Sweden. This huge area has been called the Thule province.

The fragmentation and subsidence of part of this great plateau land led to the formation of the North Atlantic Ocean.

The Thulean basalts, formed in Paleocene and Eocene times, appear to have been extruded subaerially and are not members of the deep-sea basalt association; they belong to the continental shelf.

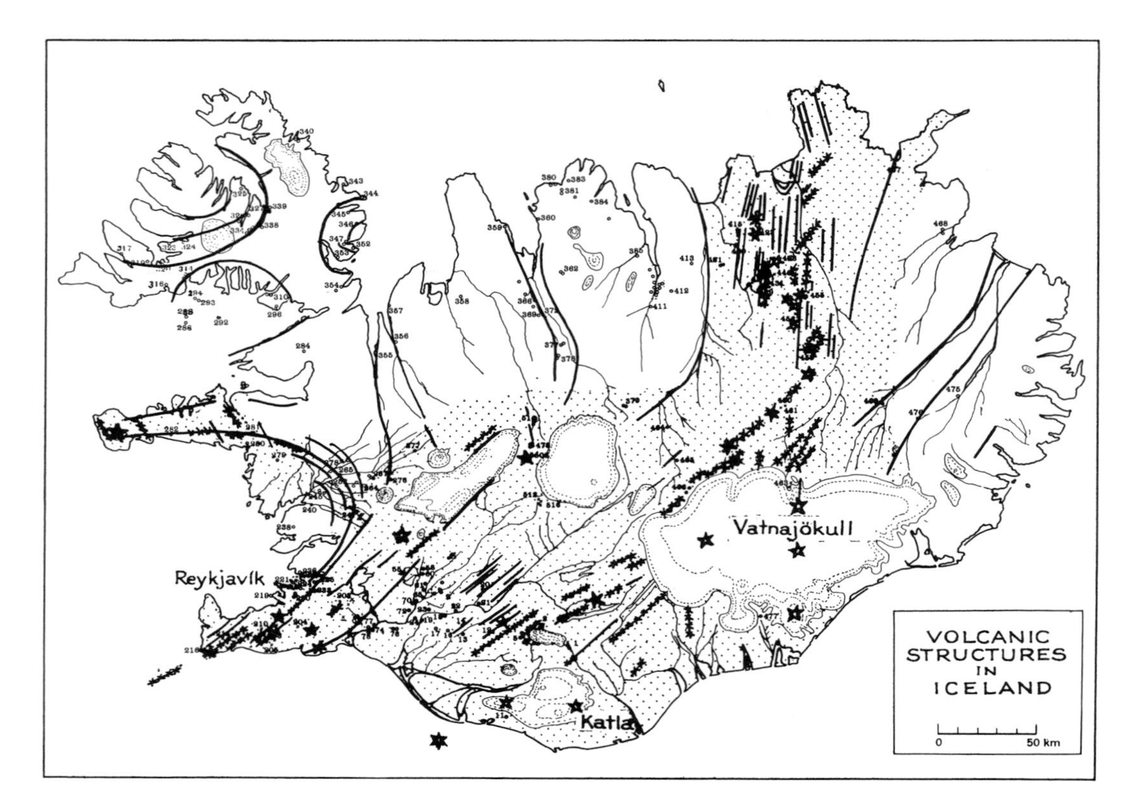

Fig. III-68. Fissure systems in Iceland. Some fissures are open, some have brought up enormous quantities of basaltic lava. The recent volcanism is restricted to the dotted areas; black stars denote individual volcanoes.

TABLE III-22

ESTIMATED AMOUNTS OF LAVA AND ASH PRODUCED BY SOME RECENT ICELANDIC FISSURE VOLCANOES

Name	Date	Lava (m^3)	Ash (m^3)
Edgjá	950	9×10^9	?
Laki	1783–1785	12×10^9	2×10^9
Askja	1875	(small amount)	2×10^9

Thicknesses of 4500 m have been estimated. It is probable that sialic material was present and available for assimilation by the magma. It is strange that Iceland has retained such a considerable mean elevation, 800 to 900 m above sea level, whereas large adjacent areas have subsided.

Modern volcanism in Iceland began less than one million years ago, is local, and has no direct relation to the Tertiary plateau basalts. It is also on a much smaller scale in spite of the impressive figures of Table III-22. The famous Laki fissure erupted in 1783 and produced more lava than any other historical eruption, flooding an area of 565 sq km.

The results of seismic investigations near Reykjavik indicate a three-layered crust—a top layer of Recent lava and ash, and two layers of (Tertiary[?]) plateau basalts; the thicknesses were 2.1, 15.7, and 10.0 km respectively, giving a total distance down to the Mohorovičič discontinuity of around 27.8 km. (Båth, 1960.)

The lavas of the Thulean province have been extensively studied, particularly on Iceland and in the British Isles. In the famous memoirs of Mull and Ardnamurchan Scottish geologists have contributed greatly to our knowledge of petrology. See Thomas and Bailey (1924), Richey (1930).

The Mull authors constructed "magma types." The conception of magma type is based upon composition alone. In this, it differs from the conception of rock type which takes into account texture as well as composition. Thus a basalt and a gabbro may belong to one magma type though admittedly representatives of different rock types.*

* Niggli (1936) elaborated on this idea and eventually proposed a complex setup of 2355 different types. Such an elaborate system is hardly called for, but the conception of "magma type" is a very useful one in petrology.

TABLE III-23

THE THREE MULL MAGMA TYPES

(E. B. Bailey, 1924)

	Plateau Type	Non-porphyritic Central Type	Porphyritic Central Type
SiO_2	45	50	45
Al_2O_3	15	13	26
Fe_2O_3 / FeO	13	13	9
MgO	8	5	4.5
CaO	9	10	15
Na_2O	2.5	2.8	2
K_2O	0.5	1.2	0.3

Three basaltic magma types are distinguished as shown in Table III-23. The plateau type is regarded as parental and gives rise to: (1) the normal Mull magma series: basalt → inninmorite → granophyre, and (2) the alkaline magma series: basalt → mugearite → trachyte. These two series correspond to cases X or A_1, respectively to A_2 as shown in Fig. III-35. Theoretically the two contrasted series can be derived in two ways:

a. By the mechanism of fractional crystallization: early olivine will not (or will) react completely with the melt which, therefore, becomes (or does not become) enriched in residual quartz yielding granophyric (respectively trachytic) end products.

b. By contamination: patches of sial would seem to be present almost everywhere under the Scottish province. By assimilation the primary alkaline magma would change to more "tholeiitic" compositions (normal Mull magma type).

It has been contended that types 1 and 2 are rather unrelated and derive from two "primary" magmas. Indeed, it was on the basis of the study of the Hebridian province that Kennedy developed his hypothesis of "tholeiite" and "alkali basalt magma" residing in two different levels in the crust. However, as explained on page 174 this hypothesis would now seem hard to uphold.

Plateau Basalts

Plateau basalts, also called flood basalts, or fissure basalts represent the most copious effusions known in the continents. They have appeared more or less at random in both space and time in non-orogenic areas as enormous floods which inundated vast areas in India, Brazil, Patagonia, Siberia, Oregon (and in the Thule province).

The great chemical similarities of the various floods are shown in Table III-24. They are always flat-lying, rather undeformed, and were, therefore, called plateau basalts by Sir Archibald Geikie; according to Tyrrell the designation *flood basalts* is preferable. Petrographically they are almost exclusively olivine-bearing basalts; it looks as if the primary magma broke through the crust and congealed

TABLE III-24

AVERAGE CHEMICAL COMPOSITION OF VARIOUS BASALTIC PROVINCES

(After Walker and Poldervaart, *Bull. Geol. Soc. Am.*, 1949)

	Trias-Lias						Cretaceous			Tertiary	Recent
	(1)	(2)	(3)	(4)	(5)	(6)	(7)	(8)	(9)	(10)	(11)
SiO_2	51.9	52.5	53.3	54.0	51.6	52.2	49.5	51.3	49.8	50.8	50.6
TiO_2	1.1	1.0	0.6	0.7	2.0	1.3	1.5	1.9	1.7	1.5	3.0
Al_2O_3	15.5	15.4	16.4	16.1	13.5	15.4	14.1	14.0	15.0	15.3	13.9
Fe_2O_3	1.0	1.2	0.5	0.8	5.3	1.6	4.8	3.3	2.7	4.5	2.4
FeO	9.7	9.3	8.3	7.4	7.7	8.7	10.1	10.1	10.2	6.9	11.8
MnO	0.3	0.2	0.2	0.1	0.2	0.1	0.2	0.3	0.2	0.2	0.2
MgO	8.2	7.1	6.7	7.0	6.3	7.3	6.1	5.5	6.5	7.5	4.7
CaO	9.7	10.3	11.5	11.1	9.8	10.0	9.9	9.8	10.9	8.6	8.3
Na_2O	1.8	2.1	1.6	1.8	2.4	2.4	2.7	2.8	2.2	3.0	3.0
K_2O	0.7	0.8	0.9	1.0	1.0	0.8	0.7	0.7	0.6	1.3	1.3
P_2O_5	0.1	0.1	....	....	0.2	0.2	0.4	0.3	0.2	0.4	0.8

(1) Average Karroo chilled basalt.
(2) Average Karroo dolerite.
(3) Average Tasmania chilled basalt.
(4) Average Antarctica dolerite.
(5) Average South American basalt.
(6) Average Palisade chilled basalt.
(7) Average Spitsbergen basalt.
(8) Average Deccan basalt.
(9) Average Siberian Trap.
(10) Average Central Victoria lavas.
(11) Average Oregon basalt.

at the surface before the processes of differentiation had had time to operate. Yet studies of the thin sections show that the plateau basalts during the period of congealing were in a process of differentiation, in the same manner as the basalts of the Pacific islands; early olivine, diopsidic pyroxene, and calcic plagioclase produce a later groundmass of trachytic composition. Small amounts of individual trachytic rocks (±quartz) are in some places associated with plateau basalts.

Flood basalts occur in many other non-orogenic areas of the world: the Algonkian region of igneous rocks, of Keweenawan age around Lake Superior; the Palisadan region of Triassic sills and flows, extending from Nova Scotia to South Carolina east of the Appalachians. Other examples are the Karroo dolerites of South Africa, the young

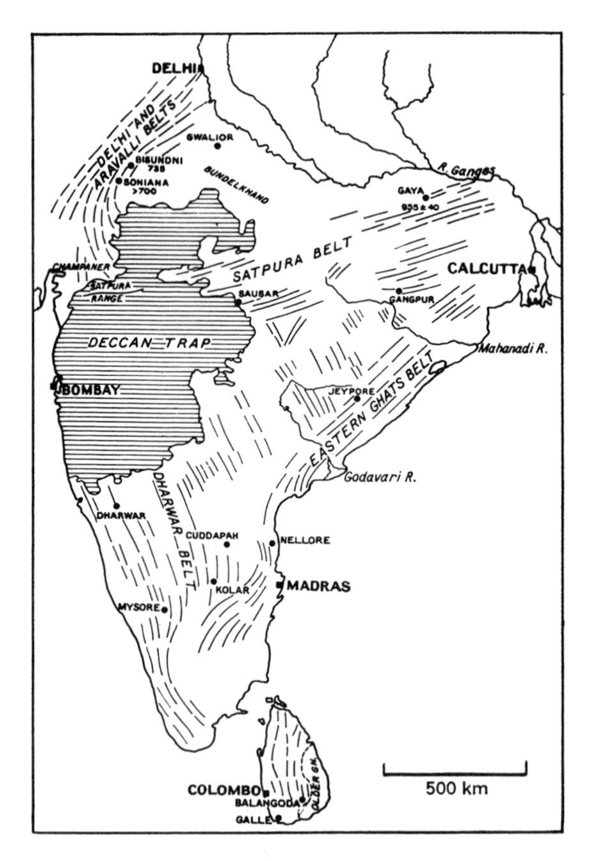

Fig. III-69. The distribution of the Cretaceous Deccan traps in peninsula India, covering an area of 600,000 km². They are surrounded by pre-Cambrian gneisses. (After A. Holmes, 1950.)

basalts of Central Victoria, the Mesozoic dolerites of Tasmania, almost perfectly matched by the Antarctic dolerites, and many more. In their general characters—physical, petrographical, and chemical—the igneous rocks of these regions conform to those described above. The flows have issued from fissures with little or no explosive, cone-building activity; the sheets are, or originally were, horizontal and of great areal extent; and the aggregate thickness of the series of sheets is very great.

The areal eruptions, so called by Reck (1930), also called *multiple vent basalts* were supposed to arise from the confluence of lava flows from a large number of small and closely spaced volcanoes. The flows coalesce into plateaux often hundreds of square kilometers in area. Later investigations on Iceland by Thorarinsson (1951) have shown that the many small vents actually represent hornitoes or pseudo-volcanoes. When basalts flood large and flat areas and flow over water-rich depressions, swamps or little lakes, the water thus trapped under the basalt flows will be heated, will boil, and in innumerable places break through the lava cover forming "multiple vents" or pseudo-craters. Examples can be found in many places in Iceland, the region of Clyde, Scotland, and in continental Europe: Auvergne, Westerwald, Bohemian Mittelgebirge, etc., of the extra-Alpine forelands.

Basalts of the Karroo System, South Africa may be classed with the plateau basalts. They are dolerites or fresh diabases dating from an early stage of the Jurassic; they are particularly abundant in Natal and Cape Province, but are represented also in the Orange Free State, Transvaal, Southern Rhodesia, Southwest Africa, and even as far north as Nyasaland. As dikes, sills, or cross-cutting sheets, these hypabyssal bodies appear at intervals throughout an area measuring at least 18 degrees of latitude and 17 degrees of longitude, and nearly 2,000,000 sq km. The extraordinary effect of this imperial invasion of the earth's crust by the basic magma has been treated in numerous publications.

Some of the sheets are locally thickened into pod-like masses 300 to 700, or more, meters in thickness. Such lenses give evidence of the complexity of a differentiation involving cooling, crystallization, and local immiscibility. They may be roughly divided into three petrological units:

1. *Roof zone* or upper acid phase of granitic composition.
2. *Central zone,* consisting primarily of dolerite or gabbro.
3. *Basal zone,* characteristic of picrite and olivine-rich hyperites.

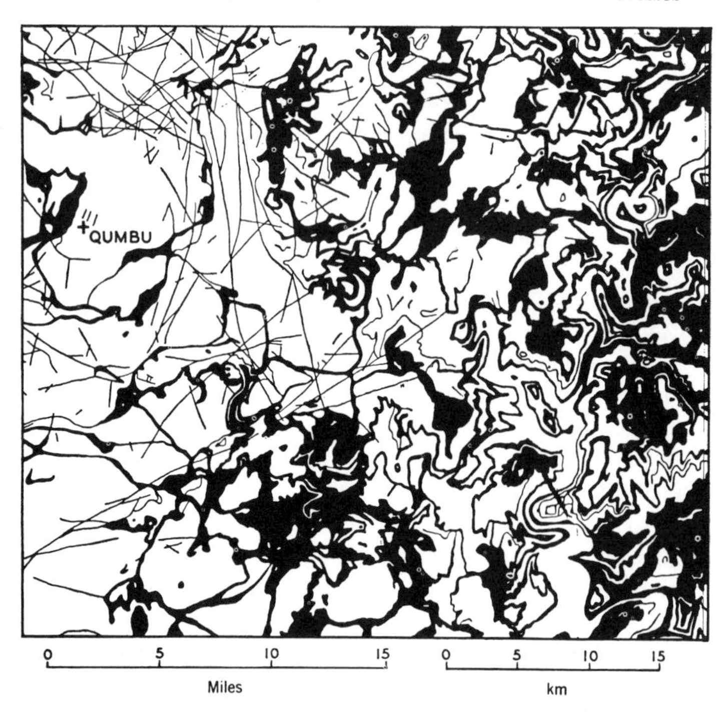

FIG. III-70. Map of complex of Karroo dolerite sheets in eastern part of Cape Province, South Africa. (After du Toit, *Geol. Surv. S. Africa,* areal map 27.)

A section is shown in Fig. III-71. Scholtz's interpretation is that during the cooling a critical combination of temperature and composition was reached which induced the sulfides to separate as an immiscible liquid (called the *ore magma*) in the form of numerous globules. But the composition of the globules was not pure sulfide. All compounds of the original magma would distribute themselves between the two immiscible liquids. In smelting operations, the

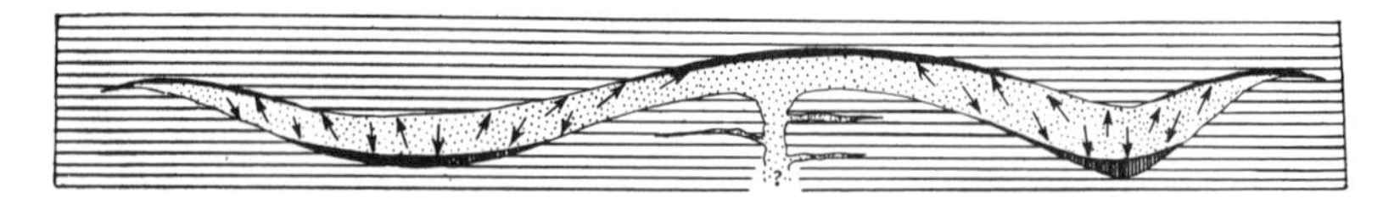

FIG. III-71. Generalized diagram to show the principal characteristics of an intrusion of a large sill of Karroo dolerite, as well as the trend of the differentiation. Horizontal ruling = invaded sediments; black = roof zone; stippled = central zone; vertical ruling = basal zone. (After Scholtz, *Publ. Univ. Pretoria,* 1936.)

liquid matte (corresponding to the sulfide phase) is well known to dissolve considerable amounts of gases. In the newly formed magmatic sulfide globules of small dimensions, the presence of such dissolved gases would have the effect of increasing their buoyancy, and hence opposing gravitative descent. When sufficiently numerous, how-

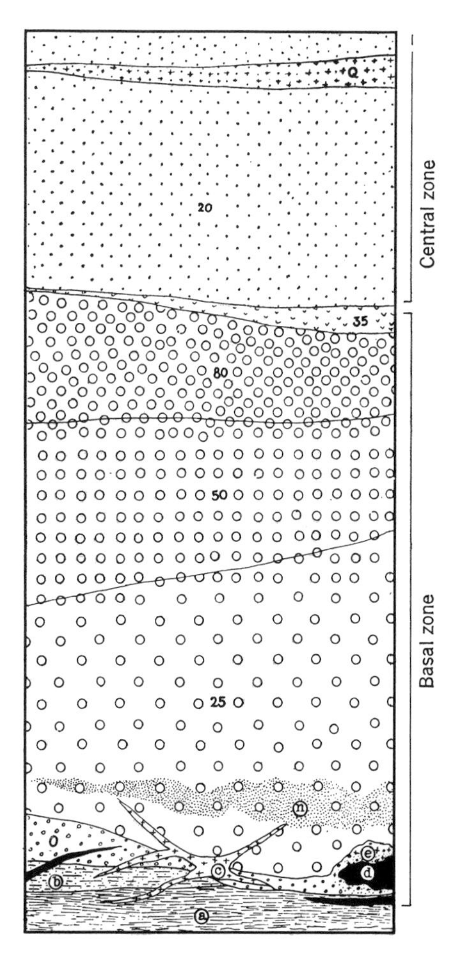

FIG. III-72. Generalized diagrammatic vertical section through the basal zone, illustrating the interrelation of the different petrological units. The junctions indicated are transitional. *a* and *b* = altered sediments; *c* = acid veins; *d* = sulfidic ore in sheets, dikes, and veins; *n* = disseminated sulfidic ore of mineralized zone; *e* = quartz-hypersthene monzonite developed in the proximity of ore bodies. Small circles = eruptive rocks of the basal zone; numbers indicate percentage of olivine in the rocks. Q = quartz-bearing hyperite. (After Scholtz, 1936.)

ever, the globules would tend to coalesce to form larger and larger droplets, which would settle more rapidly. Thus the ore which separated from the magma was concentrated in the basal zone by gravitational differentiation to form pools within hollows and irregularities present in the floor of the intrusion.

Volatiles and silicates dissolved in the ore magma would be released on cooling, and in any local accumulation, or pool, the heavy metal sulfides would tend to settle below the more siliceous fraction in a manner analogous to the separation of matte and slag in metallurgical operations.

At the same time crystal differentiation proceeded in the silicate part-magma, producing various petrological types, as diagrammatically shown in Fig. III-72. At the same time, according to a view expressed by Walker and Poldervaart (1949), small intrusions of basaltic magma without superheat were able to produce significant bodies of igneous-looking rocks of granitic composition by transfusion and sometimes by mobilization of the adjacent sediments (see page 351).

Diabase Dikes and Lamprophyres

Dikes of basaltic composition are present in all terranes, in the vast areas of pre-Cambrian gneisses as well as in the younger rock provinces in Sial. Dike complexes genetically related to larger igneous bodies usually exhibit examples of extreme magmatic differentiation. This is in sharp contrast to the diabase dikes, which are of uniform basaltic composition, strike over long distances through a variety of rock types, and, as a general rule, show no demonstrable relation to other magmatic intrusions in the same geographical area. They are younger than the surrounding rocks and usually follow tectonic lines of the same general age (Fig. III-73).

Most diabase dikes are less than 10 m across (see Fig. III-74), but occasionally they are much wider.

The so-called Great Dike of Rhodesia is one of the wonders of geology. Composed of basic and ultrabasic rocks, it measures 500 km in length by 3 to 12 km in width (Fig. III-75). Its magma was differentiated into a series of long lenses of contrasted rocks, the lenses dipping gently toward the middle of the body. This type of banding is quite unusual in dikes. Most diabase dikes display fine-grained border facies, a typical phenomenon of chilling testifying to their igneous origin. Usually no assimilation of the country rock has taken place. Whence, then, came the magma of the dikes?

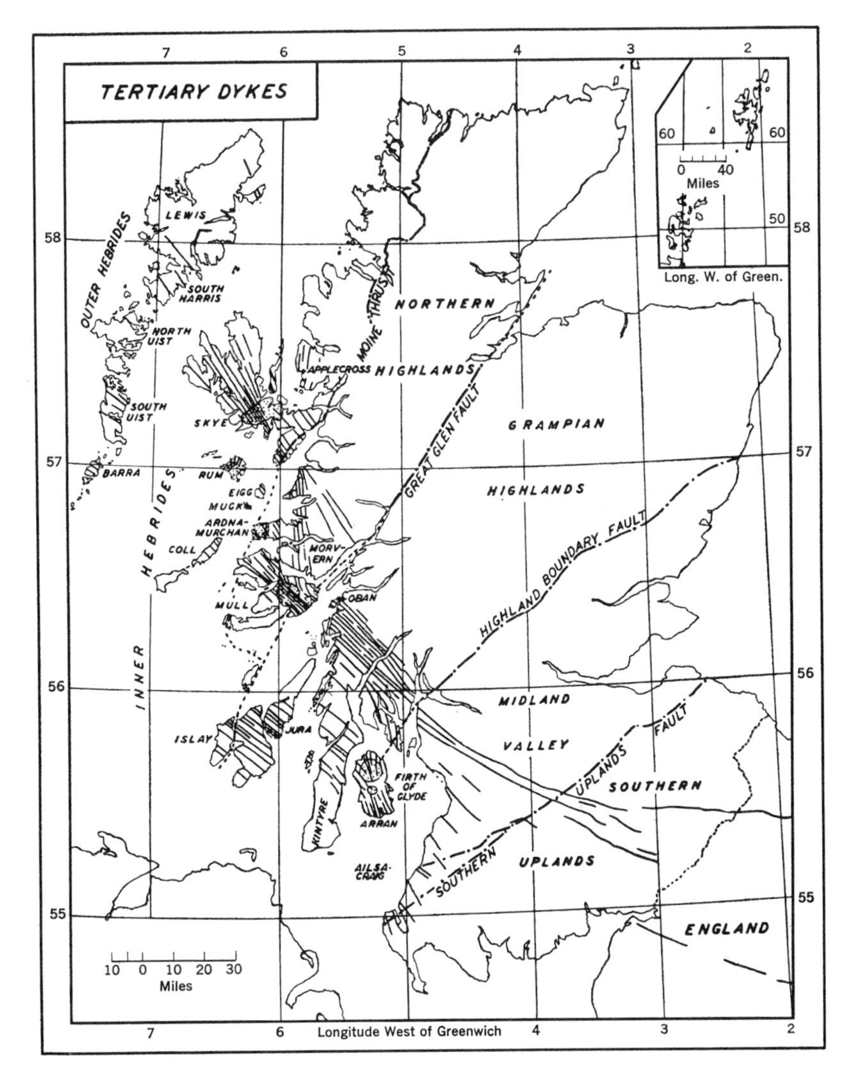

FIG. III-73. Trends of Tertiary dikes. Areas of Tertiary plutonic complexes are shown by stipple. The dikes are usually much more numerous than indicated. For example, in the case of the thickest Mull and Arran swarms, only about 2.5 per cent and 1 per cent, respectively, of the known dikes are shown. (After J. E. Richey, 1939.)

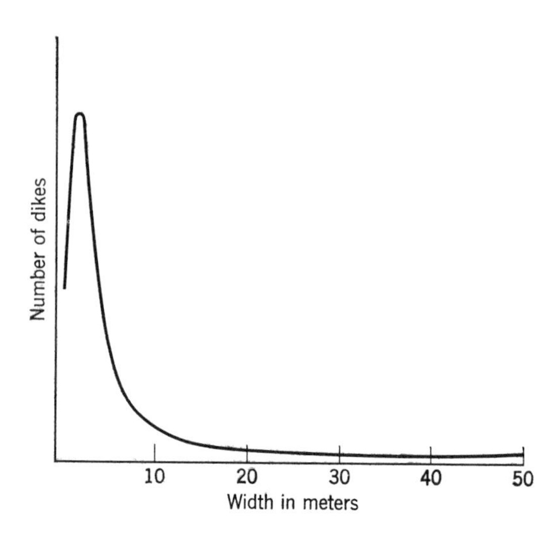

FIG. III-74. Width of individual dikes encountered in the Ullern area at Oslo. (After J. Dons, 1952.)

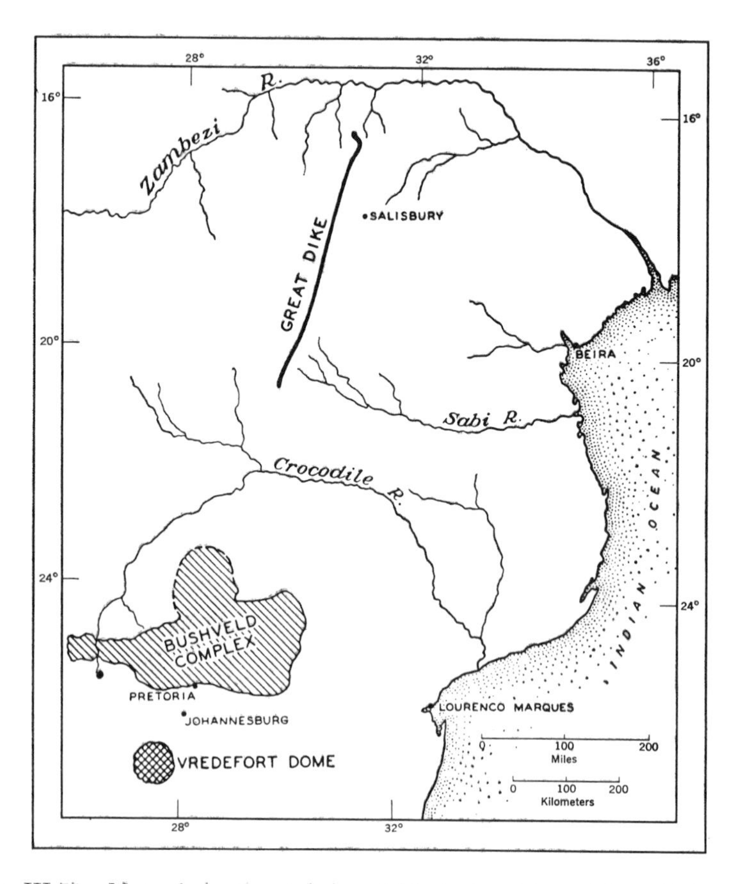

FIG. III-75. Map of the Great Dike of Rhodesia. (Copied from Daly after du Toit, *The Geology of South Africa*, 1926.)

TABLE III-25

CHEMICAL COMPOSITION OF DIABASE DIKES

	(1)	(2)	(3)	(4)	(5)	(6)	(7)
SiO_2	47.00	47.49	49.47	50.59	50.80	52.10	56.01
TiO_2	3.60	3.29	1.79	1.71	2.23	2.75	2.94
Al_2O_3	16.44	14.54	13.09	14.58	16.30	14.61	14.42
Fe_2O_3	3.31	3.36	4.18	2.91	3.91	4.24	3.94
FeO	12.34	11.80	9.95	8.22	6.95	8.72	5.93
MnO	0.04	n.d.	0.29	0.17	0.19	0.20	0.06
MgO	3.32	6.10	6.78	7.39	4.31	4.54	3.96
CaO	9.57	9.70	9.91	9.46	7.21	8.93	5.18
Na_2O	3.38	2.60	3.08	2.52	4.02	2.48	3.96
K_2O	0.67	1.12	1.07	0.17	2.17	1.22	2.95
P_2O_5	0.33	n.d.	0.24	0.25	0.94	0.20	0.64
CO_2	n.d.	n.d.	0.15	1.39	9.07	n.d.	n.d.

(1) Big diabase dike near Muray Mines, Sudbury. (Walker, 1897.)
(2) Diabase dike. Rockport, Essex County. (Washington, 1899.)
(3) Basic dikes. Skye and Mull swarms, Great Britain. Average of 5. (Harker, 1904, and Bailey, 1924.)
(4) Diabase dike. Ascutney Mountain, Vermont. (Daly, 1903.)
(5) Diabase dikes. Oslo region. Average of 11. (Brögger, 1933.)
(6) Diabase dikes. South Norway pre-Cambrian area. Average of 6. (Barth, 1939.)
(7) Diabase dikes. Silvretta, Switzerland. Average of 3. (Bearth, 1932.)

Table III-25 shows that, chemically, the diabases are practically identical with the plateau basalts. It looks as if fractures and fissures in Sial served as avenues of escape for the ubiquitous basaltic magma that in other places fed the plateau effusions. In most places the magma would reach the surface undifferentiated; nor would assimilation normally take place. The table demonstrates, however, that diabases of the Silvretta, situated within the Alpine orogeny, show a slightly abnormal chemical composition obviously approaching that of andesite. Thus it appears that the magma of these diabases had become adulterated by assimilation or by mingling with magmas indigenous to orogenic mountain chains.

Intrusions of basaltic magma in concentric structures, such as ring dikes, cone sheets, and cauldron subsidences, are common features of many volcanic districts (Fig. III-76). The classic examples occur in

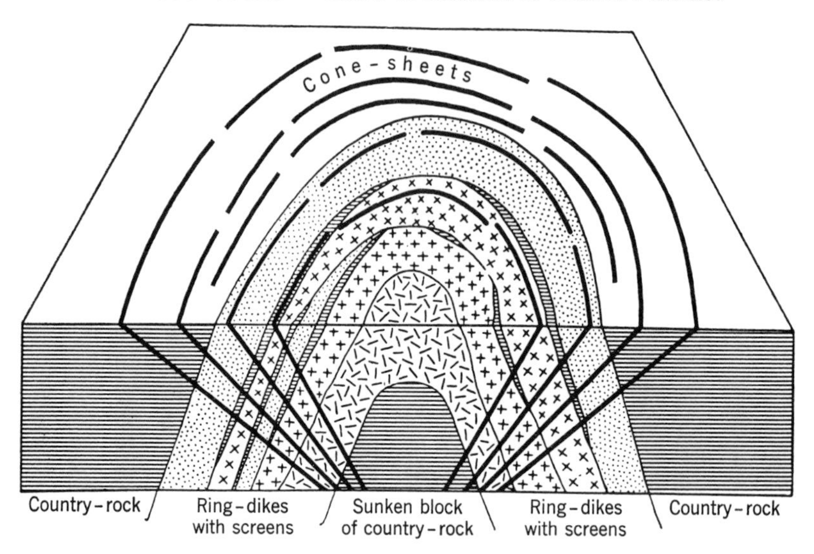

FIG. III-76. Block diagram to illustrate mutual relations of ring dikes of differing composition and dates of injection. (After Richey, 1931; Richey and Thomas, 1932.)

Scotland, especially among the Tertiary volcanics, but also in more ancient rocks. Examples also have been found in North America, on Hawaii, and elsewhere.

From the nature of cone sheets it is practically certain that their formation is due to an excess of magmatic pressure acting vertically upwards upon a relatively thin crustal covering, in a successful attempt on the part of the magma to raise its roof. On the other hand, the formation of ring-dike fissures and the concurrent intrusion of magma are independent of excessive magmatic pressure but are consequent on crustal collapse. Examples of *simple, differentiated, multiple,* and *composite* ring dikes and cone sheets are represented in the British field.

Lamprophyres. Some of the diabases of non-orogenic areas also show evidence of assimilation. A prerequisite seems to be a high water content (probably acquired from surrounding rocks). In such magmas the temperature of crystallization is depressed, and hornblende crystallizes instead of pyroxene, producing hornblende-biotite diabases that are mineralogically different from, but chemically similar to, the ordinary pyroxene diabases. However, diabase-magma usually rich in H_2O and CO_2 may develop beyond this stage; it may assimilate part of the country rock, preferentially the part rich in alkalies and "granitic" constituents. However, much evidence supports the view that lamprophyres are diabases that have been altered by metasomatic processes. Kullaites, vogesites, minettes, and, generally, lamprophyres may develop in this way.

To quote Wegmann (1948): "De nombreuse roches décrites sous le nom de lamprophyre ne sont autre chose que des filons basiques trànsformés."

Layered Sequences and the Shape of Magma Chambers

The Skaergaard intrusion at Kangerdlugsuaq, East Greenland (Wager and Deer, 1939), of Eocene age outcrops over an area of about 50 sq km and represents a rock complex the chemical variability of which must be attributed mainly to crystal fractionation.

The magma from which the rocks formed had the composition of an olivine basalt, now represented by the chilled marginal border phases (see Fig. III-77). The main portion of the complex is made up of horizontally *layered rocks* composed of clinopyroxene, orthopyroxene, olivine, and plagioclase. Each of these phases varies in composition as we progress upward in the series, the ferromagnesians becoming richer in iron, the plagioclases richer in soda. The structure is supposed to indicate gravity stratification produced by the several minerals successively separating from the magma and accumulating at the bottom, the solidification thus proceeding from the bottom up.

The following terms are used: The primary precipitate is called *cumulus,* the residual liquid being the intercumulus liquid. The enlargement of cumulus crystals by diffusion from overlying magma is called an *adcumulus* process. The upward growth of olivine is

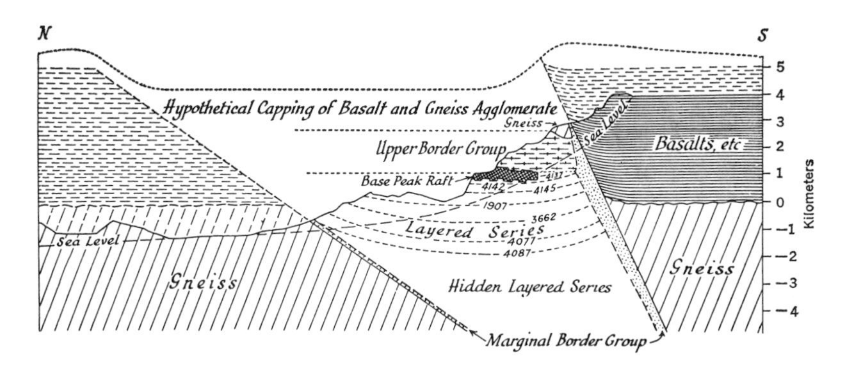

Fig. III-77. North-south section of the original Skaergaard intrusion before the flexuring and subsequent erosion. (After Wager and Deer, 1939.)

called a *harristic* cumulate (harrisite is a local name for a textural variety of peridotite). The layering is called *rhythmic* when caused by a variation in proportions of the component minerals in adjacent layers; it is called *cryptic* if caused by the gradual change in composition of the various minerals. *Igneous lamination* results from the parallelism of platy minerals.

Inasmuch as cooling and crystallization would occur predominantly along the roof, the crystals would have to reach the floor either by gravitative settling or by currents in the magma. It is believed that convective currents are set up by cooling of the magma near the roof, and crystallization in the magma increasing the bulk specific gravity. Currents of the heavy material would then descend along the walls, and move across the floor of the magma chamber toward the center, while less dense magma moved upward in the center and spread out along the roof. This process has produced two effects still visible.

1. The *fluxion banding* parallel to the contact walls indicates that crystals were deposited and subjected to a "winnowing" effect as the magma descended. The horizontal layering of the main mass bends upward on approaching the walls, as though crystals accumulating on the floor of the magma chamber were banked up against the sides.
2. The so-called *trough banding*, representing irregularities in the otherwise exceedingly regular horizontal layering, arranged radially on the floor, draining from the side walls to the center of the chamber, exhibits the appearance of shallow channels such as might be developed by a stream flowing over sand.

Probably better than anywhere else, the processes of progressive crystallization of an intrusive body can be followed by the imprints drawn in the solid rock. A series of analyses of average rocks from evenly spaced horizons through the entire vertical extent, obviously reflecting the true age relations, arranged in a variation diagram, indicates the chemical changes induced by crystal fractionation. Of particular interest is the increase in soda and ferrous iron, corresponding, respectively, to the reaction series of the plagioclases and the ferromagnesian minerals.

It is of special interest, in this case, to follow more closely the differentiation of the ferromagnesian minerals. They become, with proceeding crystallization, successively richer in iron and appear eventually as constituents in the so-called ferrogabbros made up of pyroxene and olivine of unusually high iron content. This is in complete harmony with the experimentally determined course of crystallization of a basaltic magma (see pages 95 ff.), and we might expect, therefore, that all basaltic magmas by pure crystal differentiation would yield residua very rich in iron. But most basaltic magmas do not. Why is this? Fenner (1929, 1938) seeks the solution in other processes (for instance, gaseous transfer) working in addition to, or parallel with, crystal differentiation. Bowen and Eskola argue that other residual magmas are *relatively* rich in iron, but the effect is "diluted" by a still greater enrichment in alkalies and silica.

Bowen also thinks that in this example the magma was unusually dry, consequent upon the thin and porous character of the roof of the mass. Bodies with more competent roofs are able to retain their volatile constituents and give hydrous phases, such as hornblende and biotite. It is these bodies that show the normal, natural differentiation series, basalt to rhyolite, as distinct from the laboratory (dry) series approached by the Greenland example.

In other areas the origin of layered sequences is not obvious, and it has been suggested that the layering is primarily an inherited (sedimentary) structure, and the igneous appearance of the rock is due to extreme metamorphism.

The Magma Chamber. Edwards (1942), in his study of the differentiation of the dolerites of Tasmania (close kin to those of the Karroo, see page 185), has pointed to the importance of the shape of the magma chamber. Three factors dominate the differentiation of most basaltic magmas: (1) fractional crystallization, (2) gravitational differentiation, and (3) the form of the intrusion concerned. The sinking of early-formed pyroxenes rich in magnesium gives rise to a progressive increase in the proportion of FeO:MgO in the later-formed pyroxenes.

In *closed chambers*, early-formed pyroxene and plagioclase will sink at different rates and will accumulate above the chilled bases, displacing the residual liquid into the upper parts of the chamber. Selective removal of the magnesium in the early-formed pyroxene concentrated the iron in these shrinking volumes of liquid. When the iron crystallized, the iron-rich minerals that formed could not sink past the already accumulated, early-formed minerals, and so a layer of iron-enriched rock occurs in the upper parts.

In the *subjacent chambers* without floors, on the other hand, and in the dikelike bodies, the early-formed crystals did not accumulate but, as far as the visible portion of the bodies is concerned, sank out of the chamber. The residual liquid in the visible portions of the chamber was displaced upward, and in addition a volume of fresh undifferentiated magma equal to the volume of the sinking crystals was displaced upward from the depths into the visible portion of the chamber.

Fractional crystallization, as before, caused a concentration of iron in the upper parts of the chamber. When this iron commenced to crystallize, the iron-rich pyroxenes that formed were able to sink freely because of the absence of a floor of accumulated crystals beneath them. The higher specific gravity of the iron-rich pyroxenes enabled them to sink more rapidly than the plagioclase crystallizing at the same time, and so the residual melt tended to become enriched in the feldspar constituents. The continued displacement upward of undifferentiated magma throughout the process maintained the most differentiated magma near the tops of the bodies. Convection currents may have modified this simple picture, but they do not appear to have interfered with the general trend.

In both forms of intrusion a concentration of iron must have developed in the upper part of the chamber; but, whereas in the closed chambers this concentration was maintained, in the floorless bodies it was dissipated. Presumably, iron-rich layers might be expected in the deeper parts of these bodies as layers of accumulation or remelts of accumulated iron-rich pyroxenes; but rocks so derived would differ in composition from the iron-rich rocks formed in the closed chambers.

The Glen More ring dike of Mull appears to be a striking instance on a small scale of differentiation in a subjacent chamber. An example on a larger scale is provided by the Yogo Peak stock, in the Little Belt Mountains of Montana.

This conception of closed and subjacent chambers, in which identical processes of fractional crystallization and gravitational separation can give rise to different trends of differentiation, seems of fundamental importance in petrogenesis and explains a number of otherwise irreconcilable observations. Fenner has long maintained that fractional crystallization of basaltic magma must lead to an enrichment of the residual magma in iron. Wager and Deer's study of the Skaergaard intrusive has provided a striking example of this trend.

The Skaergaard intrusive owes its unusual features to the fact that it underwent strong fractional crystallization in a closed chamber, on a larger and more impressive scale than anything hitherto recorded.

The Puzzle of Thick Stratiform Sheets of Igneous Rocks. At widely scattered localities in the world there are sheets of igneous rock which show a layered structure with certain systematic types of variation. These sheets have great thickness: for example, the Duluth, 1 to 9 miles thick; the Stillwater, 3 miles, with top not shown; the Bushveld, 2 to 3½ miles; and the Sierra Leone, about 4 miles thick, with neither base nor top exposed. The mode of formation is debated.

The Karroo basalts and the Skaergaard intrusion described above give actual examples of this structure. (See Figs. III-72 and III-77.)

For many decades geologists have pondered over these structures that seem to correspond to some universal tendency; repeatedly natural processes have operated in the same way, leading to the same conspicuous results. What are these processes? Among the numerous hypotheses, three major ones deserve special mention.

1. Fractional crystallization, with gravity sorting of crystals, and differentiation on liquefaction as a possible attendant process.
2. Successive intrusions of various mafic magmas, with or without the consequent basining of the sial under the large bodies of eruptive rocks.
3. Metasomatic replacement of a sedimentary complex. In this view the stratiform bodies are not magmatic, but the layers are relic structures indicating compositional differences in the pre-existing sediments.

The last hypothesis could not explain the systematic variations frequently observed in the layered sequences. Otherwise petrology today is not able to decide between the rivaling hypotheses; none of them is satisfactory. It seems that a promising field of research is here open for exploration.

5 · NON-OROGENIC SIALIC ROCK SERIES

But large as may loom these common universal rocks (granite and rhyolite, diorite and andesite, gabbro and basalt) we must not forget the rarae aves of petrography—which, for the study of rock geneses, rock evolution, or rock classification, may be of importance equal to the vastly greater and more numerous masses of the familiar rocks.

H. S. Washington, *Bull. Geol. Soc. Am.*, 1922.

Alkalic Rocks

According to Daly, only about one per thousand of all igneous rocks should be characterized as alkalic. By gradual transitions they are connected with the subalkalic rocks. The boundary line may be drawn according to the alkali-lime index, as proposed by Peacock (page 171). A more special definition has been given by Shand (1922), who pointed out that in the commoner kinds of igneous rocks the alkali ions are combined with aluminum and silicon in the ionic proportion of 1:1:3 (in feldspars). An alkalic rock, then, if names are to mean anything, should be one in which the alkalies are in excess of the 1:1:3 ratio, either aluminum or silicon or both being deficient. If only aluminum is deficient, quartz may crystallize, and excess alkalies combine with ferric iron to form aegirite or alkali hornblendes. This group of rock may be called ekeritic. If only silicon is deficient, the alkalies and aluminum form minerals that are unsaturated with regard to silicon, such as nepheline, analcite, or leucite, in which the atomic ratio is 1:1:1, or 1:1:2. This group was later called *miaskitic* by Fersman (1929). If aluminum and silicon both are deficient, the deficiency is again made up by the feldspathoids, and by ferric oxide, zirconia or titania, giving such minerals as riebeckite, eudialyte, or sodalite, hauyne and ussingite. These rocks were called agpaitic by Ussing (1911), and the ratio Na + K/Al the index of agpaicity.

The advantage of this definition is that an excess of alkali-feldspar or of mica does not entitle a rock to be called alkaline. It leaves us with a clear-cut natural group of rocks in which three subgroups may be distinguished.

I. Silica adequate or excessive, alumina deficient. *Ekerite,* see page 210

II. Silica deficient, alumina adequate or excessive. *Miaskitic* (v.i.)

III. Silica deficient and alumina deficient. *Agpaitic* (v.i.)

Figure III-78 shows graphically the mineralogical classification of some of the undersilicified alkalic rocks. They are mostly of miaskitic type.

Characteristically, one petrographic province is dominated by only one of the rock types: by nepheline syenite, by the urtite-ijolite series, or by litchfieldite. An example of litchfieldite is the lardalite of the Oslo region, which will be described presently. It is genetically related to syenites, quartz-syenites, and other non-alkalic rocks of the same area.

Miaskites and urtite-ijolites always occur in a foreign petrographic environment, sharply separated from the surrounding rocks. Circular or elliptical structures are common; they vary in size from the small occurrences in South Africa, less than 500 m across, to the large Khibine area on the Kola peninsula, covering more than 1300 km^2.

Calcite-bearing alkalic rocks always are of the urtite-ijolite series.

A deficiency of silica in one rock and a deficiency of alumina in another are not necessarily causatively related.

The origin of the alkalic group of rocks as a whole is multiple and difficult to survey. Some of the following mechanisms are usually contributary:

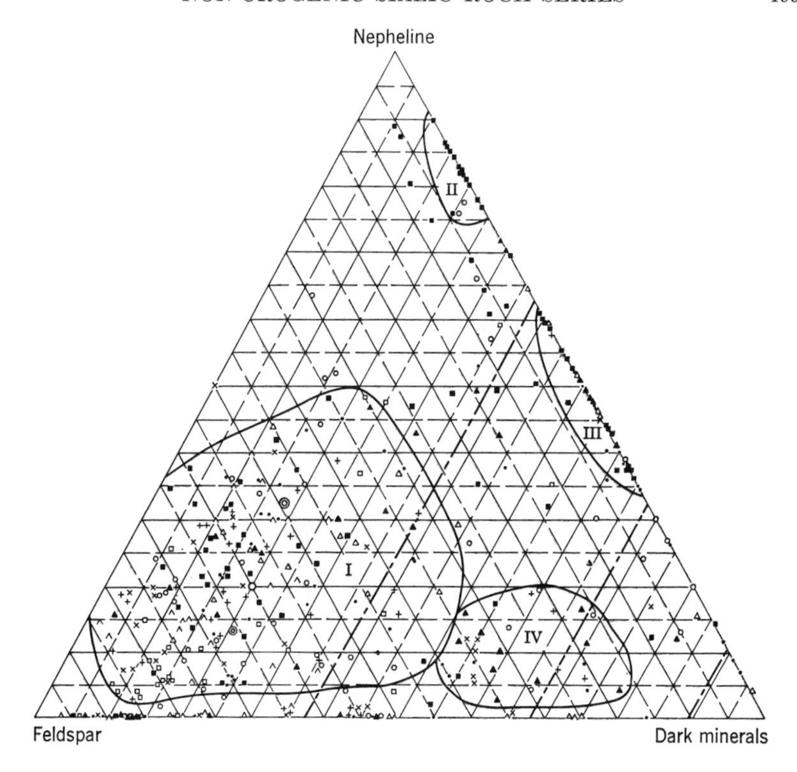

FIG. III-78. Quantitative mineralogical composition of the chief types of miaskitic alkalic rocks. (Modified after Kupletsky and Oknova, 1934.) I = nepheline syenites and litchfieldites. The feldspar of the nepheline syenites is predominantly orthoclase (often soda orthoclase). In the litchfieldites the feldspar is predominantly albitic plagioclase (or exclusively so in mariopolite and canadite). II-III = urtite-ijolite series. IV = malignites.

1. Simple fractional crystallization of primary olivine-basalt magma.
2. Gaseous transfer and thermodiffusion.
3. Differential movements of alkalies in the magma.
4. Assimilation of limestone.

Crystal differentiation of an alkali-basaltic parent magma is sufficient to explain the mildly alkalic character of the "residua" rocks of the ocean basins, and of many rocks associated with faulting and block sinking. Rock magmas of fold mountains are usually contaminated by sediments rich in silica and alumina, and thus unable to produce alkalic rocks. See Figs. III-34 and III-35 and the pertinent discussion in text.

But strongly alkalic rocks do not form by simple crystallization differentiation.

Agpaitic rocks, as emphasized by Sörensen (1960), depend for their formation on volatiles, partly moving differentially within the magma, partly as gases, carrying alkalies and other characteristic compounds to restricted parts of the magma chamber (see page 161). Agpaitic rocks are rich in typical residual elements, Zr, Ti, Nb, rare earths, and volatiles as F, Cl, and H_2O, and are in this respect related to pegmatites. Some examples follow:

Ilimaussaq, West Greenland: a number of rock types are intrusive into pre-Cambrian gneisses, they are strongly agpaitic and layered. There are three characteristic types:

Naujaite: very coarse nepheline syenite with irregular banding caused by variations in the content of eudialyte, dark minerals, and light minerals, so that red, black, and white layers develop. The chief minerals are microcline perthite, nepheline, sodalite, aegirite, arfvedsonite, and eudialyte.

Lujaurite: fine-grained, "schistose" nepheline syenite resembling crystalline schist, in particular amphibolite. The chief minerals are the same as in naujaite.

Kakortokite: coarse-grained nepheline syenite with foyaitic structure and pronounced banding of different colors: black bands rich in arfvedsonite, red bands rich in eudialyte, white bands rich in feldspar and nepheline. The banding is almost horizontal in the central parts of the body but steepens outward and becomes parallel to the vertical contact. The bands or sheets which make up the complex body are, therefore, saucer-shaped.

In the *Kola* peninsula, North-West Russia, two large bodies of agpaitic nepheline syenite are known: Khibina and Lovozero. They are similar to the main rock bodies at Ilimaussaq, but differ geochemically in containing great concentrations of Ti and P, two elements almost lacking at Ilimaussaq.

Other famous occurrences of agpaitic rocks are, for example, *Norra Kärr,* central Sweden and *Iles de Los,* French Guinea.

Miaskitic rocks frequently form through assimilation of limestone by subalkaline magmas. Assimilation of limestone results in a desilification of the magma. Carbon dioxide is liberated, and the excess CaO combines with SiO_2, which in this manner may become insufficient to form the feldspar molecules; instead feldspathoids are formed. It has been demonstrated repeatedly that alkalic rocks undersaturated in silica are formed by this mechanism: for example, the urtite group with differentiates of ijolite, malignite, jacupirangite, etc. (Daly, Shand.)

It is worthy of note that large bodies of alkalic rocks must have developed by metasomatic replacement of preexisting rock types; obviously fenites have developed in this way (page 213) but also a number of alkali syenite complexes which were believed to be truly magmatic may have been modified by metasomatic-metamorphic processes.

Tilley (1957) distinguished between *hypersolvus* and *subsolvus* nepheline syenites. In the hypersolvus, magmatic rocks, nepheline shows a great compositional range and is associated with sanidine or anorthoclase as in phonolites and other quenched lavas, or with cryptoperthite as in lardalite of the Oslo area (page 207); in low-temperature subsolvus nepheline syenite or nepheline-syenite gneiss, the nepheline has a narrow composition range converging on the formula of one K-ion in four alkali ions, and is associated with albite and/or microcline as in the Mariupol district, the Haliburton-Bancroft area, Litchfield, etc.

Some examples of contrasted occurrences of alkalic rocks will be discussed in the sequel.

The African Rift Valleys are impressive features on the face of the earth. The nature of the forces involved in the formation of these major fault zones has been summarized by Willis (1936). In the opinion of Sonder (1938) the rifts are controlled by one of the chief *zonales* of the crust, associated with effusions and great crustal instability, extending from East Africa to the Red Sea and the Jordan Valley into Asia Minor. The lavas are typically alkalic, ranging from ultrabasic (limburgite, basanite, etc.) to phonolite, trachyte, and rhyolite, including many rare types, such as melilite nephelinite, leucite basanite, and others.

In the lavas there seems to be a general tendency towards preponderance of potash in the Western Rift and preponderance of soda in the Eastern Rift.

On the other hand, closely adjacent volcanoes may exhibit strong contrasts in their chemical and petrographical characters over long periods of time (Verhoogen, 1939). The two great active volcanoes of Nyamlagira and Niragongo and the small cinder cone Nahimbi lie at the apices of a triangle whose sides are roughly ten miles long. Nyamlagira is built up entirely of potassic lavas of the leucite-basanite group. Niragongo, on the other hand, is composed of dominantly sodic melilite-nepheline basalt and potassic nepheline-leucite basalts the nepheline of which contains up to 40 per cent $KAlSiO_4$. Nahimbi has erupted nothing but sodic limburgite.

Apparently many subprovinces should be distinguished. Some have been described; for instance, the unusual series at Birunga and Toro-Ankole of Uganda, part of which is made up of the Bufumbira area forming the subject of the memoir of Holmes and Harwood (1937). The most important rock series here are: biotite pyroxenite, katungite and related melilite-rich lavas, ugandite → olivine leucitite → leucitite → potash turjaite. Some of the rocks are similar to those of the Laacher See district in the Rhine graben, and were probably formed by the action of alkaline vapors distilled from a deep-seated magma. The general hypothesis of Holmes and Harwood is that the original magma of Bufumbira showed a dominance of soda and potash and that reaction with biotite pyroxenite gave birth to the potash-rich types in the area. No other mechanism seems capable of explaining why potash should be enriched so greatly in some of these rocks. Much work is still to be done before a comprehensive survey of all the lavas of the rift valleys is completed.

The Rocks of Vesuvius are strongly alkalic because of, indirectly, an extensive assimilation of dolomitic limestone. They have been comprehensively studied by Rittmann (1960).

The oldest lavas are trachytic (Ur-Somma period). After a long time of tranquillity the second active period produced the so-called orvietites (leucocratic, leucite-bearing trachybasalts). Shortly afterwards the third active period produced ottajanites (leucocratic or mesotype basaltoid leucite-tephrites), which make up the prehistoric Young-Somma volcano. The fourth and present active period is characterized by effusions of vesuvite (mesotype plagioclase-bearing leucitite). Thus the course of differentiation is:

trachyte → trachybasalt → leucite tephrite → leucitite

This evolution cannot be explained by simple crystal fractionation. This can be seen from the fact that at tranquil times, when the vent was closed, magmatic differentiation followed another course: true gravitative and pneumatolytic differentiation, characterized by gases and light crystals (sanidine, leucite) accumulating in the upper part of the vent, heavy crystals (diopside-augite, biotite, basic plagioclase) sinking to the bottom, and a major eruption subsequently ejecting these differentiates, depositing them on the slopes of the volcano in the opposite sequence. All such Plinian eruptions have produced normal differentiation series of comagmatic rocks.

Contrasted in its differentiation is the large-scale evolution of the Vesuvius magma; each subsequent period of eruption has yielded magmas of successively lower degree of silification, with relatively high content of potash. This is shown schematically in Diagram II.

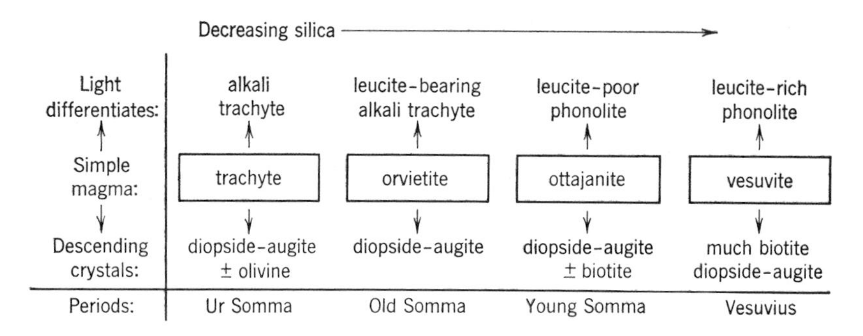

Diagram II. Vesuvian rocks. (After Rittman, 1960.)

Fractional crystallization, which obviously effected the differentiations in the Plinian effusions, cannot at the same time be made responsible for the peculiar increasing alkalinity of the chief magma. Actually this is related to assimilation of dolomitic limestone, which, as everywhere in central Italy, is present at a depth of 2000 to 3000 m. Lime and magnesia are incorporated in the magma; carbon dioxide adds to the gas content, and a thermal equivalent amount of diopside-augite crystallizes, coalesces, and sinks down. Among the products disgorged by strong eruptions are coarse-grained fragments of diopside-augite that testify to the processes occurring at depth.

Not only does the assimilation determine the petrographic character of the lavas, but also, according to Rittmann, the whole historic activity of Vesuvius depends upon it. Assimilation of limestone brings about an increase in the gas tension of the magma necessary for sustaining continued activity. It is doubtful if any new activity would have occurred after the Ur-Somma period unless the chief dolomite layer had been reached by the subvolcanic magma, had suffered assimilation, and by its content of carbon dioxide had blown new life into the old magma body.

The Petrographic Province of Central Montana has been comprehensively studied by E. S. Larsen (1940) in a most fascinating way. It includes the Tertiary igneous rocks near the eastern flank of the

TABLE III-26

ESTIMATED COMPOSITION OF THE PARENTAL MAGMAS OF THE SUBPROVINCES OF THE CENTRAL MONTANA PETROGRAPHIC PROVINCE AND OF THE SAN JUAN LAVAS OF COLORADO

(After E. S. Larsen, 1940)

	Yellow-stone	Cran-dall	Ab-saroka	Little Belt	Crazy	High-wood	San Juan	Average Plateau Basalt
SiO_2	56.0	52.8	50.6	53.8	48.8	48.7	53.8	51.2
Al_2O_3	17.7	15.5	12.9	14.5	15.0	12.7	17.2	14.7
FeO	8.1	8.4	9.3	8.0	9.1	9.0	9.8	13.9
MgO	5.6	9.0	13.3	7.8	5.2	9.2	5.8	7.0
CaO	7.6	9.2	8.1	8.5	11.1	11.0	8.4	9.8
Na_2O	3.7	2.8	2.5	3.6	5.9	4.4	3.4	2.7
K_2O	1.6	2.0	3.3	4.0	4.8	5.2	1.6	0.7
				Norm				
Or	9	11	19	23	...	...	9	4
Lc	...	...	...	...	22	23	...	...
Ab	33	25	10	17	...	...	30	24
Ne	...	...	7	9	31	23	...	...
An	27	24	14	11	...	...	26	26
Di	8	17	20	24	42	12	12	18
Hy	21	4	...	...	...	...	11	13
Ol	2	19	30	16	5	42	11	15

Rocky Mountains, from the Yellowstone National Park on the south to the Canadian boundary on the north, scattered over an area about 600 km long and 200 km wide. It is made up of a group of subprovinces. From south to north they are the Yellowstone National Park area, Absaroka Range, Crazy Mountains, Castle Mountains, Little Belt Mountains, Judith Mountains, Moccasin Mountains, Highwood Mountains, Little Rocky Mountains, Bearpaw Mountains, Big Belt Mountains, and the Sweetgrass Hills.

Some subprovinces represent the Tertiary igneous rocks of whole mountain ranges; others include only the rocks erupted during a limited time interval in a mountain range. In the Highwood Mountains four subprovinces have been recognized. The various rocks of each subprovince are closely related but are less closely related to the rocks of the other subprovinces. Chemically, the rocks of each subprovince are near a regular variation diagram (the Little Belt diagram of Fig. III-65 is one example), and the various rocks must have been derived from a common parent magma by a rather simple process. The range is from calcic rocks in the Yellowstone area to extremely potash-rich leucite-bearing rocks of the Highwood Mountains.

In Table III-26 are listed the estimates of Larsen that indicate that some of the parent magmas were derived by crystal fractionation from a common magma of olivine-basaltic composition. More accurate calculations confirm this assumption. By removing, for example, calcic plagioclase, pyroxene, and olivine from an olivine-basaltic magma we may obtain the magma of the San Juan lavas, or of the Crandall subprovince. But the laws of crystallization of silicate melts, as outlined in Section B, make it very improbable that strongly alkalic lavas (of the Highwood Mountains and the Crazy Mountains) were derived in this way, although Larsen does consider the possibility. As discussed on page 199, alkalic part-magmas may develop in consequence of thermodiffusion. Another way is assimilation of potash salt deposits. It can be shown, however, that the assimilation of, or selective reaction on, any ordinary rock is unable to produce magmas of these compositions.

The Scottish Carboniferous-Permian alkalic lavas, which are particularly abundant in the Midland Valley of Scotland, have been studied since Hutton's time by a number of geologists.

A special study of magmatic differentiation was undertaken by Tomkeieff (1937). He assumed a primary reservoir of molten basaltic magma, which by fractional crystallization produced the lava series: olivine-basalt, mugearite, trachyte, rhyolite. The relative amounts of the different lava types, plotted against silica, give a characteristic frequency distribution curve (see Fig. III-79). The continuous variation shown by all the members and the nature of the frequency distribution curve suggest that all the types originated from the parent olivine-basalt magma. The asymmetry of the curve suggests also the presence of an unaccounted for amount of basic or ultrabasic material —perhaps a product of gravitational differentiation—which was never erupted and which solidified under plutonic conditions. It represents the *accumulative phase* of the magma, whereas the portion of the curve situated to the right of the modal peak represents the *liquid fusive phase.*

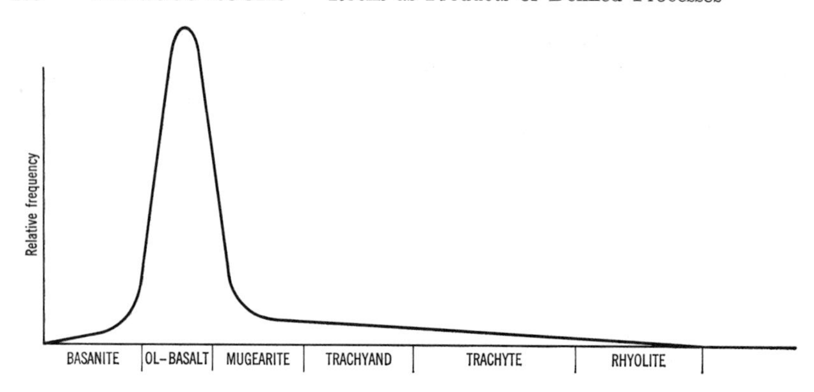

FIG. III-79. Frequency distribution curve of the Scottish Carboniferous-Permian lava types. (After Tomkeieff, 1937.)

The impression is gained that, on the whole, the series is determined by progressive fractional crystallization of feldspar and ferromagnesian minerals, the progressively diminishing amounts of the liquid fraction being analogous to the frequency distribution of the actual rock types corresponding to these liquids.

Although the main lava series is satisfactorily accounted for in this way, further difficulties arise from the fact that additional rock series (teschenite and quartz-dolerite series), which in time and space are

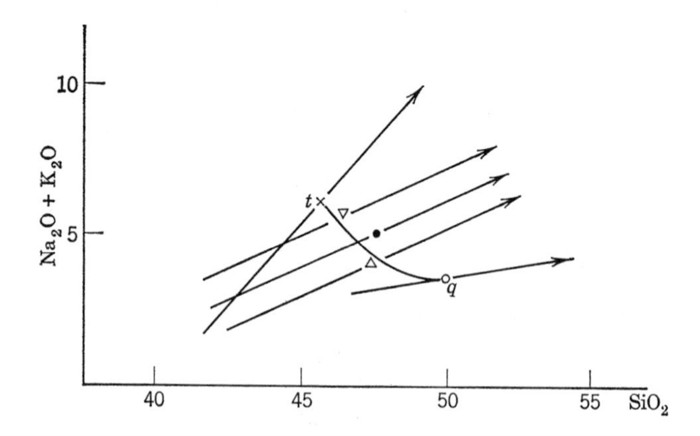

FIG. III-80. Average projection points on the alkali-silica diagrams. ○ = quartz dolerite (q), △ = olivine dolerite, ● = olivine basalt, ▽ = essexite and basanite, × = teschenite (t). The arrows indicate the trends of differentiation induced by normal crystal fractionation.

closely associated with the lava series, abound in the Midland Valley and elsewhere in central and southern Scotland. MacGregor (1948) has accounted for these facts by postulating crustal layering of one of the types suggested by Kennedy and Anderson.

Tomkeieff, on the other hand, suggested an interesting mechanism for the derivation of an alkalic (teschenitic) and a calc alkalic (quartz-doleritic) part-magma from the primary olivine-basalt magma.

A chemical classification of these magmas is shown on the alkali-silica diagram (Fig. III-80). Quartz-dolerite and teschenite are seen to be "conjugate" magmas, with olivine basalt as a center. Neither of them can be derived from olivine basalt by crystal fractionation.

The only alternative is a "transverse" differentiation, from q to t in the diagram, accomplished by a differential movement of alkalies in the magma. Volatile ions of the primary olivine-basalt magma, together with K^{+1} and Na^{+1}, may gradually diffuse through the body of the liquid magma and concentrate in the upper parts of the magma reservoir. Given enough time, the primary magma would thus split into two conjugate magma portions, the upper teschenitic, and the lower quartz-doleritic. The composition of these conjugate magmas can therefore be expressed as follows:

Teschenite magma = primary magma + alkalies and volatiles

Quartz-dolerite magma = primary magma − alkalies and volatiles

Obviously this effect is augmented in magma reservoirs of large vertical extension (for example, the Mid-Atlantic Ridge, page 178).

The Oslo Region is situated on the northern extremity of a great Permian fault zone. It represents a Permian rift valley whose rocks, dimensions, and structural pattern exhibit remarkable similarities to the African rifts. In and around the Oslo fiord large eruptions have occurred: first simatic effusions of large basalt floods, followed by sialic eruptions of rhomb porphyries (trachyandesites), associated with great subvolcanic rock bodies of alkalic affinities. The petrographical types of these rocks are shown in Table III-27.

In Fig. III-81 the frequency distribution is shown graphically. It should be compared with the distribution curve for the lava series of Midland Valley, Scotland (Fig. III-79). The dominance of olivine basalt in the Scottish rocks and the shape of the curve suggest a normal comagmatic series, with olivine basalt as the parental magma. The curve for the Oslo subvolcanic rocks is very different: (1) The gabbroic rocks are present in very small amounts and cannot possibly bear any parental relation to the syenites, which occupy a bulk of

TABLE III-27

THE OSLO ROCKS

Oslo Name	Type Name	Area, km^2
Oldest Series		
	Olivine gabbro	
Oslo-essexite	Hyperite	15
	Alkalic diorite	
Basic lavas	Alkalic basalts	500
Rhomb porphyries	Trachytes	1400
Main Series		
Kjelsåsite	Syenodiorite, augite monzonite	253
Larvikite	Augite monzonite	1670
Lardalite	Nepheline monzo-syenite	65
Nordmarkite (Granite)	Alkali syenite	1425
Ekerite	Aegirite alkali granite	821
Younger Intrusion		
Granitite	Biotite granite	840
	Total	7000

more than a hundred times that of the Oslo-essexites. If it is still maintained that the large syenitic bodies were derived from gabbroic mother magma, then it becomes necessary to postulate colossal amounts of hidden gabbro under the present surface. (2) The great preponderance of syenitic rocks distributed over a wide silica range is indicative of a comagmatic series. The origin of the chief members, larvikite-nordmarkite-ekerite, may be attributed to crystallization-differentiation of a syenite magma. The portion of the curve

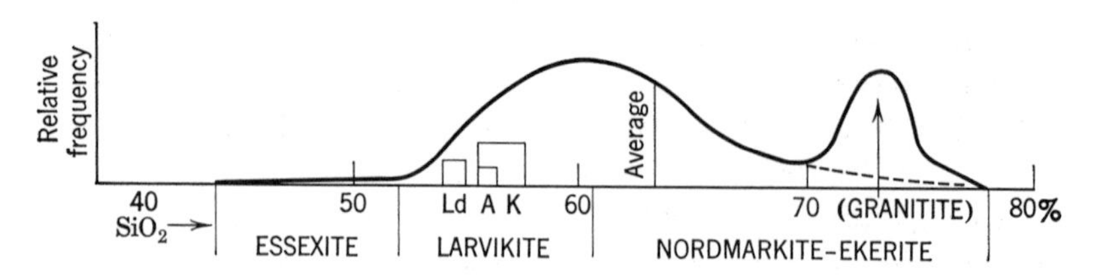

FIG. III-81. Frequency distribution of the Oslo igneous rocks. Ld = lardalite, A = akerite, K = kjelsasite.

situated to the left of the modal peak (ca. 60 per cent silica) corresponds to the accumulative phase, that to the right with the liquid-fusive phase. (3) The extra hump on the frequency distribution curve corresponding to the occurrence of granitic rocks indicates an independent position of the granite.

Whence, then, came the syenite magma? To this question there is no easy answer. Descent from a basaltic "mother magma" by fractional crystallization seems hardly possible. Naturally the idea of generation of magma by refusion suggests itself. According to Barth (1954), there are mainly two processes responsible for the production of the Oslo magma: (1) refusion of the pre-existing crust with a small introduction of "emanations" rich in alkalies, and (2) refusion and a diffusional relocation of the chemical species according to the requirement of the gravity field (see page 162 and equation 25 of the Appendix).

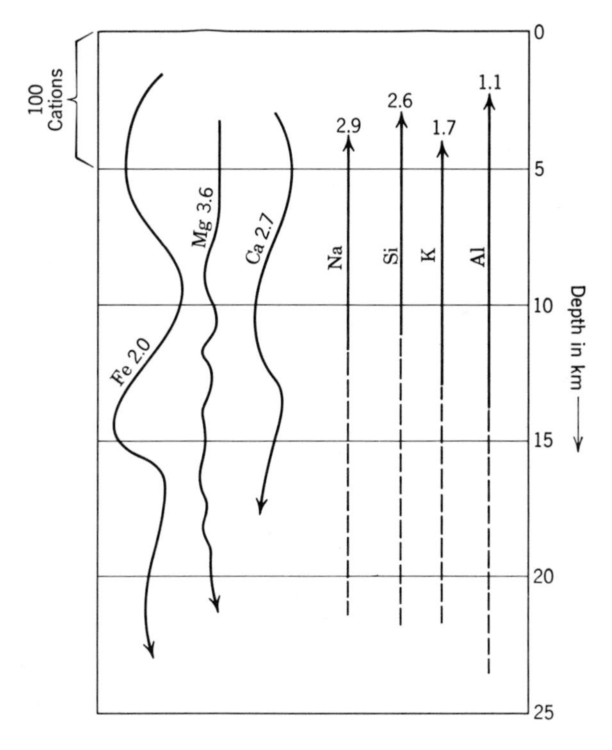

Fig. III-82. Schematic cross section through the Oslo rock prism illustrating the relocation of the cations in the pre-existing rock prism, in order to give a top layer of the composition of the average Oslo rock. (Barth, 1954.)

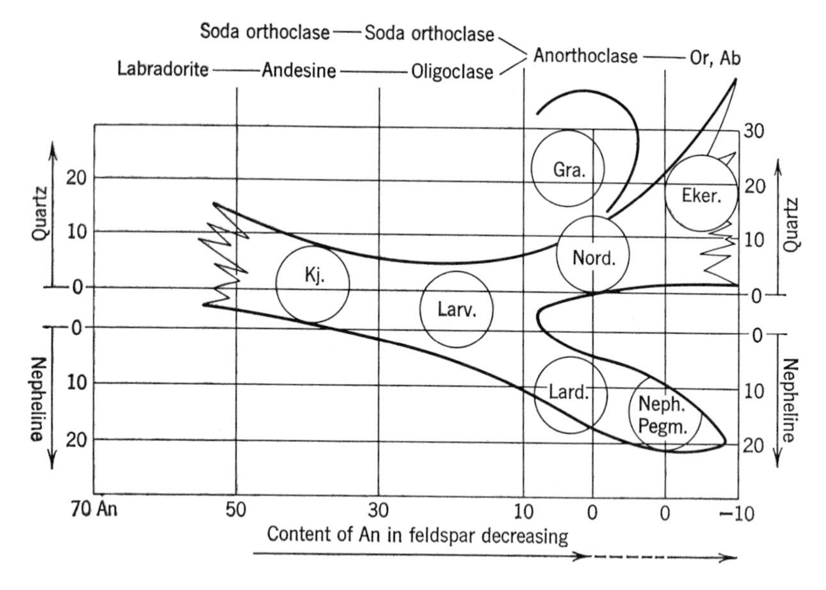

FIG. III-83. Family tree, showing the systematic position and the mutual relationship of the principal Oslo rock types. Variations in the quartz contents are read along the vertical axis. Variations in the anorthite contents are read along the horizontal axis. At the top of the diagram quartz-rich rocks are situated; downwards the quartz content decreases, and eventually becomes zero. At this stage we have, so to speak, a neutral zone in which the quartz content stays at zero while certain reactions in the colored minerals take place (for instance, augite → olivine, etc.). When the neutral zone is passed, and we still proceed farther downward, we may say that the quartz content becomes negative, which is indicated by the appearance of nepheline. From then on, the amount of nepheline will increase with decreasing silica. Let us also consider the variations in the horizontal direction. From left to right, the anorthite content in the plagioclase decreases from 70 An to zero; then we have the neutral zone where the anorthite content stays at zero. But to the right of this zone the anorthite content becomes negative. The reality of this negative content is the following: The decrease in the anorthite content is induced in the rock by an increase in the alkalies relative to alumina; when $Na_2O + K_2O = Al_2O_3$, the alkalies consume all alumina and leave nothing for the formation of anorthite. At this point the anorthite content of the feldspar thus becomes zero. If now the alkalies increase still more, we may say that the anorthite content in the feldspar becomes negative. This is indicated by the appearance of aegeirite, for the excess of alkalies now combines with ferric iron in the molecule $NaFeSi_2O_6$. From then on, the amount of aegirite will increase with increasing negativity of the anorthite content.

Figure III-59 illustrates the distribution of H_2O in different levels of a magma chamber, 30 km deep, the top of which is situated 4 km below the earth's surface, and covered by roofs impermeable to the vapor.

In the same manner the other constituents will redistribute themselves. In order for the top fifth of the column of the fused pre-existing crust to become of the composition of the average Oslo rock, a surprisingly small amount of diffusion need take place, see Fig. III-82.

By crystallization the syenite magma thus produced gave birth to various mineral reaction series. The principal rock series in its relation to the attendant mineral reaction series is as shown here:

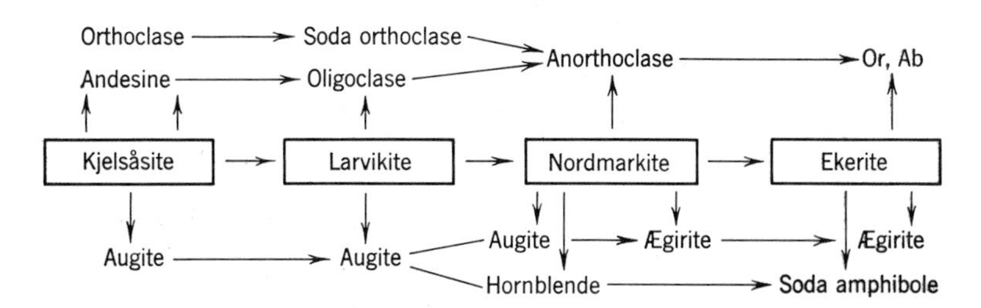

In Fig. III-83 the evolution of the Oslo magma is shown graphically. It goes from left to right. The upper branch with the ekeritic end products is quantitatively more important than the lower branch, which, except for nepheline pegmatites of agpaitic relations, stops at lardalite.

The figure shows that the branch transition nordmarkite → granite is simply effected by an increase in the quartz contents. But the analogous principal transition nordmarkite → ekerite is effected both by the feldspar reactions and by an increase of quartz.

In Fig. III-83 the systematic position and the mutual relations of all the principal rock types are displayed. Thus this diagram gives an indication of the course of differentiation followed by the Oslo magma.

Calcite-Bearing Rocks and Carbonatites

Assimilation of calcite has been much debated in the petrographic literature. By some petrographers it has been maintained that this process was necessary for the formation of alkalic rocks. The problem is related to the problem of a magmatic formation of calcite. Magmas are molten rock masses that are often connected with a vapor phase rich in water and carbon dioxide. All lava samples that have been investigated have shown carbon oxides among the volatile constituents; the presence of carbon dioxide in granitic magmas is evident from the numerous inclusions of this gas in granitic quartzes.

The magmatic carbon dioxide is, of course, in equilibrium with its products of dissociation:

$$2CO_2 \rightleftharpoons 2CO + O_2$$

The number of molecules is highest on the right side of the equation, and increasing pressure will therefore shift the equilibrium to the left. If hydrogen is introduced, we have the following reaction:

$$CO_2 + H_2 \rightleftharpoons H_2O + CO$$

which is important since there is no change in the number of molecules. At 840°C an equimolecular amount of CO_2 and H_2 shows a dissociation of about 50 per cent into CO and water vapor. At higher temperatures the amount of water vapor increases.

The solid phases that usually crystallize from such magma do not contain carbon dioxide. Not until the concentration of carbon dioxide becomes unusually high will carbonate-bearing minerals separate (calcite, scapolite, and others).

Many different types of calcite-bearing rocks have been described from various parts of the world. Also certain rocks composed exclusively of calcite, that is, marbles, have been regarded as magmatic by some authors; for example, the so-called sövite (first described by Brögger, 1914) from the Fen area, Norway. To such rocks the term carbonatite is restricted.

Carbonatites from many places in the world are now known. They occur almost exclusively as discrete masses in stocks or plugs of alkalic rocks, and often in connection with effusive rocks and rift valleys, for example, in southeast Africa.

More than fifty minerals have been reported from the carbonatites. Among the simple carbonates calcite is dominant, dolomite is next, and ferruginous and manganiferous carbonates are least in abundance. The complex carbonates are often rare-earth bearing: bastnaesite, synchisite, etc. Apatite is often abundant, monazite also is found. Barite is the principal sulfate mineral. Rather characteristic are pyrochlore, perovskite, zircon and fluorite. (Pecora, 1956.)

There are in many carbonatites exceptional concentrations of barium, strontium, rare earths, niobium, tantalum. Thus carbonatites and associated alkalic rocks are proving to be important future sources of rare commodities.

The Alnö complex, Sweden, constitutes the best studied example (von Eckermann, 1948, 1960). It occurs in pre-Cambrian gneissose granite as a stock of concentric structure with calcite carbonatite (sövite) in the core, surrounded by nepheline-bearing alkalic rocks ranging from hololeucocratic to melanocratic rich in pyroxene and melanite. In all carbonatite complexes (in the proper situation) the surrounding gneiss-granite is *fenitized*, that is, within a zone of several hundred meters it is transformed to syenite or nepheline-bearing rocks. In Alnö the carbonatite magma deprived the acid country rocks of their silicium and sodium, in return it enriched them in calcium, carbonic acid, and potassium. In consequence the gneiss-granite now grades into quartz syenite → syenite → nepheline syenite; the fenite is thus formed *in situ* and retains many of the original structural features of the gneiss-granite.

The original intrusion of Alnö consisted of a kimberlitic magma (carbonated alkali peridotite), rich in fluorine but comparatively poor

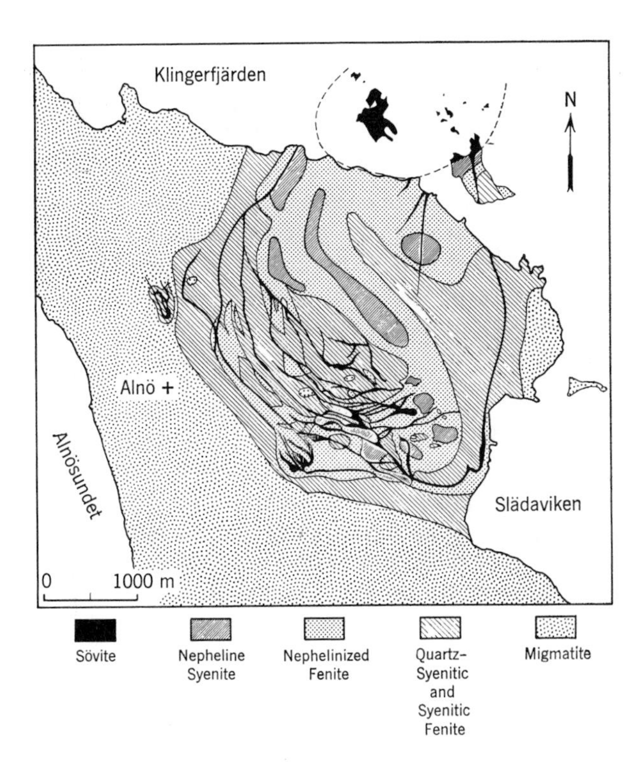

Fig. III-84. Map of the alkaline part of the Alnö Island. (After H. v. Eckermann, 1948, 1960.)

in water, which through differentiation, reaction with the surrounding rocks, and autometamorphic alterations produced an astounding variety of rock types. When the intrusion pushed upward it was headed by an almost pure dolomite-ankerite differentiate of very high internal CO_2-pressure. At a level where this pressure exceeded the external pressure of the overlying rock-roof (10 km depth) an explosion ensued, conical and radiating fractures being formed. These were immediately filled with quickly congealing carbonatite magma. The fractures thus having been sealed, the CO_2-pressure again increased, and the intrusion proceeded toward higher levels. At a depth of 5 km a new explosion followed, the events were repeated; a third explosion again happened at a depth of 3 km. At higher levels the carbonatites became increasingly calcic, the dikes issuing from the 3 km focus being sövitic in composition.

Similar intrusion patterns have been described from other carbonatite complexes. For several east African localities "ring and core," "collar structure," and "cone sheet" describe the form of the bodies.

Wyllie and Tuttle (1960) investigated experimentally the system $CaO—CO_2—H_2O$. Liquids in this system may be regarded as simplified carbonatite magmas in which CaO represents the basic oxides, and CO_2 and H_2O the volatile constituents. At 1000 bars pressure, calcite melts incongruently at 1310°C, portlandite [$Ca(OH)_2$] melts congruently at 835°C, a binary eutectic exists between calcite and portlandite at 685°C. Calcite begins to melt at 740°C on the join calcite-water vapor, and the lowest melting composition with coexisting calcite-portlandite-liquid-vapor is at 675°C. The ternary liquid exists in the pressure range 27 to 4000 bars with minimum temperatures from 685°C to 640°C. These are values easily obtainable in the crust of the earth and leave little reason to doubt a magmatic origin for those carbonatites which appear to be intrusive. They also suggest that partial melting of limestone is likely at igneous contacts, and that impure limestones may be partially melted during high-grade metamorphism.

Potassic Rocks

A quantitative definition of a potassic rock is that it contains normative leucite (in which case it is also alkalic), or that the composition of its normative feldspar plots in the field of primary potassic alkali feldspar of Fig. III-85 (v.i.).

The origin of the potassic rocks, whether deep-seated or extrusive, is still an unsolved problem. They have been explained variously as products of magmatic crystallization, of assimilation, or of metasomatism.

By direct observation Brouwer (1946) could demonstrate, on a small scale, the formation of leucite rocks around limestone inclusion in an ordinary andesitic lava from Java. In an effort to assimilate limestone the lava precipitated wollastonite, augite, plagioclase, and other lime silicates, thereby obviously becoming desilicated and alkalic (miaskitic) as explained on page 200.

But why should potassium be concentrated? The literature contains quite a number of descriptions of potash-rich leucite-bearing rocks associated with limestone. Since limestone cannot furnish potash to a magma, it seems to follow that the gases rich in CO_2 were able to redistribute the original potash content of the magma. The general question why potash should be enriched so greatly in some leucite rocks remains obscure.

It has been demonstrated that a number of devitrified lavas of fresh and unaltered aspect have suffered great changes in their chemical composition through leaching of soda and lime and a relative enrichment of potash. In the Esterel region of France devitrified obsidians and rhyolites consistently show a relative enrichment in potash as compared with the glassy lavas. Many porphyries (for example, the diabase porphyry of Mount Devon with feldspar insets containing 4.5 per cent K_2O) have been similarly altered. This is in harmony

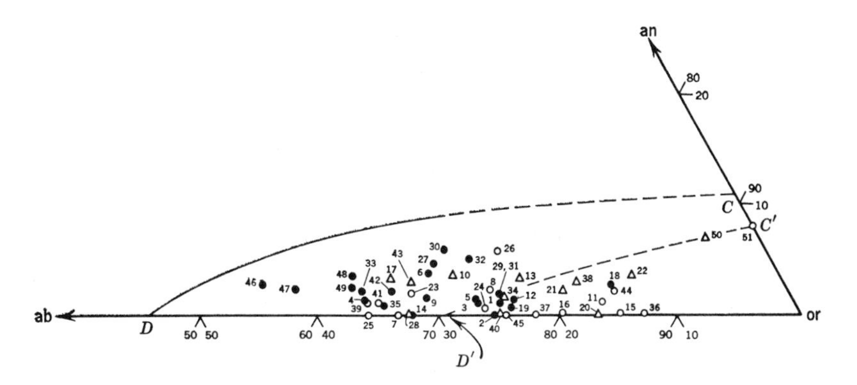

FIG. III-85. The Or:Ab:An ratio of some potassic rocks. Solid circles = granites and pegmatites; open circles = rhyolite, trachyte, and other volcanics; triangles = porphyries. (After R. A. Higazy, 1949.)

with the conclusions reached in the studies of the thermal waters and the altered rhyolites of the Yellowstone Park region. (See page 282.) Fenner (1936) thinks that there are many occurrences of similarly altered rocks in volcanic regions, which, in the absence of complete data, have been regarded as normal lavas. Analyses of such rocks have been accepted as representing original compositions.

It is of importance to define the meaning of *potassic rocks*. In order to find a useful definition of the term, it should be separated from the definition of alkalic rocks, that is, many potassic rocks are not alkalic, and vice versa.

The usual path of evolution of igneous rock series is from a basaltic, rather potash-poor magma into the "residua" systems of petrogeny, that is, systems containing alkali-alumina silicates as their chief components.

The mode of crystallization of the feldspars from a magma is shown in Fig. III-13. In the normal course of crystallization, the magmatic liquid traverses the plagioclase field, encounters the boundary curve, and, thenceforth, follows it so that in the last stages of crystallization the eutectic ratio of about 40Or:60Ab is attained. Such normal paths of crystallization are regularly observed. The important point is that in the normal course of crystallization the liquid cannot cross the boundary curve. For only if orthoclase had shown a reaction relation to calcic plagioclase could the curve be crossed.

Consequently, rocks placed on the other side of the curve, that is, in the orthoclase field, did not reach this position by the normal process of crystallization-differentiation. They are abnormal rocks, and the term potassic rocks should be restricted to them.

In order to find the position of a rock on the diagram of Fig. III-13 it is not necessary to make a plot. It can be found arithmetically in the following way. The normative feldspar of a rock must be recalculated to a sum of 100, that is, $Or' + Ab' + An' = 100$. The equation of the boundary curve in Fig. III-13 in terms of these coordinates is $(An')^2 + 2Ab' = 120$. The quantitative definition of a potassic rock is, therefore, that the composition of its normative feldspar satisfies the inequality

$$(An')^2 + 2Ab' < 120,$$

for only then will the rock plot in the orthoclase field.

Some fifty rocks from Washington's tables (1917) (of subrang dopotassic omeose) have been plotted on this diagram by Higazy (see

Fig. III-85). These are deep-seated (graphic granites, granites, pegmatites, and one monzonite), hypabyssal (porphyries), and extrusive (rhyolites, liparites, pitch, stones, comendites, and trachytes). All are potassic, and all lie in the orthoclase field.

Why, then, should the potassic rocks, if of magmatic origin, not follow the regular course of crystallization? The immediate answer is that either some less well-known magmatic processes (for example, fractional distillation or thermodiffusion) interfere with the crystallization-fractionation, or such rocks cannot result from strictly magmatic processes, that is, metasomatic alterations must have played a significant part in their formation.

6 · OROGENIC ROCK SERIES

Neither volcanism nor plutonism can be understood until we understand the formation of mountain chains.

R. A. Daly, *Proc. Am. Phil. Soc.*, 1925.

Within one orogeny, volcanism may be divided into four phases corresponding to four stages of the geotectonic development of the geosyncline.

The initial magmatism is characterized by rocks of gabbroic or ultrabasic relations. They occur as more or less steeply inclined, conformable sheets or lenticular bodies alined along the strike of the encasing rocks exhibiting the same metamorphic facies as the surroundings. The primary minerals are therefore often transformed into secondary minerals like serpentine, chlorite, and epidote, imparting a green color to the rock. Continental geologists have called them

TABLE III-28

Relation between Magma and Tectonic Stages

Magmatologic Sequence	Geotectonic Sequence	Magmatic Type
1. Initial magmatism	Geosynclinal conditions	Basaltic, ultrabasic
2. Synorogenic plutonism	Orogenesis	Granodioritic
Anatexis and Granitization		
3. Subsequent volcanism	Semi-cratonic conditions	Granitic
4. Final volcanism	Cratonic conditions	Basaltic

roches vertes, pietri verdi, or *Grüngesteine.* Rather than calling them "green rocks," which may be confused with "green schists," we shall follow an old Italian custom and call them *ophiolites.*

Ophiolites exhibit great individual variations. The great majority are basalts of subalcalic affinities and effusive of origin, with a large quantity of peridotitic rocks (usually serpentinized) occurring as a constant accompaniment. The shapes of the intrusions and certain contact relations have been interpreted to mean that the peridotites were intruded as magmas of low viscosity and moderate temperature (Bowen, 1928, Hess, 1938).

The ophiolitic lavas are, generally, of submarine eruption, showing "pillow" and allied structures. They are constantly associated with radiolarian chert and serpentinites, forming the so-called *Steinmann trinity* (see Bailey and McCallien, 1960). There is a tendency for them to become enriched in soda, whereby they change into the so-called *spilites* (albite basalt). The spilitization is a metasomatic process which in the laboratory takes place at about 300°C:

$$Na_2CO_3 + CaAl_2Si_2O_8 + 4SiO_2 \rightarrow CaCO_3 + 2NaAlSi_3O_8$$

In spite of the great amount of time given to the study of ophiolites from the earliest days of geology, it is not yet possible to present the complete story of their origin and differentiation. It is not always easy to distinguish ophiolites from the post-orogenic basic plateau basalts, with which they are to be sharply contrasted, however. For the genetic relation of ophiolites with the eugeosynclinal stage is of great importance; ophiolitic magmas characterize a geosyncline better than do the sediments. To be sure, ophiolites represent the initial magmatism of all orogenic zones: There are ophiolites following the Mediterranean orogeny from Spain on the west, through the Pyrenees, Alps, Balkans, over Iran, Himalaya, and Burma into the East-Indian archipelago; the family of the Green rocks of the Caledonian chain, the complexes of the mafic and ultramafic intrusions extending parallel to the west coast of North America from Nevada through Alaska, the 500 km belt of ultramafics and gabbros of the Ural mountains, the "Great Serpentine belt" of New South Wales, etc.

These rocks always follow tectonic zones of major movements that are also zones of negative gravity anomalies. Writes Hess (1938): "Location of a peridotite belt and dating its intrusion, locate the old tectonic axes and date the initiation of the deformation of that zone." This cannot be fortuitous, and must be explained further. A reasonable conclusion is that ophiolites regularly develop in consequence of special geologic conditions always present at the beginning of an orogeny. Generally speaking, it is possible that the primary olivine-basalt magma of the depths of the geosynclines was activated by the initial orogenic movements, squeezed and intruded into the sedimentary strata, filtered and strained, with consequent separation of olivine and other early precipitates, which subsequently were made over into peridotites and serpentines. The strained-off magma in the usual way, through fractional crystallization under assimilatory control, gave birth to one or the other of the various rock series belonging to the heterogeneous ophiolite family.

The particular problem of the genesis of the ultramafic types merits special mention.

Peridotite is a general name for olivine-rich rocks; the most extreme among them are *dunites* (or olivinites), containing chiefly magnesian olivine to the near exclusion of all other minerals. They occur usually in even-grained, coarse, massive rock bodies of not very large dimensions (up to about 5 km across). Also sills, sheets, dikes, and necks are known, but no effusives. A remarkable porphyritic dunite, with large crystals of olivine set in a groundmass of smaller olivine grains and small octahedra of picotite (spinel), is known in the Island of Mull. A dunite from Seiland in northern Norway contains eighty-eight parts of olivine and eleven parts of spinel.

The home of the dunites is, generally speaking, the folded mountain chains. Bodies of serpentine occur in the same geological environment, and have been variously explained as altered olivinite bodies or as crystallized directly out of a serpentine magma. Both dunites and serpentines have been regarded as products of the orogenic magmatism, and have received special mention under the discussion on ophiolites. Their mode of formation is doubtful; some of them may not be of magmatic origin.

Many peridotites are local, layered differentiates within gabbroic and noritic bodies. Such occurrences, often adjoined by pyroxenites as well as schlieren and dikes associated with larger gabbro intrusions, are explained as early accumulative crystal differentiates. But stocks, dikes, and sills are harder to explain. Wyllie (1960) discussed the origin of peridotites in terms of the system $CaO—MgO—FeO—SiO_2$; and also (1961) determined the temperature of the intrusion of a picrite sill into Torridonian sandstone to 1175°C.

There is a considerable body of geological evidence that, in submarine environment, extrusions of lavas with a composition similar to peridotite have taken place. But laboratory investigation has cast much doubt on the existence of such rocks. Known surface lava temperatures do not exceed 1200°C, while the melting range of peridotite is around 1700°C. But peridotites are often associated with submarine volcanic activity, and under moderate pressure of water the melting points of rocks and minerals are normally depressed several hundred degrees. When a lava is extruded at the surface, it cannot contain volatiles at pressures much in excess of atmospheric pressure. When the same material is extruded in a submarine environment volatiles can be retained at pressures corresponding to the water pressure at the given depth. It follows that submarine lavas will occur at much lower temperatures than do their surface equivalents.

Laboratory experiments conducted under water pressures corresponding to the deeper parts of the oceans indicate that rocks of serpentine composition melt almost completely at 1300–1400°C.

Pure magnesian serpentine has a maximum temperature of existence at approximately 500°C; magnesian olivine is stable in contact with water vapor down to about 430°C, and iron olivines are stable to still lower temperatures.

It thus appears that peridotite lavas could flow on the deep ocean floors. At greater depths intrusion of such materials should be possible at even lower temperatures.

But there is also another way in which peridotite dikes or sills may form: In the experiments of Bowen and Tuttle it was found that silica was abstracted from the charges by water vapor, the loss occurring even when no free SiO_2 was present. Thus enstatite was transferred to forsterite. It was also shown that SiO_2 and other substances (for example, FeO) can be transferred to the charge by water vapor. (See page 250.) In the light of these observations Barth (1950) suggested that, if a crack formed in a mass of pyroxenite, and if water vapor unsaturated with SiO_2 streamed through the crack, the rock adjacent to the crack could be converted to a type consisting mainly of olivine, which might appear to be a dike in the pyroxenite. And it seems also possible that water vapor charged with SiO_2 and other substances and streaming through a crack in peridotite could convert the adjacent rock to pyroxenite. This mode of formation is not logically limited to bodies having a dike-like relation; any other form is conceivable.

Thus bodies of peridotite may develop simply by pneumatolytic (hydrothermal, metasomatic) transport of material. Large-scale material transport by metasomatism is well known in geology. Indeed, ultrabasic rocks of orogenic zones (New Caledonia, Indonesia, and the Alps) are now explained as products of metamorphic processes

(Avias, 1949; van Bemmelen, 1950; Perrin and Roubault, 1951). Again, olivinites, peridotites, and pyribolites (rocks composed of pyroxene and amphibole) of well-exposed pre-Cambrian areas of Greenland are regarded as siliceous dolomites and other supracrustal rocks having recrystallized during high-grade regional metamorphism (Sörensen, 1953, Berthelsen, 1960).

The Synorogenic Plutonism. It seems reasonable to assume that orogenic magmas are produced by remelting of the deeper parts of a geosyncline. Experiments by Wyllie and Tuttle (1960) have shown that most sediments in geosynclines (granitic rocks, arkoses, shales, and limestones) will begin to melt in the depth range 20–25 km. The first magma to form is granitic in composition (page 111), as the temperature rises the composition will become more basic, but by folding and downwarping additional sediments may be assimilated thus consuming the heat introduced and preventing further basification. Unusual heat conditions or much deeper burial would be required, therefore, for the formation of basaltic liquids.

The biotite diorite series of rock is typical of all large orogenic zones. For instance, the so-called opdalite-trondhjemite series of the Caledonian intrusives of southern Norway are for all practical purposes identical to the tonalite series of the Alps, or the diorites of the South American Andes. They all belong to the mica diorite type of rock series. Also the geological mode of occurrence is identical: in time and space they are connected with large folded mountain chains and intrusive into the folded geosynclinal sediments.

The normal case of crystal differentiation of a gabbroic primary magma was mentioned briefly on page 111. The differentiation diagram is repeated below (Diagram III).

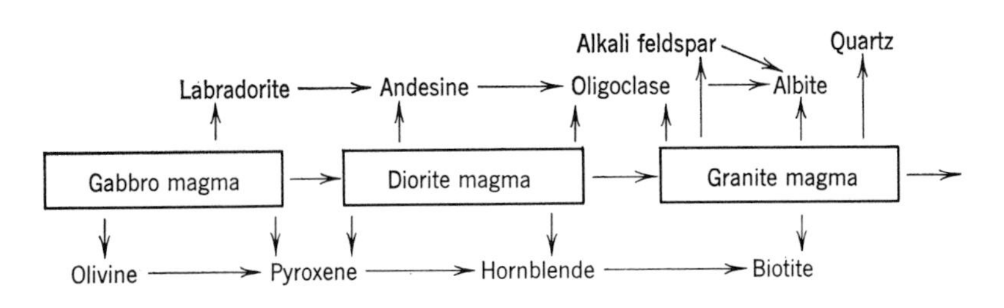

DIAGRAM III. General case.

With this "normal" diagram we shall compare an analogous diagram of the biotite diorite series of rock (V. M. Goldschmidt, 1922). (Diagram IV.)

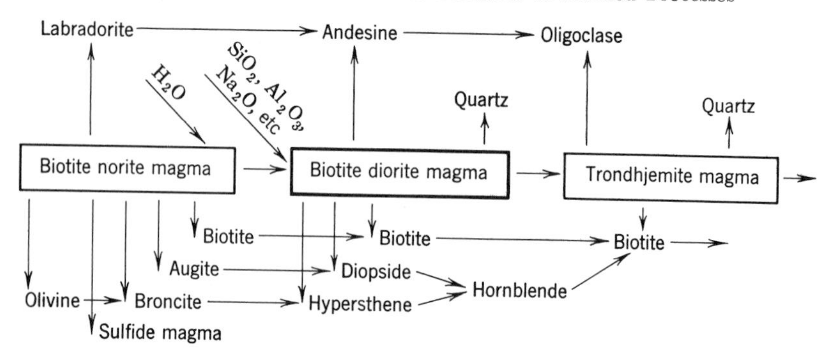

DIAGRAM IV. The differentiation of the biotite diorite series of rock. The development is actually from right to left. The arrows should, therefore, be turned around. See text.

The most distinctive difference in this kind of series as compared with the normal one is the small amount of gabbroic liquid. The mother magma is biotite dioritic in composition and yields early and copious precipitates of *biotite* with which the paucity or even complete lack of potassium feldspar is causally related. The early-formed biotites extracted so much potassium from the magma that nothing was left for the formation of potassium feldspar in the later stages. The preference of biotite over potassium feldspar is evidently conditioned by a relatively high content of water in the magma.

Whence, then, comes this water? These rock series developed in the folded mountain chains of the geosynclines, into which large quantities of water-rich sediments were put, the magmatic intrusions being contemporaneous with the folding. These rock series follow the orogenic belts over distances as long as a thousand kilometers without ever penetrating into the foreland of the mountains. It would be a rather bold hypothesis to maintain that long and narrow reservoirs containing magma rich in water happened to exist. It would seem much more likely that the water content of the magma of the folded mountains was derived from the water-rich clayey sediments of the geosynclines, that the sediments by local and regional metamorphism lost part of the water, and that just this water became dissolved in the silicate magma.

Other than water, the usual geosynclinal sediments are particularly rich in silica and alumina, which therefore will likewise become incorporated in the magma. Hence the preponderance of dioritic rocks which greatly outweigh rocks of gabbroic (basaltic) affinities.

Rock series of this kind are thus formed by differential anatexis. There is no primary magma; solid sediments represent the "primary" substance from which the rock series developed. Thus, by turning around the arrows in Diagram IV the best graphical picture of this process is obtained.

The charnockite-anorthosite series develops at great depths. It was regarded by V. M. Goldschmidt (1922) as truly magmatic. The distinctive feature of this series is that both biotite and hornblende are scarce or missing, the pyroxene minerals (diopside and hypersthene) extend into the late granitic fractions of the crystal differentiates, and the separation of potassium feldspar starts at an early stage.

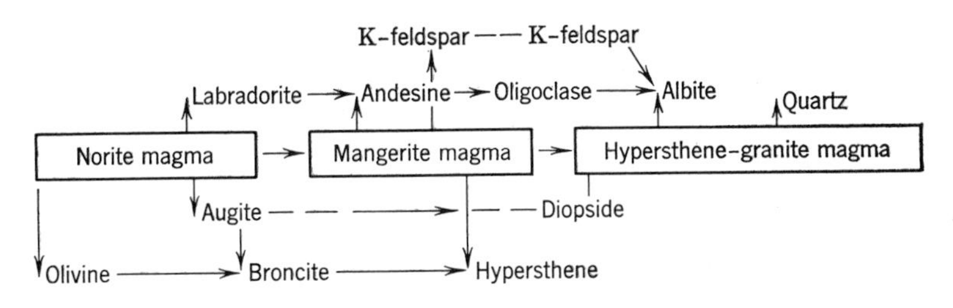

Diagram V. The development of the charnockitic series of rock.

Clearly these features may be explained by an unusually low content of water preventing the formation of biotite in the magma. The geological mode of occurrence is suggestive. This rock series is always intrusive into old granites or dry desiccated gneisses, the geological environment thus being poor in water. For these reasons Goldschmidt regarded them as products of particularly dry magmas.

Later investigations, however, indicate that the rocks have suffered complete metamorphism (and metasomatism). They may represent plutonic igneous rocks which have undergone slow recrystallization in the solid state under influence of high temperature and high, mainly uniform pressure at great depths in the earth's crust. They are typically associated with anorthosite, another representative of deep orogenesis and ultrametamorphism.

Anorthosite, like granite, frequently forms very large batholiths covering tens of thousands of square kilometers. Anorthosite is a *monomineralic* rock consisting almost exclusively of plagioclase (usually andesine or labradorite, but oligoclasites and bytownite rocks are not unknown).

An anorthositic magma could exist only above a rather high temperature at which plagioclase melts. But there is no field evidence indicating such high temperatures in or around anorthositic bodies.

It is probable that smaller bodies of anorthosite, occurring in strongly differentiated stratiform sheets with strongly contrasted rock facies in layers, represent accumulative differentiates of early plagioclase from a gabbroic magma. The plagioclases of accumulative anorthosite are commonly >60 per cent An. But the plagioclases of the very large bodies are always <65 per cent An.

Buddington (1939) has shown that, generally speaking, anorthosite and associated gabbroic and noritic anorthosite occur in one of these two contrasted sets of relationships reflecting two somewhat contrasted modes of origin.

An igneous origin has always been assumed. Most authors have favored a gabbroic parent magma from which enormous amounts of plagioclase crystals separated.

In the *Adirondack province* the parent magma was dioritic in composition (Balk, 1931). Large amounts of plagioclase and much smaller amounts of ferromagnesian minerals (especially pyroxene) separated and remained suspended in the mother liquor as it moved obliquely upward through the crust of the earth. During a long period of intrusion the magma gave birth to three rock types: gabbro, anorthosite, and syenite. Gabbro and anorthosite were regarded as crystal differentiates (accumulation of solid minerals), syenite as the mother liquor. Through movements in the crust the syenitic liquid was squeezed out (filter pressing), while the solid crystals remained and developed into large anorthositic rock masses, including small spherical bodies of gabbro. During the process of intrusion the mother liquor acted as a lubricant, but in places of strong squeezing, particularly along the edges of the anorthositic body, the solid crystals were crushed and broken. Splinters and fragments broken off from the plagioclase crystals were caught by the moving liquid, and are now seen at various places in the solid syenite which surrounds the anorthositic rock masses.

According to Balk's interpretation, the Adirondack anorthosites were not intruded as a magma but as a mush of solid plagioclase crystals lubricated by a syenitic mother liquor, the plagioclase crystals having been formed in conformity with Bowen's hypothesis by accumulation of plagioclase precipitated from an original dioritic magma. Buddington (1939) has emphasized that the Adirondack rocks were reworked by metamorphic agents.

In the *Egersund-Sogndal district* (southern Norway) no evidence of squeezing can be seen, but observations indicate rather that these rocks have gone through complicated processes of a metasomatic nature. The evidences are:

(1) Giant crystals of homogeneous andesine in places attain a length of 120 cm (4 feet). (2) Big, stumpy crystals of hypersthene, 20 cm long, form trails and trends in the anorthosite. (3) Homogeneous and undeformed crystals of "porphyric" andesine are in a foliated or gneissose groundmass. They are evidently porphyroblasts of non-magmatic origin. (4) Large folds and gneissic structures in some places are very distinct, but in other places are almost obliterated by recrystallization of coarse andesine porphyroblasts. (5) A relatively large number of dikes of anorthosite dissect the main anorthosite. The plagioclase in the dikes is more sodic than in the main body. The dikes might represent the "restmagma," but, if they do, the monzonitic and granite rocks around do not, and lose their genetic relation to the anorthosite.

Michot (1957) concluded that part of the Norwegian anorthosites is of magmatic origin and derived from basaltic magma by assimilation of pelitic sediments in huge quantities. This is inferred to take place in the deepest part of a geosyncline during an orogeny. Michot interprets "block structures" as a result of the development of a leuconoritic magma by anatexis. The purer anorthosite is inferred to be a post-anatexis residual material. Still another process resulted in the formation of beds of anorthosite in association with a norite-granite series of gneisses of variable composition, viz.: the arrival of emanations of K, Na, and Si transforming a norite-anorthosite series in conjunction with expulsion of Mg and Fe which went to form a basic front (anorthositization of norite).

Buddington (1957) thinks that the mineralogical and geological observations as well as the new experimental evidence are all consistent with high temperatures essential for the formation of anorthosites from a magma.

Conclusions

Tentative and uncertain as the theories may be, they do explain in a general way that, in addition to rock series keeping their distinctive character throughout, there are also those of mixed character that laterally and temporally change into other types. It is easy to see that at one and the same place very different rock series may develop at different times, depending on the variation of the geological environment with time. It is also clear that in different places of the earth similar rock series develop as reflections of similar geological conditions.

E. SELECTED REFERENCES

PART III-A

At the end of the eighteenth century, the first tentatives of rock classification appeared in the works of K. Haidinger and A. G. Werner in Germany, followed by R. J. Haüy in France and J. Pinkerton in England. A classification based on chemical analyses was first used by P. L. Cordier in his "Mémoires sur les substances minérales" in *J. de physique,* Paris, 1815. Various systems of petrochemistry were subsequently developed: (F. Schröckenstein, *Silicatgesteine und Meteorite,* Prague, 1897. F. Loewinson-Lessing, *Études sur la composition chimique des roches éruptives,* Bruxelles, 1890; *The Problem of the Origin of Magmatic Rocks* (in Russian), Leningrad, 1934. A. Osann, *Versuch einer chemischen Klassifikation der Eruptivgesteine,* 1900; *Elemente der Gesteinslehre,* Stuttgart, 1923. F. Becke, "Darstellung von Gesteinsanalysen," *Tchermak Mitt.,* 1925. A. Michel-Lévy, "Classification des magmas des roches éruptives," *Bull. soc. géol. France,* 1897; *Bull. serv. carte géol. France,* 1902–1904. A. Lacroix, *Minéral. de Madagascar,* Paris, 1922; *Bull. serv. géol. Indochine,* Hanoi, 1933. B. Choubert, "Géochimie des magmas," *Soc. géol. France mém.* 54, 1947. C. Burri, *Petrochemische Berechnungsmethoden,* Basel, 1959. T. F. W. Barth, Oxygen in rocks: a basis for petrographic calculations, *J. Geol.* **56,** 1948.)

Additional textbooks and monographs are:

Balk, R., Structural behavior of igneous rocks, *Geol. Soc. Am. Mem.* **5,** 1937.

Bowen, N. L., *The Evolution of the Igneous Rocks,* Princeton, 1928.

Burri, C., *Petrochemische Berechnungsmethoden auf äquivalenter Grundlage,* Basel, 1959.

Cross, W., J. P. Iddings, L. V. Pirsson, and H. S. Washington, A quantitative chemico-mineralogical classification and nomenclature of igneous rocks, *J. Geol.* **10,** 1902.

Daly, R. A., *Igneous Rocks and the Depth of the Earth,* New York, 1933.

Johannsen, A., *Descriptive Petrography of the Igneous Rocks,* 4 vols., Chicago, 1931–1938.

Niggli, P., *Gesteins- und Mineralprovinzen,* Berlin, 1923.

Nockolds, S. R., Average chemical composition of some igneous rocks, *Bull. Geol. Am.* **65,** 1954.

Rittmann, A., *Vulkane und ihre Tätigkeit,* 2nd ed., Stuttgart, 1960.

Shand, S. J., *Eruptive Rocks,* 3d ed., London and New York, 1947.

Wahlstrom, E. E., *Introduction to Theoretical Igneous Petrology,* New York, 1950.

Washington, H. S., Chemical analyses of igneous rocks, *U.S. Geol. Survey, Profess. Paper* **99,** 1917.

(For further references see also page 2.)

PART III-B

Synthetic studies on silicate melts were begun in a systematic way by the Geophysical Laboratory of the Carnegie Institution of Washington, D. C. A great number of the known phase diagrams of silicates have been worked out by workers in many countries and by none more than the present and earlier staff members of said laboratory.

All the phase diagrams have been collected and published by F. P. Hall and H. Insley: "Phase Diagrams for Ceramists," issued as Part II of the November, 1947 issue of the *J. Am. Ceram. Soc.*

Another great work of compilation is: W. Eitel's *Physical Chemistry of the Silicates,* Chicago, 1952.

Most of the original papers are to be found in *Am. J. Sci., J. Geol.,* and in the Annual Reports of the Director of the Geophysical Laboratory. Worthy of special mention are:

Bowen, N. L., *The Evolution of the Igneous Rocks,* Princeton (Princeton University Press), 1928; New York (Dover), 1958.

Tuttle, O. F., and N. L. Bowen, Origin of granite in the light of experimental studies, *Geol. Soc. Am. Mem.* **74,** 1958.

PART III-C

Barth, T. F. W., Volcanic geology, hot springs, and geysers of Iceland, *Carnegie Inst. Wash. Publ.* **587,** 1950.

Cloos, H., Bau und Tätigkeit der Tuffschloten, *Geol. Rundschau,* 1941.

Drever, H. I., Immiscibility in picritic intrusion, west Greenland, *XXI Int'l. Geol. Congress,* Pt. 13, Copenhagen, 1960.

Einarsson, T., Eruption of Hekla, *Visindafélag Islandinga,* **202,** 1949.

Fenner, C. N., Immiscibility of igneous magmas, *Am. J. Sci.,* **246,** 1948.

Greig, J. W., Immiscibility in silicate melts, *Am. J. Sci.,* **13,** 1927.

Holgate, N., Roles of liquid immiscibility in igneous petrogenes, *J. Geol.* **62,** 1954.

Krauskopf, K. B., Heavy metal content of magmatic gases, *Econ. Geol.* **52,** 1957.

Kuhn, W., and S. Vielhauer, Propagation of earthquake waves, *Zeit. phys. Chemie* **202,** 1953.

Lacroix, A., *La Montagne Pelée et ses eruptions,* Paris, 1904.

Larsen, E. S., Batholith of S. California, *Geol. Am. Mem.* **29,** 1948.

Neumann, H., Hydrothermal differentiation, *Econ. Geol.,* **43,** 1948.

Perret, F. A., *The Eruption of Mt. Pelée (1935)*, Carnegie Institute of Washington Publication, No. 458.

Rittmann, A., *Vulkane und ihre Tätigkeit,* 2nd ed., Stuttgart, 1960.

Roedder, E., Low temperature liquid immiscibility in the system K_2O-FeO-Al_2O_3-SiO_2, *Am. Mineralogist* **36,** 1951.

Shimazu, Y., A thermodynamical aspect of volcanic gas, *J. Earth Sci. Nagoya Univ.*, 1960.

Steiner, A., Origin of ignimbrites; A new petrogenetic concept, *New Zealand Geol. Surv. Bull.* **68,** 1960.

Verhoogen, J., *Exploration du Parc National Albert,* Bruxelles, 1948.

PART III-D

Backlund, H. G., Der Magmaaufsteig in Faltengebirge, *Bull. comm. géol. Finlande* **115,** 1936.

Bailey, E. B., and others, Tertiary and post-Tertiary geology of Mull, Lock Aline and Oban. *Mem. Geol. Survey Scotland,* 1924.

Bailey, E. B., and W. J. McCallien, Some aspects of the Steinmann trinity, *Quart. J. Geol. Soc. London* **116,** 1960.

Balk, R., Structural geology of the Adirondack anorthosite, *Min. Petr. Mitteilungen* **41,** 1931.

Barth, T. F. W., The igneous rock complex of the Oslo Region, *Norske Videnskaps-Akad. Oslo* **9,** 1944; **4,** 1954.

Becke, F., Die Eruptivgebiete der böhmischen Mittelgebirge und der amerikanischen Andes. Atlantische und pazifische Sippe der Eruptivgesteine, *Tschermak's mineral. u. petrog. Mitt.,* XXII, 1903.

Bemmelen, R. W. van, The origin of igneous rocks in Indonesia, *Geol. en Mijnbouw,* 1950.

Benson, W. N., Basalts. *Am. J. Sci.* **239,** 1941; *Royal Soc. New Zealand Trans.,* **68,** 1937; **71,** 1941; **72,** 1942; **74,** 1944.

Buddington, A. F., Adirondack igneous rocks. *Geol. Soc. Am. Mem.* **7,** 1939.

———, The origin of anorthosite re-evaluated, *Rec. Geol. Survey India* **86,** 1957.

Dewey, H., and J. S. Fleet, Some British pillow-lavas and rocks associated with them, *Geol. Mag.* **8,** 1911.

Eckermann, H. von, The alkaline district of Alnö, *Sveriges Geol. Undersk. Ser. Ca No. 36,* 1948; *Bull. Geol. Inst. Uppsala* **40,** 1961.

Edwards, A. B., Differentiation of the dolerites of Tasmania, *J. Geol.* **50,** 1942.

Eskola, P., Origin of granitic magma, *Min. Petr. Mitteilungen* **42,** 1932.

Ganser, A., Ausseralpine Ophiolitprobleme, *Eclogae Geol. Helv.* **52,** 1959.

Goldschmidt, V. M., Stammestypen der Eruptivgesteine, *Norske Videnskaps-Akad. Oslo* **11,** 1922.

Harker, A., *The Natural History of Igneous Rocks: I, Their Geographical and Chronological Distribution,* Science Progress, 1896.

———, *The Natural History of Igneous Rocks,* London, 1909.

Hess, H. H., A primary peridotite magma, *Am. J. Sci.* **35,** 1938.

———, Stillwater igneous complex, *Geol. Soc. Am. Mem.* **80,** 1960.

Holmes, A., and Harwood, Petrology of the volcanic area of Bufurubira, *Geol. Surv. Uganda Mem.,* No. IV-Pt. II, 1937.

Kuno, H., High-alumina basalt, *J. Petr.* **1,** 1960.

Larsen, E. S., Some new variation diagrams, *J. Geol.* **43,** 1938.

———, The petrographic province of central Montana, *Bull. Geol. Soc. Am.* **51,** 1940.

Macdonald, G. A., Hawaiian petrographic province, *Bull. Geol. Soc. Am.* **60,** 1949.

———, Dissimilarity of continental and oceanic rock types, *J. Petr.* **1,** 1960.

Murata, K. J., A new method of plotting chemical analyses of basaltic rocks, *Am. J. Sci.* **258A,** 1960.

Niggli, P., Die Magmentypen, *Schweiz. Min. Petr. Mitt.* **16,** 1936.

Peacock, M. A., Classification of igneous rock series, *J. Geol.* **39,** 1931.

Pecora, W. T., Carbonatites, *Bull. Geol. Soc. Am.* **67,** 1956.

Richey, J. E., and H. H. Thomas, The geology of Ardnamurchan, etc. *Mem. Geol. Survey,* Scotland, 1930.

———, The Tertiary ring complex of slieve gullion, *Quart. J. Geol. Soc. London,* **88,** 1932.

Rittmann, A., *Vulkane und ihre Tätigkeit,* Stuttgart, 1960.

Stearns, H. T., and W. O. Clark, *Geology and Water Resources of the Kau District,* U.S. Geol. Surv. Water-Supply Paper, 1930.

Steiner, A., Origin of ignimbrites, *New Zealand Geol. Surv. Bull.* **68,** 1960.

Steinmann, G., Die ophiolitischen Zonen in den Mediterranen Kettengebirgen, *XXIV C.R. Int'l. Geol. Congress* **2,** Madrid, 1926.

Sörensen, H., The agpaitic rocks, *XXI Int'l. Geol. Congress,* Pt. 13, Copenhagen, 1960.

Tilley, C. E., Problems of alkali rock genesis, *Quart. J. Geol. Soc. London* **113,** 1957.

Tomkeieff, S. I., Petrochemistry of the Scottish Carboniferous-Permian igneous rocks, *Bull. Volcan. Ser. II* **1,** 1937.

Ussing, N. V., Geology around Julianehaab, *Medd. Grönland* **14,** 1911.

Tyrrell, G. W., Flood basalts and fissure eruptions, *Bull. Volcan. Ser. II* **1,** 1937.

Verhoogen, J., Les volcans Viruna et l'eruption du Nuamlogira, *Soc. géol. Belg. Ann.* **62,** 1939.

Wager, L. R., and W. A. Deer, The petrology of the Skaergaard intrusion, *Medd. Grönland* **105,** No. 4, 1939.

Wegmann, C. E., Geological investigations of Southern Greenland, *Medd. Grönland* **113,** 1938.

Wyllie, P. J., and O. F. Tuttle, Hydrothermal melting of shales, *Geol. Mag.* **98,** 1961.

———, Experimental investigations of silicate systems containing two volatile components, *Am. J. Sci.* **259,** 1961.

Part IV · Metamorphic Rocks

*l'application des règles de la chimie . . . de la thermodynamique . . . suppose l'existence de systèmes fermes et d'autres conditions, généralement non réalisées dans l'ecorce terrestre.**

E. C. Wegmann, Métasomatisme et analyse tectonique, *Int. Geol. Congr.*, London, 1948.

1 · INTRODUCTION

In the earth's crust of the continents the most important group of rocks, quantitatively, are those called metamorphic (including gneisses and migmatites). Any part of the crust that is affected by pressure or other kinds of stress and/or a simultaneous elevation of temperature, will undergo metamorphism.

Obviously the igneous rocks as well as the sediments will fall victim to the alteration. The former represent products of high T, the latter of low T. None of them is adjusted to moderately elevated temperatures paired with pressure (including shearing stress), varying from low to very high, but will respond to such conditions by reaction and recrystallization.

In the latter half of the eighteenth century Abraham Gottlieb Werner, who was professor at the School of Mines at Freiberg in Saxony, rendered many great services to geology, especially by improving the methods of distinguishing the mineralogical characters of rocks. Werner's main geological theory was called the Neptunian theory, and for many years it enjoyed great popularity.

Meanwhile, James Hutton (1726–1797) of Edinburgh, rightly regarded as the founder of modern geology, advanced the opinion that certain crystalline strata, which Werner called primitive and had declared were precipitated from a hot primeval ocean, were ordinary sedimentary strata subsequently altered by heat.

The Huttonian theory was adopted by Sir Charles Lyell, who in his *Principles of Geology* (1833) proposed "the term Metamorphic for the altered strata, a term derived from μετα, *trans*, and μορφη, *forma.*"

* The application of the rules of chemistry—of thermodynamics—requires the existence of a closed system, and of other conditions not generally realized in the crust of the earth.

In modern language metamorphism applies to transformations in rocks effected without melting and exclusive of the processes of weathering and sedimentation. In special cases the transformations may be purely mechanical, resulting in a deformation of the existing crystals and a change in the structure and texture of the rock. Usually the deformation is accompanied by a chemical recrystallization of the minerals (which also, of course, results in structural and textural changes).

The recrystallized minerals may be of the same kind as the old ones. They usually are different, however, and then a relocation of the chemical constituents has taken place. But the total chemical composition is not affected, that is, metamorphism is normal or isochemical.

Generally, metamorphism is accompanied by a change in the total chemical composition of the rock, that is, export and import of chemical matter result in metasomatic metamorphism, or simply metasomatism (change of body; σῶμα = body).

The geological conditions creating the free energy gradients driving the metamorphism are: (1) Increasing temperature induced by adjacent intrusions of magma or magmatic exhalations (*thermometamorphism* or, with reference to the geological environment, *contact metamorphism*). (2) Pressure gradients (stress) following dislocations in the crust along which the rocks are sheared and rolled out (*dislocation-metamorphism* or *dynamo-metamorphism*, often resulting in the formation of marked zones of extremely fine-grained, crushed rock densely cemented together, that is, the so-called mylonites). This metamorphism, like contact metamorphism, is local and restricted. (3) The most widespread metamorphism is induced by increase in both temperature and pressure in orogenic regions (vast segments of the crust represented by the folded mountain ranges). Heat is produced partly by downwarping and deep burial, partly by regional magmatism. Changes in rock induced by various combinations of the above factors are grouped together under *regional metamorphism*.

Additional terms used in the study of metamorphic rocks are *automorphism*, caused by decrease in temperature in newly congealed igneous rocks in which residual hydrothermal solutions are able to react with the igneous minerals. *Diaphtoresis*, or *retrograde metamorphism*, is a general term used for any metamorphism that takes place in consequence of decreasing temperature and pressure. *Static metamorphism* (∼ *load metamorphism* ∼ *geothermal metamorphism*) has been used to describe changes in rocks supposedly brought about by a regular increase in temperature and hydrostatic pressure through deep burial.

A distinction must be made between happenings in the superstructure (Oberbau) of the orogenic mountain chains and the migmatic infrastructure (Unterbau) lying below the usual depth range of regional metamorphism.

Migmatization takes place in consequence of the physico-chemical conditions of the infrastructure. In the absence of a better classification, migmatites will be grouped with metamorphic rocks.

2 · METAMORPHIC MINERALS

A general principle in scientific progress is to work from the simple to the more complex. In igneous geology the individual minerals and their melting relations were first studied, then simple mineral combinations, and finally the complex mineral combinations representing the igneous rocks. Thus were gradually established the general laws governing the complex equilibrium relations crystal $\rightleftharpoons$ melt in a polycomponent magma. These laws form the basis of modern igneous petrology.

Metamorphic geology is less advanced, the main reason being that rather complicated laboratory techniques are necessary to negotiate the difficulties inherent in the slow-working subsolidus reactions of chief importance in the genesis of metamorphic rocks. Consequently, the conditions of formation, the stability regions, and the thermodynamic properties of a great many metamorphic minerals are not known. We can only guess at them from circumstantial evidence furnished by field geology.

Again it is necessary to study the equilibrium relations of the individual metamorphic minerals and simple mineral combinations before the complex mineral combinations representing natural metamorphic rocks can be understood.

Minerals of special importance in contact metamorphism will be discussed later (pages 257 ff.). First we shall look into the relations of the regional-metamorphic minerals.

In regional metamorphism quartz and feldspar, on the one hand, and the colored minerals, pyroxene, amphibole, mica, chlorite, on the other, represent, quantitatively, the most important metamorphic minerals. They also may serve as examples of the most important types of structure in silicates and explain the relations between crystal structure and rock structure as met with in practically all regional metamorphic rocks.

In all crystalline silicate lattices, silicon is surrounded by four oxygens, forming tetrahedra $[SiO_4]^{4-}$. Aluminum is, after silicon, the most important cation and has a double role; either it is surrounded by four oxygens and acts, crystallo-chemically, like silicon, or it is surrounded by six oxygens and acts like the larger cations, magnesium, iron, etc. It is interesting to note that, generally speaking, at higher temperatures alumina prefers 4-coordination; at higher pressures and lower temperatures it also accepts 6-coordination. Tetrahedra in which the central Si-ion may be partly replaced by Al is symbolized by writing ZO_4.

Oxygen in silicates may belong to two tetrahedra $[ZO_4]$ at the same time. Thus distinct anions are formed, the most important of which form endless bands, chains, or sheets of the general type Z_xO_y.

Olivine-pyroxene-amphibole-biotite-chlorite, representing the common, colored rock-making silicates, belong to the important reaction series of Bowen, and reflect in their crystal structure interesting relations to the geological mode of occurrence.

In olivine (and other orthosilicates) individual tetrahedra of $[SiO_4]^{4-}$ are present. The pyroxene structure is characterized by infinite chains, $\infty[ZO_3]^{2-}$; amphibole forms silicate bands,

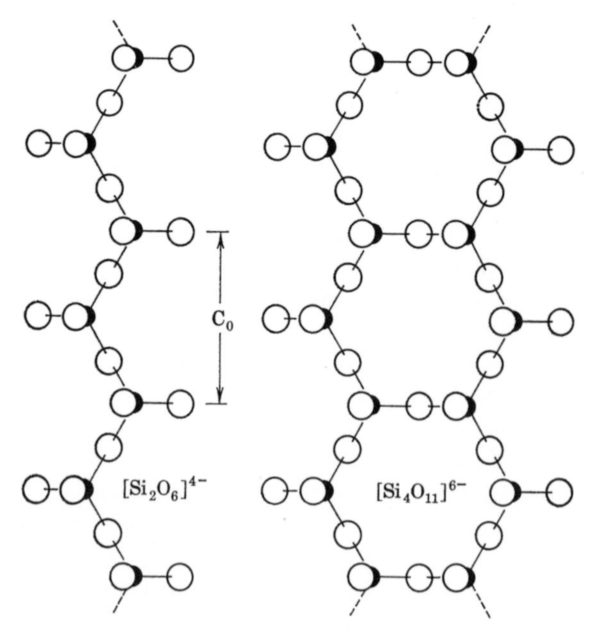

FIG. IV-1. $(SiO_3)^{2-}$-chain in pyroxene and $(Si_4O_{11})^{6-}$-bands in amphibole.

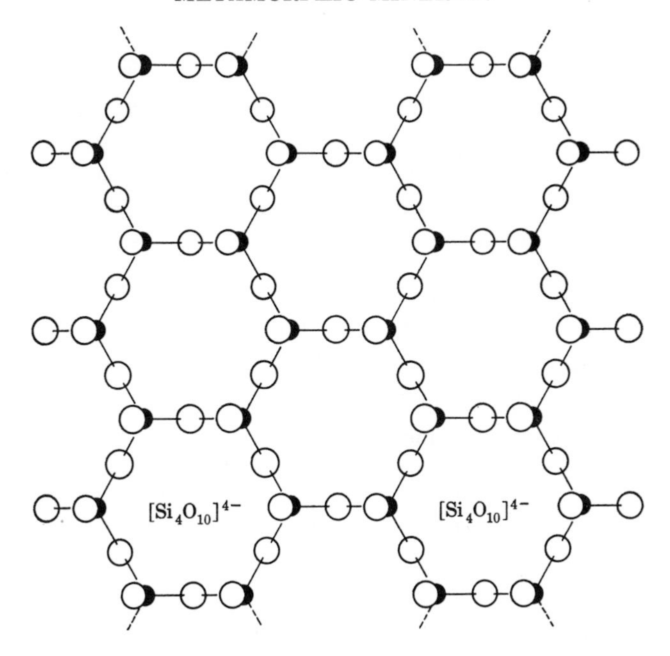

FIG. IV-2. Sheet structure in mica-like minerals.

$\infty[Z_4O_{11}]^{6-}$; and mica and chlorite form silicate sheets, $\infty[Z_4O_{10}]^{4-}$. (See Figs. IV-1 and IV-2.)

For better comparison, the three types of anion may be written thus:

$[Z_4O_{12}]$	pyroxene chain
$[Z_4O_{11}]\cdot[OH]$	amphibole band
$[Z_4O_{10}]\cdot[OH]_2$	mica sheet
$[Z_4O_{10}]\cdot[OH]_8$	chlorite sheet

The letter Z stands for both Si and Al; the fraction taken by Al increases regularly in the series from zero in certain pyroxenes to at least 25 per cent in micas.

The relative stability of these lattices increases with decreasing temperature in the direction pyroxene $\rightarrow$ chlorite (corresponding to increasing hydration). Conversely, if these minerals are subjected to increasing temperature in the rocks, first chlorite, then in turn mica and amphibole become unstable; pyroxene persists until it melts. That all minerals *per se* are stable over a wide range is indicated, for instance, by the fact that muscovite has been synthesized at 200°C and at 900°C (under high confining H_2O pressure), the latter temperature being well beyond the upper limit of any normal metamorphism.

In pyroxene and amphibole the chains and bands extend in the direction of the *c*-axes and account for the prism and needle shape of these minerals. Likewise the (pseudo) hexagonal sheet-like structure in mica and chlorite is explained. In chlorite the consecutive (Si_4O_{10}) sheets are more widely spaced than in mica and separated by additional layers of hydroxide.

List of Important Regional Metamorphic Minerals Ordered According to Crystal Structure Properties

In the subsequent mineral formulas the letters W, X, Y, Z designate ions of the following volume relations:

Ion	Si^{4+}	Al^{3+}	Fe^{3+}	Mg^{2+}	Fe^{2+}	Na^{+}	Ca^{2+}	K^{+}
Radius	0.39	0.50	0.67	0.65	0.82	0.98	1.06	1.33

←——Z——→ (Si–Fe³⁺) ←————X————→ (Mg–Ca)
←————Y————→ (Al–Fe²⁺) ←————W————→ (Na–K)

(A) Nesosilicates (νῆσος = *island*), *with individual* (SiO_4) *tetrahedra*

All minerals are subjected to structural control, but some of the nesosilicates are weakly affected.

Olivine $(Mg,Fe)_2SiO_4$. Characteristic of high temperature and deep metamorphism (eclogite facies). Also as contact mineral. At very high pressures olivine changes into a high-pressure modification of spinel structure. At similar temperatures iron-rich olivines require a smaller pressure than magnesium-rich olivines (Fig. IV-3). At great depths, of the order of magnitude of 500 km (depending upon the temperature) there is a zone in which Mg-rich olivine and iron-rich spinel are in equilibrium. See Fig. IV-9.

Garnet Group

Garnet group, mixed crystals of the general formula $X_3Y_2(SiO_4)_3$, with some $(HO)_4$ (in grossularite) substituting for (SiO_4). The first garnet to form (at the lowest temperature) is rich in *spessartite*, $Mn_3Al_2(SiO_4)_3$ (greenschist facies). Most abundant in regional metamorphic rocks are solid solutions of *almandite*, $Fe_3Al_2(SiO_4)_3$, pyrope, $Mg_3Al_2(SiO_4)_3$, and smaller amounts of *grossularite*, $Ca_3Al_2(SiO_4)_3$. In grossularite Fe^{3+} may proxy for Al forming *andradite* which is typical of limestone contacts. With increasing pressure the mutual solubility of pyrope in almandite increases and becomes complete above ca. 800°C.

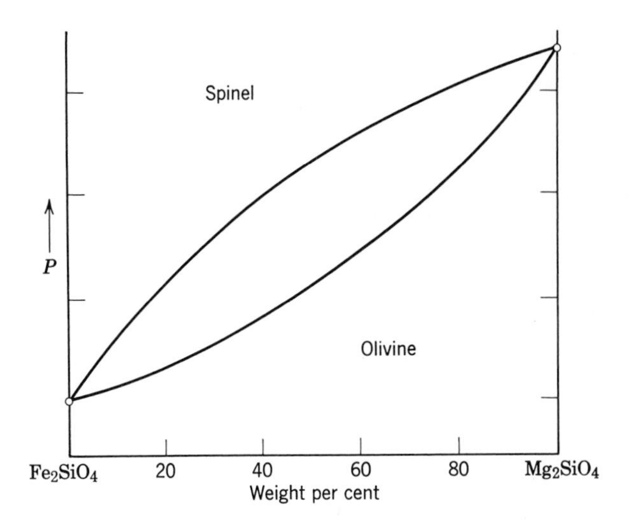

FIG. IV-3. Schematic diagram of the transition olivine ⇌ spinel with increasing pressure. The transition is formally similar to that of olivine ⇌ melt with increasing temperature. See Fig. III-19.

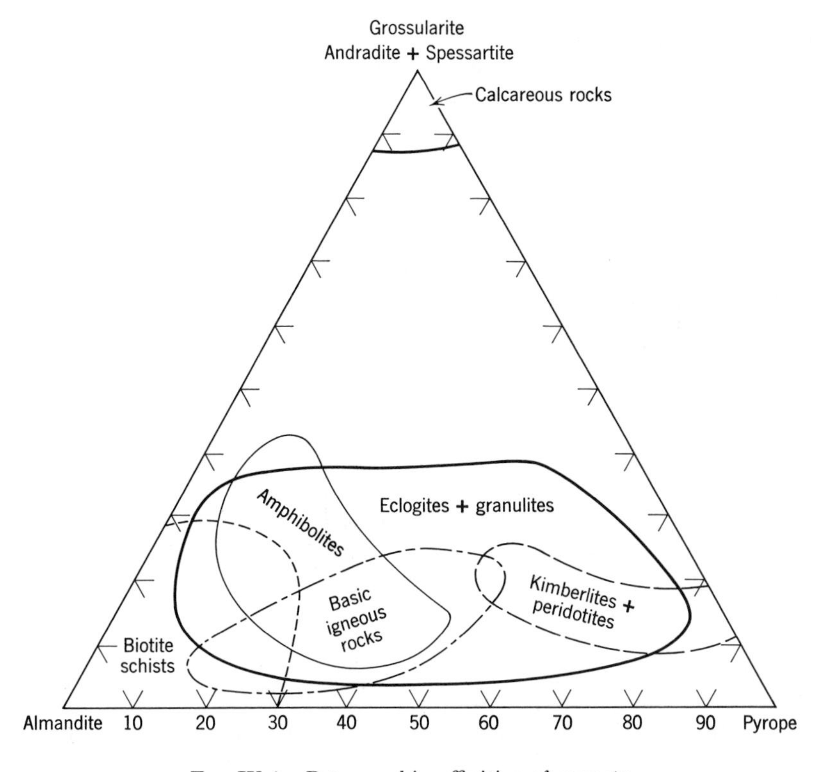

FIG. IV-4. Petrographic affinities of garnets.

Andalusite and *kyanite* are two different modifications of Al_2SiO_5. A third modification is *sillimanite* which, however, contains tetrahedral chains [$AlSiO_5$], and may be classified with the inosilicates. Andalusite is characteristic of contact rocks (chiastolite, long, thin pigmented porphyroblasts, usually undirected, in shale etc.), but it occurs also in mica schists. It does not develop in stressed rocks. The growth of kyanite (often as porphyroblasts) appears to be favored by stress (gneisses, alpine schists). Sillimanite represents the highest temperature stage (aluminum-rich gneisses). It always shows a directed growth (parallel fibers) corresponding to its crystal structure. According to the phase rule, the several aluminum silicates should never occur side by side. Actually it is not very rare to find two of them in the same rock; even all three have been found in the same paragenesis.

There has been much speculation about the mode of occurrence and the stability regions of these three minerals. Harker suggested that kyanite is a *stress mineral,* whereas andalusite represents an *anti-stress mineral.* To be sure, it is true that the lattice of andalusite is easily destroyed by stress forces. But, although stress may favor the development of kyanite, it is not necessary; kyanite can grow in pegmatite where no stress seems to have been present.

In referring to the equilibrium diagram, Fig. IV-5, a word of caution should be given. The curves of the diagram probably rightly

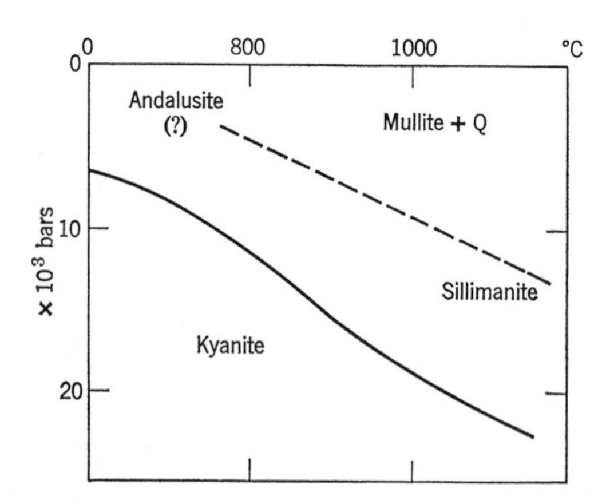

Fig. IV-5. Tentative equilibrium diagram for Al_2SiO_5. The kyanite-sillimanite curve is probably correct; the dashed curve may actually correspond to a band of solid solutions between sillimanite and mullite. The position of andalusite is uncertain, it may be unstable, and under equilibrium conditions always monotropically changing into either sillimanite or cyanite.

reflect the equilibrium under the conditions of the experiment. But this does not mean that kyanite in rocks always requires pressures of 10 kilobars or more (corresponding to a depth of 35 km). Rocks are far from being pure aluminum silicate plus water; appreciable amounts of ferric iron and of other elements enter into the lattices of the natural aluminum silicates. The SiO_2 liberated by the sillimanite-mullite transformation may react with other rock constituents in various ways, etc. All such things will affect the position of the curves, and it is entirely possible that the kyanite-sillimanite curve under favorable conditions could be displaced toward lower pressures so much that a natural kyanite becomes stable at room pressure. See also Fig. IV-6 (jadeite) and text.

Staurolite $\left\{\begin{matrix} Al_2SiO_5 \\ Fe(OH)_2 \\ Al_2SiO_5 \end{matrix}\right\}$, structurally, is a kyanite with intercalated layers of $Fe(OH)_2$. It occurs with kyanite (often in parallel growths) in tectonically deformed mica schists, etc., usually as porphyroblasts.

Chloritoid is a phyllosilicate of related chemical composition $\left\{\begin{matrix} Al_2SiO_5 \\ Fe(OH)_2 \end{matrix}\right\}$, often crystalloblastic in fanshaped or sheaf-like groups in tectonically deformed schists of relatively low metamorphism. Manganoan chloritoids may be called *ottrelites*. Chloritoid crystallizes in both monoclinic and triclinic polymorphs. Above certain temperatures chloritoid decomposes at high pressures to almandine, staurolite, hercynite, and vapor; at low pressures to iron cordierite, hercynite, and vapor. Precise determinations of the conditions for these decompositions require control of partial pressure of oxygen.

(B) Sorosilicates (σωρός = group), with finite clusters of tetrahedra, e.g., $[Si_2O_7]^{6-}$

Epidote-Zoisite Group

General formula $X_2Y_3Z_3O_{12}(OH)$. Epidote itself has the formula $Ca_2(Al,Fe''')_3(SiO_4)_3 \cdot OH$ and is common in rocks of the lower stages of regional metamorphism, substituting here for the anorthite component of plagioclase (the corresponding iron-free member is zoisite, see further under plagioclase, page 254). Epidote and albite (or oligoclase) are, therefore, frequently associated in rocks. Epidote is, moreover, a common contact mineral. It has been synthesized at 2–8 kilobars and 500–600°C. This mineral group is remarkable for the many and variegated chemical substitutions, for instance, in *allanite* (orthite) tervalent cerium and univalent sodium substitute for bivalent calcium.

Pumpellyite is possibly a hydrated clinozoisite, $Ca_2(Al,Mg \cdot Fe'')_3$-$Si_3O_{12}(OH) \cdot H_2O$.

Lawsonite, $CaAl_2Si_2O_8 \cdot 2H_2O$, corresponds in its water-free part to anorthite, but it is much heavier (density = 3.09) and is a critical mineral of highly stressed rocks (glaucophane schists of California, the Apennines, Piedmont, etc.).

Ilvaite (lievrite), $CaFe_2''Fe'''Si_2O_8(OH)$, shows structural relations to epidote. Perfect crystals in limestone contacts at Elba, Seriphos, etc.

Prehnite is a low-temperature mineral substituting for epidote. Structurally it is an inosilicate, $Ca_2Al_2Si_3O_{10}(OH)_2$, and in its mode of occurrence it is similar to a zeolite.

(*C*) *Cyclosilicates* (συκλός = *ring*) *with finite rings of the type* $[Si_6O_{18}]^{12-}$

Cordierite, $Mg_2Al_3(AlSi_5O_{18})$, occurs in many variants. Some Fe may substitute for Mg, and in the laboratory pure Fe-cordierite has been made. But in natural cordierite the amount of Fe is rather small. It is a common constituent of many gneisses (often with garnet, sillimanite, etc.), also a contact mineral. The formula contains no water but, like *beryl* ($Al_2Be_3Si_6O_{18}$), the $(Al,Si)_6O_{18}$-rings are stacked on top of each other, making a central passage, the inner diameter of which is about 2.7 Å, thus permitting water, (OH), fluorine, and alkalies (up to 5 per cent) to be stuffed into the hole in the center of the rings. Cordierite has been found to break down at high pressures in the laboratory. A number of variants were recently recognized. They are listed in Table IV-1.

Tourmaline; the structure shows $(Si_6O_{18})^{12-}$-rings and flat BO_3 islands, which are kept together by Al, Mg, and Na. The ideal for-

TABLE IV-I

POLYMORPHOUS PHASES OF $Mg_2Al_4Si_5O_{18}$

Name	Symmetry	Structure	Occurrence	Synthetic
High indialite	hexagonal	Single rings beryl type	para-lavas	>830°C
Low indialite	hexagonal	Single rings beryl type	(?)	<830°C
High cordierite	pseudo-hex	Single rings beryl type	volcanics	. . .
Low cordierite	pseudo-hex	Single rings beryl type	metamorphics	. . .
μ $Mg_2Al_4Si_5O_{18}$	(?)	α $LiAlSiO_4$	(?)	850–800°C
Osumilite	hexagonal	milarite	volcanics	. . .

mula is $NaMg_3Al_6(BO_3)_3(Si_6O_{18})(OH)_4$, but complicated substitutions with K, Ca, Li, Fe, Ti, Cr, F, etc., are common. It is a frequent pneumatolytic mineral (luxullianite = tourmalinized granite). It forms very early (at low temperature) during regional metamorphism (small crystals in phyllites and aluminous schists).

(D) Inosilicates (ἴς, *genitive* ἰνός = *fiber*), *with chains or bands as shown in Fig. IV-1.*

Pyroxene Group

Pyroxenes are important metamorphic minerals. *Enstatite-hypersthene* $(Mg,Fe)SiO_3$ are common contact minerals also found in regional metamorphic rocks. Under high pressure and temperature, enstatite can dissolve up to 19 weight per cent of Al_2O_3. A useful relation between pressure and the solubility limit of Al_2O_3 in enstatite can possibly be established. At still higher pressure MgFe-pyroxenes will invert to spinel, liberating coesite:

$$\begin{array}{llll} 2MgSiO_3 \rightarrow & Mg_2SiO_4 + & SiO_2, & \Delta v = -3.3 \text{ cm}^3/\text{mole} \\ \text{enstatite} \rightarrow & \text{spinel} & + \text{coesite} & \end{array}$$

The effect of $FeSiO_3$ in solid solution is analogous to that in olivine. A transition zone is formed which probably overlaps the olivine-spinel transition zone.

Clinohypersthene has not been encountered in metamorphic rocks; it probably requires too high temperatures for its formation there. The subsolidus relations of this and some other pyroxenes are shown in Fig. III-21.

Diopsidic pyroxene, $CaMgSi_2O_6$ with some iron and alumina, occurs at contacts and in lime and magnesia-rich crystalline schists; augite, containing more aluminum, iron, titanium, sodium, etc., occurs at contacts and in some high-temperature gneisses.

The typical metamorphic pyroxenes (that is, pyroxenes not known in igneous rocks) are aluminous. The following have been recognized:

Jadeite	rich in $NaAlSi_2O_6$
Omphacite	$(Ca,Na)(Mg,Fe^{+2},Al)(Si,Al)_2O_6$
Chloromelanite	$(Ca,Na)(Mg,Fe^{+3},Al)(Si,Al)_2O_6$
Fassaite	$(Ca)(Mg,Al)(Si,Al)_2O_6$

Fassaite is a contact mineral. Omphacite and chloromelanite are characteristic of deep metamorphism (eclogite facies). Jadeite likewise is a high-pressure mineral.

Petrographically important are the thermodynamic relations of the reactions.

$$NaAlSi_3O_8 + NaAlSiO_4 \rightarrow 2NaAlSi_2O_6 \quad \text{and}$$

$$2NaAlSi_3O_8 \rightarrow 2NaAlSi_2O_6 + 2SiO_2$$

$$\text{(albite)} \rightarrow \text{(jadeite)}$$

The equations are better to survey by using the one-cation equivalent unit (see page 67).

$$5Ab + 3Ne \rightarrow 8Ja, \tag{III-1}$$

$$10Ab \rightarrow 8Ja + 2Q: \tag{III-2}$$

for these reactions the following constants have been determined:

$$\text{for (1)} \quad \Delta S^\circ = -14.6, \quad \Delta H^\circ = -6670, \quad \Delta V = -33.6$$

$$\text{for (2)} \quad \Delta S^\circ = -14.8, \quad \Delta H^\circ = -1230, \quad \Delta V = -34.6$$

By equation 13 in the Appendix, $\Delta G = \Delta H - T\,\Delta S$, one obtains the changes in free energy

$$\text{for (1)} \quad \Delta G_{298} = -2300 \text{ cal/mole}$$

$$\text{for (2)} \quad \Delta G_{298} = +3200 \text{ cal/mole}$$

Thus the free energy decrease of the first reaction indicates that a nepheline syenite (Ab + Ne) is unstable at room temperature and pressure, and should invert into jadeite. By the approximate equation 16 in the Appendix, $\Delta G_T \backsim \Delta H^\circ - T\,\Delta S^\circ$, we find the inversion temperature, T_i, at room pressure

$$\Delta G_{T_i} = 0 = \Delta H^\circ - T_i\,\Delta S^\circ, \text{ or}$$

$$T_i = 456°K = \backsim 180°C$$

This result indicates that at approximately 180°C natural nepheline syenite and jadeite rock are in equilibrium. Above this temperature nepheline syenite is stable, below this temperature jadeite is stable.

If quartz is present, however (reaction 2) it takes a much higher pressure at all temperatures to change albite into jadeite.

The phase diagram for these reactions is given in Fig. IV-6. But the occurrence of jadeite in rocks is much more complicated. Natural albite is not pure, but may take both K and Ca into solid solution; likewise the jadeite lattice will easily accommodate ions like Mg,

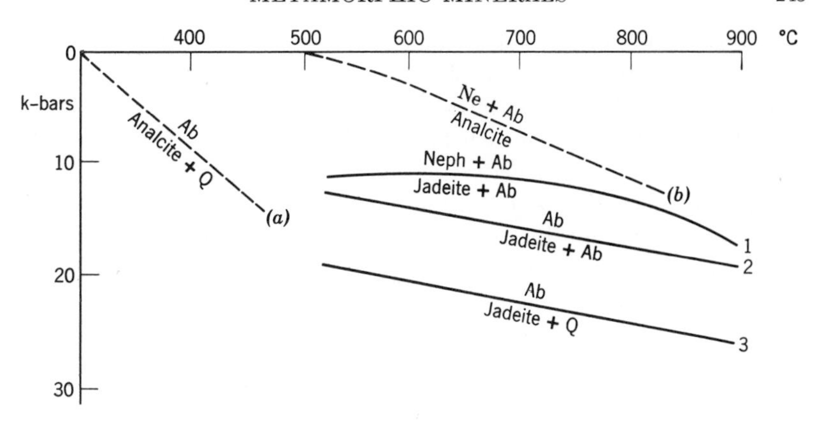

FIG. IV-6. Dashed lines refer to pressure of H_2O, and give the equilibrium conditions for (*a*) $NaAlSi_3O_8 + H_2O \rightleftharpoons NaAlSi_2O_6 \cdot H_2O + SiO_2$, and, (*b*) $NaAlSiO_4 + NaAlSi_3O_8 \rightleftharpoons 2\ NaAlSi_2O_6 \cdot H_2O$. Solid lines refer to "dry" pressure. Nepheline and albite are mutually soluble at high temperatures and high pressures. Nepheline saturated with albite changes into silica-deficient jadeite with albite remaining unchanged: curve 1. Albite dissolves in the jadeite as pressure is raised. Pure albite exsolves jadeite along the boundary: curve 2. At still higher pressures albite breaks down completely to jadeite plus quartz: curve 3.

Fe, Ca, etc., making it extremely difficult to determine the *PT* conditions under which natural jadeites actually have formed.

Amphibole Group

General formula $X_{2-3}Y_5Z_8O_{22}(OH,F)_2$. *Common hornblende* has a very wide distribution in metamorphic rocks of medium to high regional metamorphism (amphibolites, etc.). Varieties rich in Al_2O_3, often called *hastingsites,* are supposed to indicate a higher degree of metamorphism.

Tremolite (grammatite), $Ca_2Mg_5Si_8O_{22}(OH)_2$, and *actinolite*, $Ca_2(Mg,Fe)_5Si_8O_{22}(OH)_2$, are characteristic of crystalline schists of relatively low temperature, and form individual rocks (actinolite schists). They are also present as contact minerals. They may contain some Na_2O and/or Al_2O_3. *Smaragdite* is a grass-green actinolite in intergrowth with omphacite in eclogite rocks. *Nephrite* is a dense aggregate of actinolite. In the form of jade it occurs in ornaments and utensils of prehistoric man; it is obtained at various points in central Asia, in New Zealand, and in serpentinized gabbros in brecciated zones in crystalline schists of Grisons, Switzerland, and the Ligurian Apennines.

Grünerite, $Fe_7Si_8O_{22} \cdot (OH)_2$, and *cummingtonite,* $Mg_7Si_8O_{22}(OH)_2$, are present in many metamorphic rocks, for example, at Collobrières, France, in central Sweden, in the Mesabi Range, Mich., at Cummington, Mass., and elsewhere. Chemically identical are *anthophyllite* and the Al_2O_3-rich variety *gedrite* (snarumite), which, in contradistinction to all other hornblendes, are orthorhombic. Their stability region as compared to that of the grünerite-cummingtonite series is not well known; about 50 per cent of magnesium may be replaced by iron, but at higher iron content the monoclinic grünerite occurs.

Glaucophane, $Na_2(Mg,Al)_5Si_8O_{22}(OH)_2$, in solid solution with riebeckite, is supposed to be a high-pressure mineral occurring in tectonically highly deformed schists in the Alps, the Greek Archipelago, in the Coast Ranges of California, etc. Perhaps more widely distributed is the so-called *gastaldite,* containing much of the actinolite molecule.

Riebeckite-arfvedsonite, solid solutions of $Na_2Fe_3{}^{2+}Fe_2{}^{3+}Si_8O_{22}(OH)_2$ – $Na_3Fe_4{}^{2+}Fe^{3+}Si_8O_{22}(OH)_2$, have been synthesized, and the experimental data agree with the observations that arfvedsonites are stable at higher temperatures and lower oxidation states in alkalic igneous rocks; in contrast, riebeckites are stable at low temperatures and high oxidation states in metamorphic environments; more common is *crossite* containing some aluminum.

(E) Phyllosilicates (φύλλον = *leaf*), *with* Si_4O_{10}*-sheets, as shown in Fig. IV-2.*

Mica Group

The constituent sheets of the mica structure can be illustrated as follows:

Z	$(Al,Si)_4O_{10}$	Tetrahedral sheet (fourfold coordination)
Y	Al, Mg, Fe, etc.	Octahedral group (sixfold coordination, binds sheet together)
Z	$(Al,Si)_4O_{10}$	
W	K, Ca, Na, etc.	Large cation group (twelvefold coordination, binds ZYZ groups)

Hydroxyl groups alternating with oxygen lie parallel to the sheets. According to the number of Y-ions one can distinguish between dioctahedral and trioctahedral micas. Dioctahedral are muscovite, paragonite, and margarite:

Muscovite, $KAl_2(AlSi_3)O_{10}(OH)_2$, is stable from the very beginning of metamorphism up to rather high grades. Artificially it has been made to crystallize from gel-like substances at temperatures down to about 100°C; finely crystalline, scaly muscovite formed at low temperatures are often called *sericite*. But it will also crystallize under high vapor pressure up to and above 800°C.

Paragonite (sodium muscovite), $NaAl_2(AlSi_3)O_{10}(OH,F)_2$, is a characteristic mineral of, for example, paragonite schists in Tessin, Switzerland, often associated with kyanite and staurolite, with corundum (Delaware County, Pa., and elsewhere in the Appalachians) with actinolite (Tirol, etc.). There is solid solubility between muscovite and paragonite. The ratio K/Na in mica, compared to the ratio K/Na in coexisting feldspar could be developed into a geological thermometer.

Margarite (calcium mica), $CaAl_2(Al_2Si_2)O_{10}(OH)_2$, is rare; it occurs in some eclogite-anorthosite provinces, in chlorite schists (for example in Tirol), is commonly associated with corundum, and in many cases formed from it.

Still other variations in the large cation group are possible; particularly Ba has a tendency to concentrate (*oellacherite*). In the octahedral group Cr may concentrate (*fuchsite*).

The following micas are trioctahedral:

Biotite, $K(Mg,Fe)_3(AlSi_3)O_{10}(OH,F)_2$, shows much variation in chemical composition, and there are many variants of crystal structure corresponding to more or less regularity in the stacking of the $(Si,Al)_4O_{10}$ sheets. It requires a certain stage of metamorphism to grow; possibly low-grade biotites are high in MnO. The pure Mg-member is called *phlogopite*, and is characteristic of contact metamorphism, the Fe-end member is *siderophyllite*. The melting relations are shown in Fig. IV-7.

The possible substitutions by rare metals in biotite are very great. In the octahedral group large amounts of manganese may be present, *manganophyllite* with 18 per cent MnO. Lithium may occupy up to two of the three octahedral positions: *lepidolite*, $KLi_2AlSi_4O_{10}(OH,F)_2$, which is particularly interesting also because it is the only mica for which the tetrahedral positions are completely occupied by Si.

Zinnwaldite, $KLiAlFe^{2+}AlSi_3O_{10}(OH,F)_2$. Appreciable amounts of Ti may be present and relatively large amounts of Ca, Zn, Sc, V.

Into the large cation group appreciable amounts of Rb and Cs may enter.

The stability relations of the micas may be discussed with reference to the curves in Fig. IV-7. Under the condition of the experiments, biotite, and in particular magnesian biotite, is stable at much higher

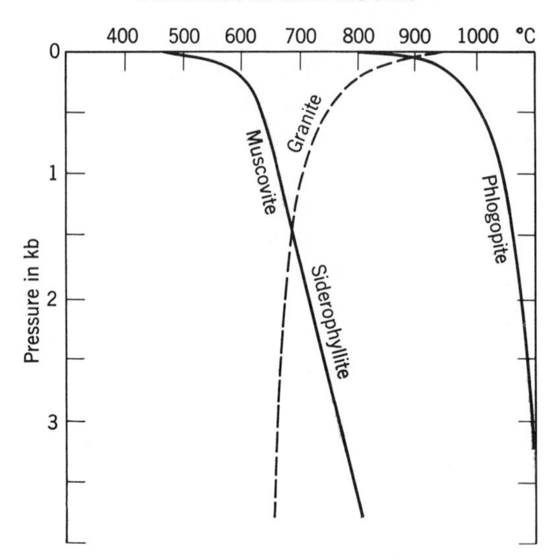

FIG. IV-7. Experimental decomposition curves for micas. (After Yoder and Eugster, 1954, 1955.) The temperature of the breakdown of muscovite is, in the experiment, only dependent on the water pressure. But the decomposition of the iron-bearing biotite is also dependent on the partial pressure of oxygen in the experiment. The upper limit is given by the curve of iron-free phlogopite; the lower limit is near to the muscovite curve, if oxygen pressure corresponds to the quartz-magnetite-fayalite assemblage. Superimposed on these curves is the minimum melting curve of granite, after Bowen and Tuttle.

temperature than that of molten igneous rocks. This is consistent with the occurrence of biotite as an igneous mineral. For muscovite this is different; it takes a high vapor pressure, according to the figure at least 1500 atmospheres, to make muscovite stable at a temperature as high as that of molten granite.

However, nature is not as simple as the experiments. If muscovite exists in equilibrium with a granitic magma, it must also be in equilibrium with quartz. Now it is seen from the reaction:

$$KAl_3Si_3O_8(OH)_2 \rightleftharpoons KAlSi_3O_8 + Al_2O_3 + H_2O,$$

that the decomposition products containing corundum are incompatible with quartz; presence of quartz therefore "contaminates" the reaction and depresses the reaction temperature. Because of the low angle of intersection of the curves "muscovite"–"melting of granite," a small depression of the muscovite curve would eliminate its intersection with the melting curve. It is conceivable, therefore, that muscovite under no conditions will crystallize as a primary mineral from a granitic magma.

Pyrophyllite-Talc Group

The structure is analgous to that of mica.

Pyrophyllite, $Al_2Si_4O_{10}(OH)_2$, occurs in low-grade schistose metamorphic rocks, in places as a very fine-grained, dense variety called *agmatolite*.

Talc, $Mg_3Si_4O_{10}(OH)_2$, is commonly associated with serpentinite, chlorite schists, and similar low grade rocks; it is also a product of low contact metamorphism of dolomitic limestone. A fine-grained, dense variety is called *steatite*.

Minnesotaite, $(Fe,Mg,H_2O)_3(Si,Al)_4O_{10}(OH)_2$, found in ore deposits in Minnesota, and easily obtainable artificially by hydrothermal synthesis together with pyrophyllite and clay minerals.

Serpentine Group ("Trioctahedral"). $Y_6Z_4O_{10}(OH)_8$.

The (Si_4O_{10}) sheets are analogous to those in kaolinite. True serpentines are: *Antigorite*, $Mg_6Si_4O_{10}(OH)_8$, occurring as thin and lamellar crystals, separated into translucent folia, usually compact or granular massive. *Chrysotile* of the same composition has a fibrous structure often occurring in seams. *Asbestos* is usually made up of chrysotile.

Also belonging to the serpentine group are minerals containing aluminum and iron, ranging from *Amesite* $(Mg_4Al_2)(Al_2Si_2)O_{10}\cdot(OH)_8$, to *Cronstedtite* $(Fe_4{}^{2+}Fe_2{}^{3+})(Fe_2{}^{3+}Si_2)O_{10}\cdot(OH)_8$.

Serpentines occur in crystalline schists in the form of layers, dikes, or necks, or constituting rock masses often formed metasomatically after olivine-rich rocks.

$MgO—SiO_2—H_2O$

The system $MgO—SiO_2—H_2O$ has been investigated experimentally. The stable phases occurring are periclase, forsterite, enstatite, quartz, brucite, talc, serpentine, and vapor.

Figure IV-8 shows the stability regions of these minerals on the *P-T* diagram. It is seen that, whereas enstatite and forsterite (as well as periclase and quartz) are stable throughout the whole range, the minerals with sheet lattices become unstable at higher temperatures.

The significance of curve I is that the low-temperature combination brucite-serpentine becomes unstable, and olivine becomes stable in contact with water vapor:

$$\text{Brucite} + \text{serpentine} \rightleftharpoons \text{forsterite} + \text{vapor}$$

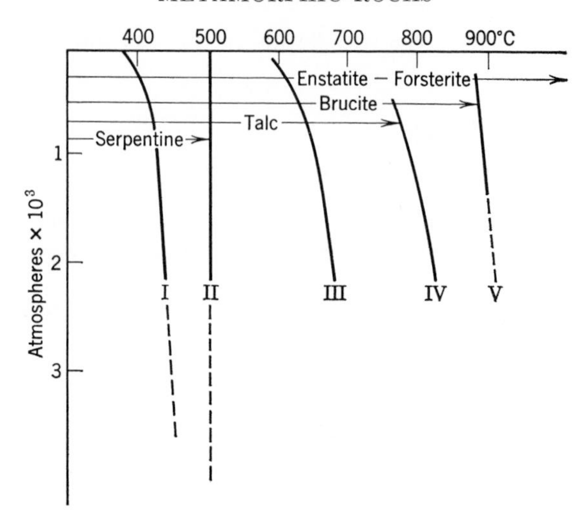

FIG. IV-8. Stability regions in the system $MgO—SiO_2—H_2O$. (After Bowen and Tuttle, *Bull. Geol. Soc. Am.* **60,** 1949.)

At curve II (ca. 500°C) serpentine itself becomes unstable. Curve III represents the temperature and pressure at which the combination talc-forsterite becomes unstable; at higher temperatures talc is stable only in combination with enstatite, which, from now on, is stable in contact with water:

$$\text{Talc} + \text{forsterite} \rightleftharpoons \text{enstatite} + \text{vapor}$$

Curve IV shows the upper limit of the stability of talc, and curve V gives the upper limit of the stability of brucite.

Natural olivines are easily subjected to serpentinization. If the olivine has about 12 per cent FeO, curve I is lowered some 80°. Olivine with about 30 per cent FeO is not serpentinized even at 295°C and 1000 atm.

Finally it should be noted that the experiments do not always correspond to the natural conditions. Frequently the pressure acting on the vapor phase will be less than that acting on the minerals. Then water will be squeezed out of the serpentine crystal lattice at lower temperatures than those given by the curves I and II.

Chlorite Group

There are a great many different varieties. The crystal lattice is built up of sheets of Z_4O_{10} of the talc structure alternating with sheets of brucite $Mg(OH)_2$. A great many structural variants result from various stacking patterns of these sheets.

The chemical variations are also great. The following "end members" form complete mutual solid solutions.

Talc-chlorite	Mg_6	Si_4	O_{10}	$(OH)_8$
Clinochlore	(Mg_5Al)	$(AlSi_3)$	O_{10}	$(OH)_8$
Corundophilite	(Mg_4Al_2)	(Al_2Si_2)	O_{10}	$(OH)_8$

Intermediate members are *pennine* and *prochlorite* (rhipidolite). There is some Fe substituting for Mg; the end members are:

Daphnite	$(Fe^{\cdot\cdot}_{4.5}Al_{1.5})$	$(Al_{1.5}Si_{2.5})$	$O_{10}(OH)_8$
Pseudothuringite	$(Fe^{\cdot\cdot}_4 Al_2)$	(Al_2Si_2)	$O_{10}(OH)_8$

The following have both two and three-valent iron:

Chamosite	$(Fe^{\cdot\cdot}_3 Fe^{\cdot\cdot\cdot})_6$	$(AlSi_3)$	$O_{10}(OH)_8$
Thuringite	$(Fe^{\cdot\cdot}_4 Fe^{\cdot\cdot\cdot}_2)$	(Al_2Si_2)	$O_{10}(OH)_8$

The decomposition temperatures of some important artificial chlorites are known. Among the decomposition products are fayalite, forsterite, spinel. If quartz was added, it would react with these products and thereby depress the temperature of the reaction. In quartz-chlorite schists, clinichlore would therefore not be stable at as high temperature as indicated by the experiments.

Synthesis always proceeds by the growth of a crystal of antigorite (serpentine) structure which slowly inverts to chlorite. The chlorite minerals are of considerable geologic importance, and occur in many different rocks of the early stages of regional metamorphism. Metasomatism, for example, in wall-rock alterations around ore bodies, may also be indicated by the resulting chlorite zones.

Iron-rich chlorites occur in chamositic iron formations, and in vein deposits. But some chamosites being products of chemical sedimentation exhibit (unstably (?)) kaolinite structure and will be converted to true chlorites by low-grade metamorphism.

Stilpnomelane, $(K,H_2O)(Fe,Mg,Al)_{4-6}(Si_4O_{10})_2 \cdot (OH)_4 4H_2O$, has a position between clay minerals and chlorites. It occurs in certain chlorite schists of low-metamorphic grade (New Zealand, Swiss Alps, Norwegian Caledonides, and elsewhere). It is hard to recognize in thin sections, and probably has a wider distribution than previously believed.

Clay Mineral Group

This group has been treated before, see pages 33 ff. Hydrothermal syntheses at low temperatures in the system $Al_2O_3—SiO_2—H_2O$ yield clay minerals and also pyrophyllite.

(F) Tectosilicates (τεκτονεία = *framework*) *with three-dimensional frameworks of* ZO_2 *chains as shown in Fig. III-4*

SiO_2. *Quartz* is an important metamorphic mineral. Its properties have been described on page 75, it is not necessary to repeat it here. Experiments have shown that even at comparatively low temperatures and pressures the solubility of SiO_2 in steam is appreciable. At 400–500°C the solubility increases rapidly with pressure; at 1000 atm the gas contains 0.25 per cent silica. This corresponds to a partial pressure of silica of 2.5 atm, and silica is one of the least volatile of substances. Reaction takes place rapidly. Even when the rate of passage is such that the steam is in contact with quartz for less than 10 minutes, the vapor phase becomes three-quarters saturated. At this easily attainable pressure, the solubility of silica in steam is ample to account for the formation of quartz in pegmatite deposits.

At lower temperatures silica gel is deposited which develops into opal, agate, chalcedony, and other subcrystalline, hydrated phases of silica. In such situations the unstable and "stuffed" *silica O* may also form.

Coesite and *stishovite* are the high-pressure forms of silica. See page 76. The stability relations are shown in Fig. IV-9. Quartz is the only form of silica known to develop under the conditions of regional metamorphism. See also Fig. III-6.

Feldspars are of the general formula WZ_4O_8, showing that the two to one ratio for oxygen to $Z(Z = Si + Al)$ is maintained for all feldspars.

The mineralogical classification and the melting phenomena have been discussed (pages 80 ff.). In metamorphic geology the subsolidus relations become of special importance and will be treated here.

The alkali feldspars, $(K,Na)AlSi_3O_8$, form a series of complete solid solutions at elevated temperatures, but exsolve into perthites at lower temperatures. The exsolution area is shown in Fig. IV-10 and demonstrates that at, say, 500°C the two feldspars in equilibrium, 1 and 2 in the figure, must have the composition 80Or and 3Or respectively. Furthermore, if a mixture of potassium and sodium feldspar grows at 500°, two phases will develop, one of composition 80Or, another of composition 3Or. Consequently, in a rock containing alkali feldspar and no plagioclase, the composition of the two feldspar phases serves as a geological thermometer. However, in most rocks one of the feldspar phases is plagioclase, the other is an alkali feldspar; both phases contain three constitutent "molecules", Or, Ab, and An. At constant temperature and pressure Ab will distribute itself between the two

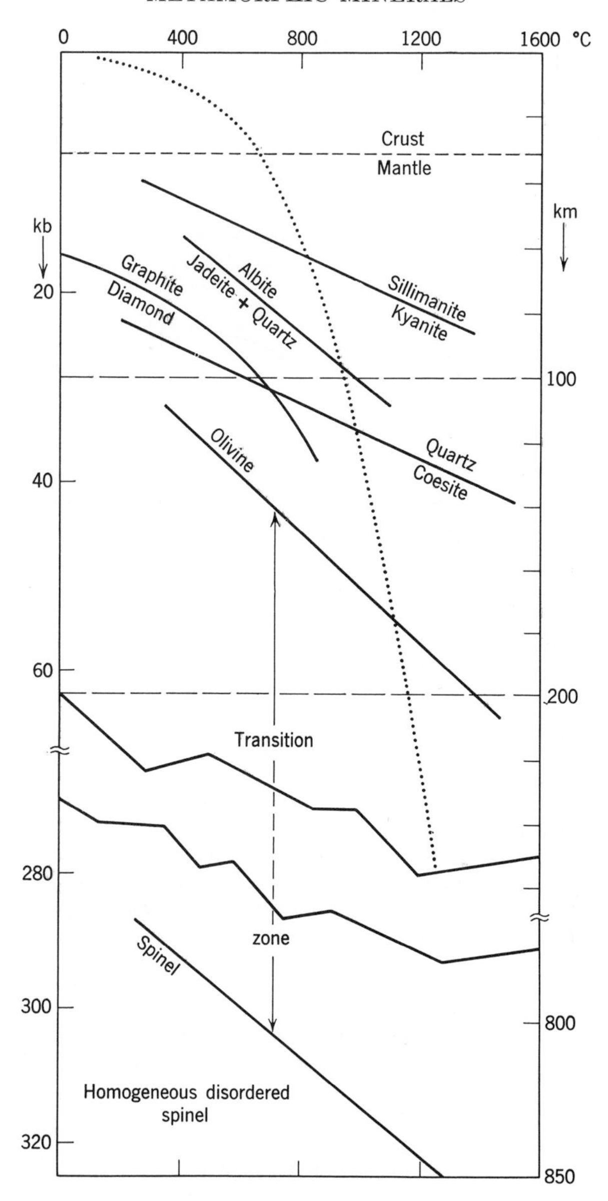

FIG. IV-9. Stability relations of a number of high-pressure minerals. Stishovite becomes stable at a depth of about 400 km, but is probably unable to form by lack of free SiO_2. In the transition zone, extending over a depth range of approximately 500 km Mg-rich olivine coexists with Fe-rich spinel (see Fig. IV-3). The dotted line gives the normal geothermal gradient.

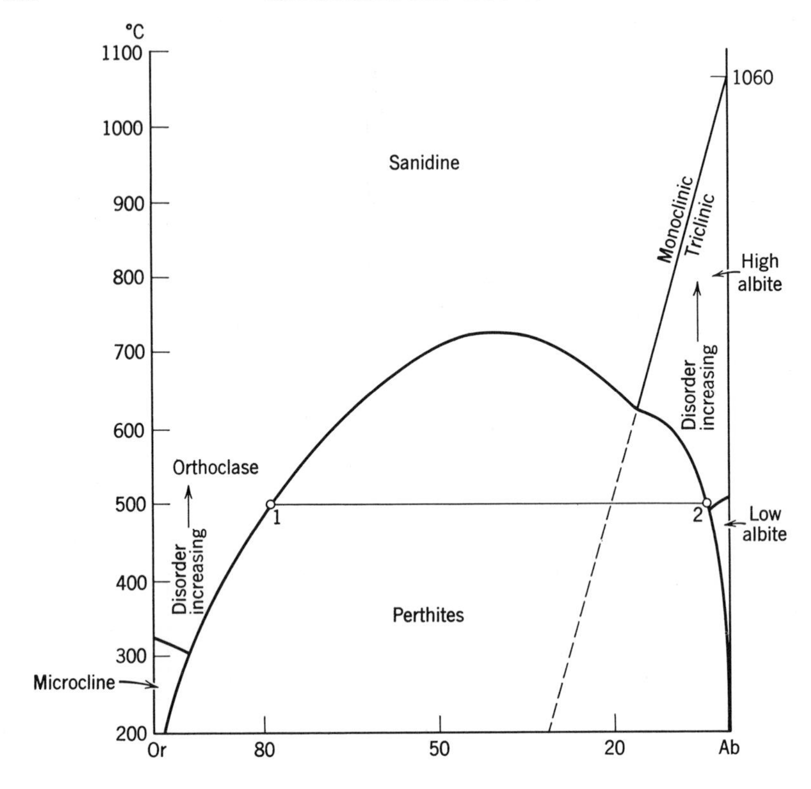

FIG. IV-10. The subsolidus relations in the system of alkali feldspars (Barth, 1959). The low-temperature forms, microcline and low albite are fully ordered with Al and Si in discrete positions. But the high-temperature forms, sanidine and high albite are disordered in regard to the Al/Si positions; this may be symbolized by the formula $K(Al,Si)_4O_8$. The transition order → disorder takes place over a temperature range of approximately 500°C.

phases with a nearly constant ratio of distribution (this holds approximately for rocks of usual chemical composition, Barth, 1956):

$$\frac{\text{Mole fraction of Ab in Or}}{\text{Mole fraction of Ab in An}} = K_{(T,P)}$$

The ratio has a temperature coefficient, that is, K varies from 0.1 to 0.6 with temperatures increasing from 360° to 860°C. Thus the composition of two feldspars coexisting in equilibrium will indicate the temperature of (re)crystallization of the rock. See Fig. IV-11.

Plagioclase feldspars, (Na,Ca)(Al,Si)$_4$O$_8$, were believed to form complete solid solutions at all temperatures. However, a submicroscopic exsolution takes place in the range 5An–20An with formation of *peristerites.* This phenomenon is of importance when plagioclase associates with epidote (v.i.). Plagioclases in the range 20An–70An possess an "intermediate structure" when in low-temperature equilibrium. This structure is generally interpreted in terms of a kind of superlattice or unmixing on a unit-cell scale. At around 300°C and 175 atm water pressure, both microcline and albite will slowly break up into muscovite and paragonite respectively. By weathering they hydrolyze and go into clay minerals or lateritic minerals. See page 35.

Albite is reasonably stable at low temperature, but anorthite and calcic plagioclases will at low temperatures react with the hydrous pore solutions in rocks, or with some other disperse phase, break up, and change into zoisite, scapolite or other reaction products depending on the composition of the rock.

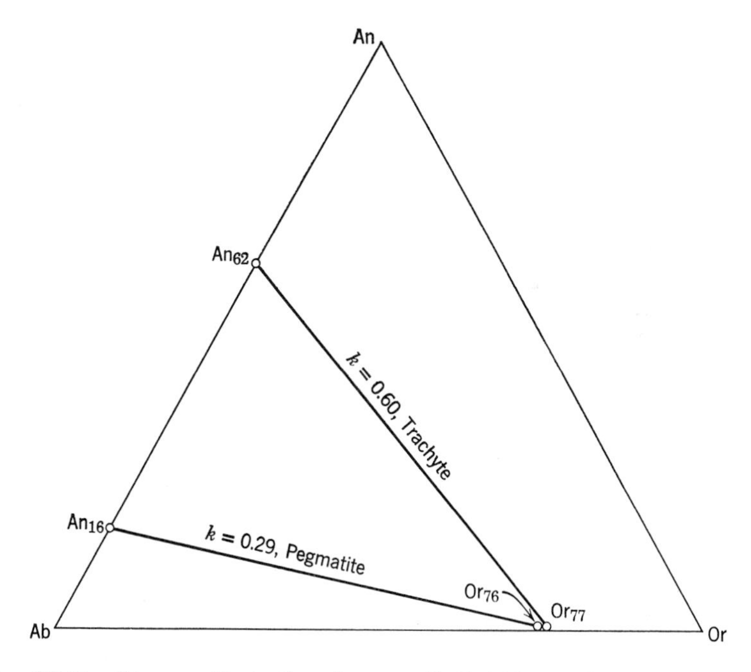

Fig. IV-11. Diagram illustrating that an alkali feldspar of composition Or$_{77}$ coexists with a plagioclase of An$_{62}$ in a trachyte (860°C), whereas an alkali feldspar of the same composition coexists with a plagioclase of An$_{16}$ in a pegmatite (560°C).

Artificially, anorthite can be changed into zeolite according to the equation:

$$\underset{\text{anorthite}}{CaAl_2Si_2O_8} + \underset{\text{water}}{2.4H_2O} = \underset{\text{thomsonite}}{CaAl_2Si_2O_8 \cdot 2.4H_2O} \quad \text{(Goldsmith and Ehlers, 1952)}$$

the temperature of the reaction varies with the water pressure and ranges slightly above 300°C. If excess silica is present other zeolites may form, thus Fyfe et al. (1958) found Ca-analcite as a breakdown product at low partial water pressure and at temperatures slightly above 400°C.

Saussuritization is the alteration of zoisite (or epidote) from anorthite. It is regularly produced by hot residual magmatic solutions acting on their own precipitates. Gabbros, for instance, attacked by their own mother liquors, fall victim to saussuritization (autometamorphism). From anorthite, zoisite may form by the following reaction:

$$\underset{\text{anorthite}}{4CaAl_2Si_2O_8} + \underset{\text{orthoclase}}{KAlSi_3O_8} + \underset{\text{water}}{H_2O} \rightleftharpoons \underset{\text{zoisite}}{2Ca_2Al_3Si_3O_{12} \cdot OH} + \underset{\text{muscovite}}{KAl_3Si_3O_{10} \cdot OH} + \underset{\text{quartz}}{2SiO_2} \quad (3)$$

or

$$\underset{\text{anorthite}}{4CaAl_2Si_2O_8} + \underset{\text{water}}{H_2O} \rightleftharpoons \underset{\text{zoisite}}{2Ca_2Al_3Si_3O_{12} \cdot OH} + \underset{\text{kyanite}}{Al_2SiO_5} + \underset{\text{quartz}}{SiO_2} \quad (4)$$

or, if lime-bearing solutions are present,

$$\underset{\text{anorthite}}{3CaAl_2Si_2O_8} + \underset{\text{lime}}{Ca(OH)_2} \rightleftharpoons \underset{\text{zoisite}}{2Ca_2Al_3Si_3O_{12} \cdot OH} \quad (5)$$

If ferric iron is present, epidote rather than zoisite will form. In this sense zoisite is the low-temperature "modification" of anorthite. Albite has no equivalent low-temperature modification, but is itself stable at low temperatures.

Hydrothermal solutions in contact with anorthite will extract $Ca(OH)_2$. It is probable, therefore, that the pore solution in natural rocks will be alkaline, and that reaction 5, rather than 3 or 4, pictures the true conditions during saussuritization.

Water will act on plagioclase by transforming the anorthite end member into zoisite without affecting albite. Thus calcium is extracted, and the remaining crystal becomes gradually more sodic with decreasing temperature. However, at low temperatures there is no miscibility in plagioclase with less than 20An (peristerite range). At this point, therefore, plagioclase changes discontinuously into almost pure albite. See Fig. IV-12. It is clear, therefore, that albite

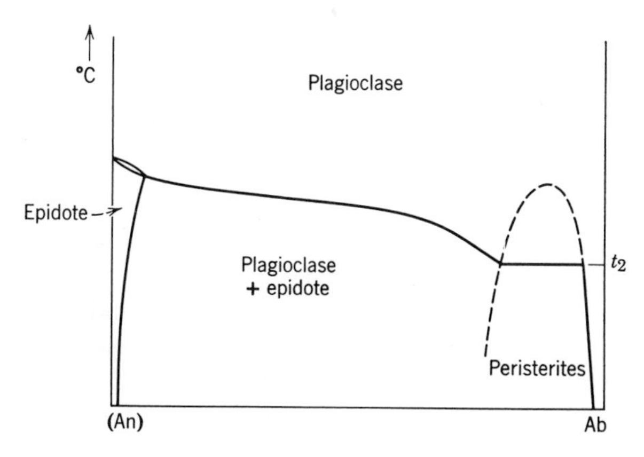

Fig. IV-12. Stability relations of plagioclase and zoisite (epidote) in presence of water vapor pressure (O. H. J. Christie, 1960).

and zoisite (epidote) will be in stable equilibrium over a long range of low temperatures up to t_2. At this temperature (t_2) albite changes abruptly into plagioclase of 20An; thenceforth becomes gradually more calcic as the temperature increases.

The range of temperature, however, is influenced by various other factors. High pressure will favor the formation of epidote; the equilibrium will shift so as to decrease the number of gas molecules (rule of Le Châtelier, page 212). Likewise the high ratio of iron to aluminum favors its formation (because ferric iron enters into the epidote lattice but not into the anorthite lattice). In rocks of common chemical composition and under pressure of around 1000 atm (corresponding to about 3 to 4 km depth) the upper limit of saussuritization may be about 450°C.

Scapolite, mixed crystals essentially of $3(NaAlSi_3O_8)NaCl$ and $3(CaAl_2Si_2O_8)\ CaCO_3$, can be synthesized from the constituents at 800°; typical of limestone contacts, but also found in regionally scapolitized gabbros (ödegaardites from Bamble, S. Norway).

Cancrinite, $(Na,Ca)_{3-6}Al_6Si_6O_{24}(CO_3,H_2O)_{1-3}$, forms inter alia in certain rocks of primary nepheline having suffered a cancrinitization which is analogous to scapolization. Potassium may partly substitute for Na, and (SO_4) and Cl for CO_3.

Zeolites and some other rarer silicates, for example, *datolite*, $CaBSiO_4(OH)$, whose chief mode of occurrence is from hydrothermal solutions in vesicles and fissures of lava rocks, have usually formed through cooperation of emanations and gases emitted by the lavas themselves. Zeolites also form in low-temperature metamorphic rocks under the conditions of the zeolite facies (page 312). Here laumontite, $CaAl_2Si_4O_{12} \cdot 4H_2O$ is usual, also heulandite $CaAl_2Si_6O_{16} \cdot 5H_2O$ may be present.

Most zeolites have the general formula

$$W_{5-12}(Al,Si)_{40}O_{80} \cdot sH_2O$$

W = Na, Ca, K, Sr, Ba; also rarely, Mg, Mn.

Unlike what is usually found in silicates of a common general formula, the various zeolites are not isomorphously related to each other; there are triclinic, monoclinic, rhombic, trigonal, hexagonal, and cubic members. Apophylite $KCa_4(Si_4O_{10})_2F$ is chemically different from all other zeolites in that it contains no alumina.

The framework-like space lattice of zeolites has open holes and passages in which water molecules are accommodated. The water content is a function of temperature; it decreases gradually with rising temperature and increases again upon cooling. Synthetic studies indicate that the temperature of zeolite formation is in the range of 100° to 300°C.

Non-Silicates

In almost all metamorphic rocks silicates are the essential constituents. Among non-silicates *graphite* may occasionally form concentrated deposits, but usually it is derived from metamorphic coal beds; in other places its formation may be referred to carbon introduced by ascending hydrocarbons or carbonyl-bearing gases. *Spinel*, $MgAl_2O_4$, has been mentioned under contact metamorphism; *pleonaste*, containing some Fe^{3+}, is probably the most common variety. *Corundum*, Al_2O_3, is present in contact and regional metamorphic rocks. The most concentrated deposits (for example, on the island Naxos and on other islands in the Aegean Sea, at Smyrna [Asia Minor], and at Chester, Mass.) are the so-called emery deposits, consisting of fine-grained corundum with *magnetite*, Fe_3O_4, *hematite*, Fe_2O_3, and *quartz*.

Hydrargillite, brucite, diaspore, and other hydroxides of iron, manganese, and alumina have been discussed in Part II of this book.

Calcite, $CaCO_3$, *dolomite*, $CaMg(CO_3)_2$, *magnesite*, $MgCO_3$, *siderite*, $FeCO_3$, and *ankerite*, $Ca(Fe,Mg)(CO_3)_2$, are important metamorphic minerals and essential constituents of many rock masses. At elevated temperature calcite takes up an appreciable amount of Mg in solid solutions. This has been used as a geologic thermometer. Marbles (attaining a purity of 99 per cent $CaCO_3$) are often formed by simple recrystallization of sedimentary limestone. Dolomites (dolomite marbles) are usually formed by metasomatism. At elevated temperature, contact-metamorphic and hydrothermal dolomitization takes place. Likewise magnesite rock masses are of metasomatic origin. These carbonates also form constituents of many other metamorphic rocks, but will not be further discussed here (see pages 259 ff.).

Other carbonates (*siderite, malachite,* etc.) as well as sulfates (*gypsum, anhydrite, alunogene,* etc.) are important and may form large parts of the rock in which they occur. They are, however, too special to warrant further description in this book. (See pages 41 ff.)

Still another large group of minerals is of great importance. Sulfides, arsenides, sulfosalts, etc., command special interest as widely distributed and economically important ores. Some of the ores, for instance *pyrite,* FeS_2, *chalcopyrite,* $CuFeS_2$, are present in all kinds of rocks. During contact metamorphism at rather high temperature pyrite changes into *pyrrhotite* (FeS) if the gas pressure of sulfur is low. Other ore minerals are typical for metamorphic or metasomatic deposits: *galena,* PbS, *sphalerite,* ZnS, etc.

Calc Silicate Minerals

Quartz and calcite constitute a stable mineral assemblage in many marbles. However, during high-temperature metamorphism it becomes unstable, for at elevated temperatures these minerals react to form wollastonite and carbon dioxide:

$$\begin{array}{ccccccc} CaCO_3 & + & SiO_2 & \rightleftharpoons & CaSiO_3 & + & CO_2 \\ \text{calcite} & + & \text{quartz} & \rightleftharpoons & \text{wollastonite} & + & \text{gas} \end{array}$$

When this reaction proceeds to the right, a gas molecule is produced; the temperature of the reaction point will, therefore, increase with increasing pressure. At one atmosphere the formation of wollastonite requires a temperature of about 450°C. The change in temperature with pressure was first calculated by Goldschmidt, who used this reaction in his scheme of classification of metamorphic rocks. In Fig. IV-13 the stippled line shows Goldschmidt's curve. It is seen that, when the pressure exceeds 5000 atm (corresponding to about 20 km depth), the temperature of the reaction is almost 900°C. However, Goldschmidt's computation was made under the assumption that the pressure acting on the carbon dioxide is the same as that acting on

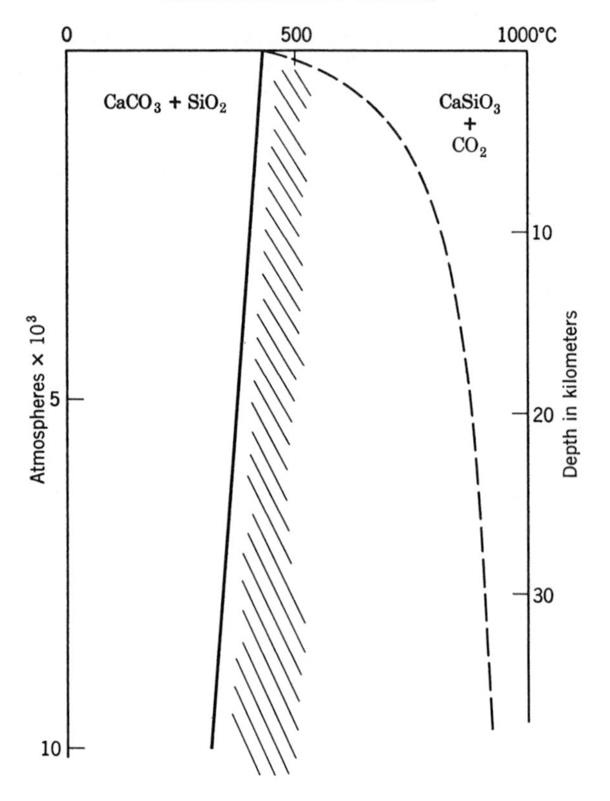

FIG. IV-13. The change of the temperature of the reaction $CaCO_3 + SiO_2 = CaSiO_3 + CO_2$ with pressure. Dashed line: curve after V. M. Goldschmidt (1912), valid only in a closed system of constant pressure. Full line: curve for an open system from which carbon dioxide escapes. In most rocks the conditions will lie close to the full line, probably in the hatched area.

the rock. In most natural situations this is not true. The rock cover is usually semi-permeable to carbon dioxide gas, which therefore escapes from the locale of the reaction and percolates to the surface. If the gas finds easy egress, the molecular volume of the gas can be neglected, and the change in the temperature of the reaction with pressure can be calculated from the Clapeyron formula (equation 19 in the Appendix), taking into consideration only the molecular volumes of the solid phases. The molecular volumes in cubic centimeters are as follows:

$$\begin{array}{ccccc} CaCO_3 + SiO_2 & \rightarrow & CaSiO_3 + CO_2 \uparrow \\ 37 \quad + \quad 22 & \rightarrow & 41 \end{array}$$

and $\frac{dT}{dP} = \frac{\Delta VT}{\Delta H} = \frac{-18 \times 720}{10^6} = -0.013$ degree per atmosphere. This shows that, if carbon dioxide is effectively removed as soon as it forms, the differential pressure will squeeze carbon dioxide out of calcite at a lower temperature than is otherwise possible. Instead of being impeded, the formation of wollastonite will be facilitated by increasing pressure.

If an impure carbonate rock is subjected to slowly increasing temperature at constant pressure, a series of reactions involving progressive elimination of carbon dioxide takes place, each of which is determined by the temperature, the pressure, and the bulk composition of the reacting system. In natural rocks, water and traces of other fugitive compounds are usually present that facilitate the metamorphism and permit certain reactions to take place at low temperatures. The pressure usually has no great influence (see Fig IV-13).

We start with a dolomitic limestone at low temperature composed of the phases calcite + dolomite + quartz (+traces of water, etc.). Dolomite reacts easier than calcite. If CO_2 escapes, the decomposition of pure dolomite can take place at about 375°C. And in the presence of other components capable of reaction with dolomite, decomposition takes place at still lower temperatures.

If the temperature is raised only a moderate amount, talc will form:

$$\underset{\text{dolomite}}{3CaMg(CO_3)_2} + \underset{\text{quartz}}{4SiO_2} + \underset{\text{water}}{H_2O} \rightarrow \underset{\textit{talc}}{Mg_3Si_4O_{10}(OH)_2} + \underset{\text{calcite}}{3CaCO_3} + \underset{\text{gas}}{3CO_2} \quad (1)$$

At higher temperatures tremolite forms:

$$\underset{\text{talc}}{5Mg_3Si_4O_{10}(OH)_2} + \underset{\text{calcite}}{6CaCO_3} + \underset{\text{quartz}}{4SiO_2} \rightarrow \underset{\textit{tremolite}}{3Ca_2Mg_5Si_8O_{22}(OH)_2} + \underset{\text{gas}}{6CO_2 + 2H_2O} \quad (2)$$

The next step is the formation of forsterite:

$$\underset{\text{tremolite}}{Ca_2Mg_5Si_8O_{22}(OH)_2} + \underset{\text{dolomite}}{11CaMg(CO_3)_2} \xrightarrow{250°} \underset{\textit{forsterite}}{8Mg_2SiO_4} + \underset{\text{calcite}}{13CaCO_3} + \underset{\text{gas}}{9CO_2 + H_2O} \quad (3)$$

followed by the formation of diopside:

$$\underset{\text{tremolite}}{Ca_2Mg_5Si_8O_{22}(OH)_2} + \underset{\text{calcite}}{3CaCO_3} + \underset{\text{quartz}}{2SiO_2} \xrightarrow{270°} \underset{\textit{diopside}}{5CaMgSi_2O_6} + \underset{\text{gas}}{CO_2 + H_2O} \quad (4)$$

At still higher temperatures, brucite will form if water is present:

$$\underset{\text{dolomite}}{CaMg(CO_3)_2} + \underset{\text{water}}{H_2O} \longrightarrow \underset{\textit{brucite}}{Mg(OH)_2} + \underset{\text{calcite}}{CaCO_3} + \underset{\text{gas}}{CO_2} \quad (5)$$

This is the highest step attained by siliceous dolomite during regional metamorphism.

$$\underset{\text{calcite}}{CaCO_3} + \underset{\text{quartz}}{SiO_2} \xrightarrow{450°} \underset{\textit{wollastonite}}{CaSiO_3} + \underset{\text{gas}}{CO_2} \quad (6)$$

With steadily increasing temperature, the remaining successive steps are:

$$\underset{\text{dolomite}}{CaMg(CO_3)_2} \longrightarrow \underset{\textit{periclase}}{MgO} + \underset{\text{calcite}}{CaCO_3} + \underset{\text{gas}}{CO_2} \quad (7)$$

$$\underset{\text{diopside}}{CaMgSi_2O_6} + \underset{\text{forsterite}}{Mg_2SiO_4} + \underset{\text{calcite}}{2CaCO_3} \xrightarrow{560°} \underset{\textit{monticellite}}{3CaMgSiO_4} + \underset{\text{gas}}{2CO_2} \quad (8)$$

$$\underset{\text{diopside}}{CaMgSi_2O_6} + \underset{\text{calcite}}{CaCO_3} \xrightarrow{600°} \underset{\textit{akermanite}}{Ca_2MgSi_2O_7} + \underset{\text{gas}}{CO_2} \quad (9)$$

$$\underset{\text{calcite}}{3CaCO_3} + \underset{\text{wollastonite}}{2CaSiO_3} \longrightarrow \underset{\textit{tilleyite}}{Ca_3Si_2O_7 \cdot 2CaCO_3} + \underset{\text{gas}}{CO_2} \quad (10)$$

$$\underset{\text{tilleyite *}}{Ca_3Si_2O_7 \cdot 2CaCO_3} \xrightarrow{650°} \underset{\textit{spurrite}}{2Ca_2SiO_4 \cdot CaCO_3} + \underset{\text{gas}}{CO_2} \quad (11)$$

$$\underset{\text{wollastonite}}{4CaSiO_3} + \underset{\text{spurrite}}{2Ca_2SiO_4 \cdot CaCO_3} \longrightarrow \underset{\textit{rankinite}}{3Ca_3Si_2O_7} + \underset{\text{gas}}{CO_2} \quad (12)$$

$$\underset{\text{akermanite}}{Ca_2MgSi_2O_7} + \underset{\text{calcite}}{CaCO_3} \xrightarrow{750°} \underset{\textit{merwinite}}{Ca_3Mg(SiO_4)_2} + \underset{\text{gas}}{CO_2} \quad (13)$$

$$\underset{\text{rankinite}}{Ca_3Si_2O_7} + \underset{\text{calcite}}{CaCO_3} \xrightarrow{800°} \underset{\textit{larnite}}{2Ca_2SiO_4} + \underset{\text{gas}}{CO_2} \quad (14)$$

* Tilleyite is a rare mineral which is expected to occur between akermanite and spurrite. However, experimental studies by Tuttle and Harker (1957) show that calcite + wollastonite react directly to spurrite.

The production of the mineral phases in progressive metamorphism of siliceous dolomitic limestone has been studied in great detail by N. L. Bowen (1940), and C. E. Tilley (1951). The estimated temperatures of most of the reactions are taken from Bowen's paper. The minerals are listed below in the order of their production at the fourteen steps:

1. Talc
2. Tremolite
3. Forsterite
4. Diopside
5. Brucite
6. Wollastonite
7. Periclase
8. Monticellite
9. Akermanite
10. Tilleyite
11. Spurrite
12. Rankinite
13. Merwinite
14. Larnite

The arrangement is in the order of their production with rising temperature, but Bowen strongly emphasizes that there is nothing inherently of a high-temperature character in the minerals with the high numbers. All indications are that any of these minerals is stable at ordinary temperatures (larnite inverts to a low-temperature form, but the compound Ca_2SiO_4 is stable). It is only with respect to their production in metamorphism of carbonates that these minerals can be thus arranged in an ascending temperature series.

Examinations of natural occurrences show that at the relatively low temperatures of regional metamorphism the diopside stage is rarely passed. At the somewhat higher temperatures of the contact zones about granites, which come from the coolest of magmas, the wollastonite stage may be attained but not passed. It is only at the hotter contacts of syenitic or monzonitic masses that we begin to get members higher in the series; and the highest members are associated only with basic, for the most part basaltic, rocks.

3 · CONTACT METAMORPHISM

This kind of metamorphism is local and restricted and sooner or later . . . will have to be removed by geologists from metamorphism proper and considered apart.

H. H. Read, *Geology,* 1949.

Fundamentally there should be no difference in the assemblages of minerals obtained under similar conditions, whether they result of regional or of contact metamorphism.

H. S. Yoder, *Am. J. Sci.,* Bowen Vol., 1952.

Igneous magma of high temperature may penetrate into sedimentary rocks; it may reach the surface or it may solidify in the form of intrusive bodies. Heat from such bodies spreads into the surrounding sediments; and, since the mineral assemblages of the sediments are adjusted to low temperatures, the heating-up will result in a mineralogical and textural reconstruction known as contact metamorphism.

The width of the thermal aureole of contact metamorphism surrounding igneous bodies varies from almost complete absence in the case of small intrusions (basalt dikes or diabase sills) to several kilometers in the case of large igneous bodies.

The effects produced do not depend only upon the size of the intrusive. Other factors are amount of cover and the closure of the system, composition and texture of the country rock, and, not least important, the abundance of gaseous and hydrothermal magmatic emanations. The heat conductivity of rocks is so low that gases and vaporous emanations become chiefly responsible for the transportation and transfer of heat into the country rock.

Stratified rocks are altered in the contact zone to what is commonly called hornfels or hornstone. They are hardened, often flinty, rocks, usable for road material and so fine-grained that the mineral components can be discerned only with the microscope. Hornfels used to be regarded as "silicified" sediments. Not until Kjerulf, in the later half of the nineteenth century, analyzed sedimentary shale and "silicified" shale of the Oslo region and to his surprise found that, chemically, they were identical (except for H_2O and CO_2) did geologists realize that the "hardening" of the shale took place without appreciable change in the chemical composition. Kjerulf summarized his results by saying that the composition of a hornfels depended on the original sediment (the shale) and was independent of the kind of adjacent igneous rock.

Rosenbusch later arrived at the same conclusion (studies in Alsace) and pronounced that no chemical alterations accompany the formation of hornfelses except for the removal of fugitive constituents. The Kjerulf-Rosenbusch rule is useful but needs modification, in that chemical changes may ensue from hydrothermal and pneumatolytic action.

The next problem, then, is to see how the mineral assemblages of the hornfelses depend upon the chemical composition of the original sediments. An approximate solution was given by V. M. Goldschmidt in his great treatise, "Kontaktmetamorphose des Kristiania-Gebietes," published in 1912.

The chief types of sedimentary rocks are sand, clay, and limestone. Among the varieties of hornfelses which may develop from different mixtures of these components, the continuous series from shale to limestone is the most interesting.

Most shales contain some iron and magnesia-bearing constituents in addition of feldspar and clay, and quartz is always admixed. Consequently, sufficient SiO_2 is usually present in the hornfelses to form highly silicified minerals. Other than SiO_2, we thus have the following four chief chemical constituents: Al_2O_3, CaO, FeO, and MgO. As usual in petrology we group the last two together and make a triangular plot, as shown in Fig IV-14. We thus have a system of three components: alumina, lime, and ferromagnesia.

Applying now the mineralogical phase rule, announcing that the number of stable minerals in a rock shall not be larger than the number of components, we readily see that, except for quartz (and some alkali-bearing minerals, see below), no more than three additional minerals should occur in any one of these hornfelses. Observations have verified this: from alumina, lime, and ferromagnesia, seven

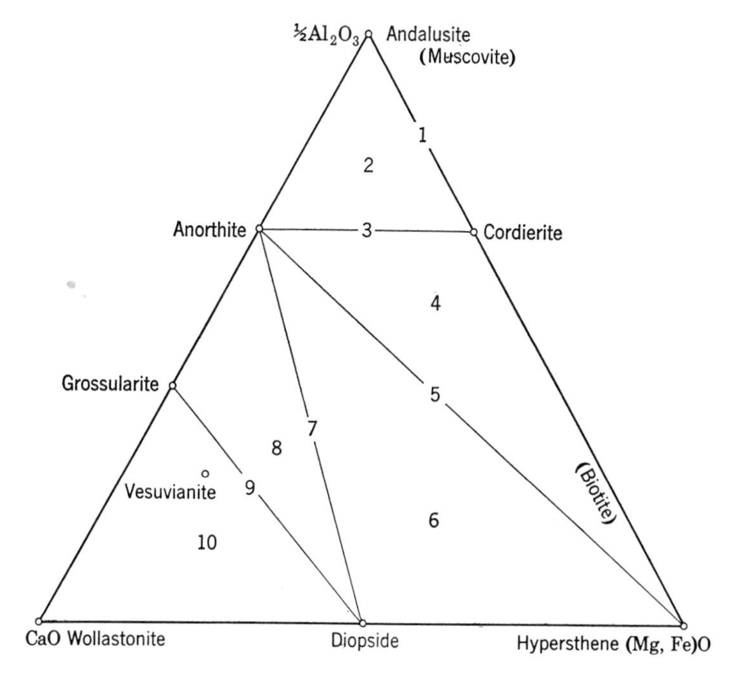

Fig. IV-14. Diagram illustrating the mineral assemblages of the ten Goldschmidt classes of hornfelses. Other mineral constituents, in addition to those shown in the diagram, are quartz, orthoclase, and accessories like apatite, etc.

minerals will form that are stable under the conditions of contact metamorphism: *andalusite,* Al_2SiO_5; *cordierite,* $Mg_2Al_4Si_5O_{18}$; *anorthite,* $CaAl_2Si_2O_8$; *hypersthene,* $(Mg,Fe)SiO_3$; *diposide,* $Ca(Mg,Fe)$-Si_2O_6; *grossularite,* $Ca_3Al_2Si_3O_{12}$, *wollastonite,* $CaSiO_3$. But only three (or less) of these minerals can occur together. In this way different mineral combinations develop, each combination (plus quartz and an alkali-bearing mineral) representing a natural hornfels. Figure IV-14 shows that there are ten such combinations, corresponding, according to Goldschmidt's terminology, to hornfelses of classes 1 to 10.

Class 1. Pure shale contains, in addition to quartz, alumina and magnesia which form andalusite and cordierite.

Class 2. A small addition of lime produces anorthite (in the actual hornfels represented by plagioclase, since some Na_2O is present) in addition to andalusite and cordierite.

Class 3. More lime produces more anorthite, whereas andalusite disappears.

Class 4. Next hypersthene will form in addition to cordierite and anorthite.

Class 5. With still more lime cordierite disappears, with formation of more anorthite and hypersthene.

Class 6. Lime is now so dominating that it combines with magnesia to form diopside in addition to anorthite and hypersthene.

Class 7. Hypersthene disappears.

Class 8. More lime reacts with anorthite to form grossularite in addition to diopside.

Class 9. Anorthite is completely used up in the grossularite reaction.

Class 10. Pure lime silicate, that is, wollastonite occurs with diopside and grossularite.

Variations from this scheme are easily explained. Some of the Oslo shales carry a good deal of feldspar. The primary alkali content will appear in an additional mineral phase; the original microcline may recrystallize into orthoclase, but usually enough water is present to make a mica. As seen in Fig. IV-14, the chemical composition of biotite, $KMg_3AlSi_3O_{10}(OH,F)_2$, makes it possible for it to substitute for hypersthene and develop in hornfelses of classes 3, 4, 5, and 6. Presence of biotite in these rocks is revealed by a characteristic chocolate-brown color. In potash-bearing hornfelses of classes 1, 2, and 3 muscovite will appear and partly or completely substitute for andalusite. In hornfelses of class 10 some lime-rich hydrous silicates may develop, for example, vesuvianite (idocrase), $Ca_{10}(Mg,Fe)_2Al_4Si_9O_{34}(OH)_4$, and the presence of ferric iron will produce andradite, $Ca_3Fe_2Si_3O_{12}$, which will form mixed crystals with grossularite.

In a great many places, silica-rich pelitic sediments alternate with calcareous sediments. By subsequent contact metamorphism hornfelses will develop that are rather similar to those of the Oslo region. In certain regions, however, limestone is missing as a component in the sediments, and the corresponding hornfelses will be poor in lime, with a composition approaching the ternary system SiO_2–Al_2O_3–(Mg,Fe)O. Figure IV-15 shows the mineral associations encountered in such hornfelses. Each of the seven triangular areas of Fig IV-15 corresponds to a possible mineral assemblage, two of them, quartz cordierite-andalusite and quartz-cordierite-hypersthene, being representatives of the usual silica-rich pelitic hornfelses, the others occurring in hornfelses of, for instance, the Hartz Mountains, Germany, of the Combrie area, Perthshire, Scotland, and elsewhere.

Sapphirine (synthetic $Mg_4Al_{10}Si_2O_{23}$, with MgSi substituting for Al_2) will develop only under special conditions (contact at Peekskill, New York). *Periclase,* MgO, develops during high-temperature con-

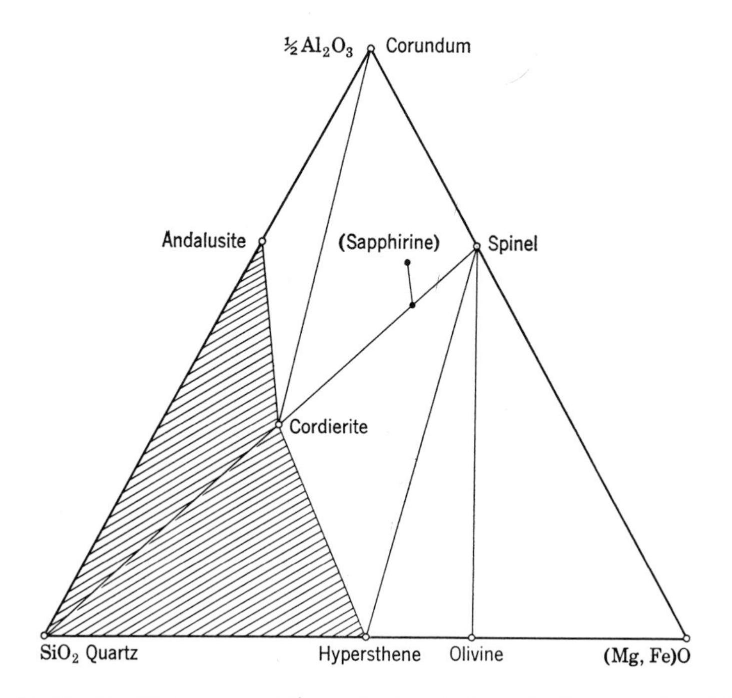

Fig. IV-15. Equilibrium assemblages of mineral phases in lime-poor hornfelses. Shaded area: field of normal pelitic sediments. (After C. E. Tilley, 1923.)

tact metamorphism and especially through dedolomitization. Some dedolomitized rocks have been called pencatite (calcite and brucite in equal-molecular proportions), predazzite for varieties richer in calcite, and, generally, brucite marble. Brucite or, more frequently, *ferro brucite* $(Mg,Fe)(OH)_2$ usually substitutes for periclase.

The natural hornfelses very closely follow the theoretical scheme of mineral associations, showing incontestably that a state of chemical equilibrium was approached in the rocks and that the laws of physical chemistry may be applied to metamorphic processes.

Nature, however, is always more complicated than theory. Although the thermochemical laws are never to be neglected and, indeed, are the only adequate guides in our research, the conditions of their validity should always be carefully investigated. Non-adjustment of equilibrium does occur in connection with special geological conditions. In contact with effusive rock or with small basic sills or dikes, that is, in places where high temperature prevailed for a short time only and cooling was rapid, vitrification, that is, melting of argillaceous or arenaceous sediments, may occur, the effect being restricted to a few centimeters from the contact (*buchites* are alumina-rich glasses). In the same situation quartz may convert into tridymite, and impure calcareous sediments may yield anomalous mineral assemblages and rare mineral species, some of which are to be regarded as metastable forms and some as special high-temperature forms. Examples are *melilite* and *fassaite* (aluminous pyroxene) from contact-metamorphic dolomite of Monzoni, Tirol, and the remarkable association of pleonaste and melilite with

merwinite	$Ca_3Mg(SiO_4)_2$
larnite	Ca_2SiO_4
scawtite	$Ca_4Si_3O_8(CO_3)_2$
spurrite	$2Ca_2SiO_4 \cdot CaCO_3$
rankinite	$Ca_3Si_2O_7$

in metamorphic chalk in contact with dolerite at Scawt Hill, Antrim (Tilley, 1942). In similar assemblages *monticellite*, $CaMgSiO_4$, and other rare calcium silicates or unusual minerals like *perovskite*, $CaTiO_3$, may be encountered.

Metasomatism in contact metamorphism is a factor of importance and will be discussed here, although the general discussion of metasomatism and its role in regional metamorphism will be deferred until we have dealt with the effects of normal metamorphism in different classes of rocks and studied the general laws governing the metamorphic processes.

Igneous intrusion is regularly accompanied by a magmatic gas phase of high temperature and high penetrating power supplying the surrounding rocks not only with heat but also with water and other volatile compounds, resulting in the so-called pneumatolytic or hydrothermal contact metamorphism.

Thus the chemical composition of the surrounding rocks is changed, and the Kjerulf-Rosenbusch rule does not hold. In most hornfelses the effect is seen for a few feet from the contact, and in places large rock masses are affected.

The primary magmatic gases are acid (page 154) and show in consequence high reactivity. If the contact rock is basic, especially limestone, the acid gases will react effectively with it. Limestone, therefore, acts as a filter capturing the escaping gases, with the formation of a great variety of reaction minerals; the corresponding rocks are known as skarns. Reaction rocks at the contact of limestone and composed of lime silicates form mainly garnet and pyroxene, often accompanied by fluorite and phlogopite, and with sulfides of iron, zinc, lead, or copper; in other occurrences magnetite is formed.

Magmatic gases may contain heavy metals (see pages 150 ff.). The metals are then captured by the limestone and retained in the skarn rock.

Tactite has been proposed as a general term for pneumatolytic contact-metamorphic rock. Special terms are garnetite, porcellanite, etc.

Contact metasomatic deposits include a number of important deposits of garnet, emery, and graphite, and metalliferous deposits, as Banat in Hungary, copper ores of Utah, Arizona, New Mexico, and Mexico. Zinc ores of Hanover, N. H., Trepca, Yugoslavia, Broken Hill, Australia; and others.

The following equations illustrate what happens when gases rich in iron fluoride meet with limestone:

$$2FeF_3 + 3CaCO_3 = Fe_2O_3 + 3CaF_2 + 3CO_2$$

Hematite and fluorite are formed.

If silica is present in the limestone, andradite garnet will form instead of hematite:

$$2FeF_3 + 3SiO_2 + 6CaCO_3 = Ca_3Fe_2Si_3O_{12} + 3CaF_2 + 6CO_2$$

Similar equations can be drawn for the formation of many other of the pneumatolytic minerals listed in Table IV-2.

Of economic value are many ore deposits connected with the intruding magma. The genetic association of specific kinds of mineralization with specific kinds of igneous rocks presents a complex and intricate problem. It depends upon the distribution of the mineralizing elements in the various types of igneous rocks, and in the various phases (solid, liquid, gaseous) formed by the magma. It is a geochemical problem, affecting not only (1) contact-metamorphic deposits *sensu stricto,* but also (2) deposits in or near roof pendants of older rocks surrounded by intrusive masses, (3) lodes in the periphery of intrusives, and (4) normal zonal distribution of mineralization with respect to igneous masses.

TABLE IV-2

SUBSTANCES CONCENTRATED IN LIMESTONE BY METASOMATIC CONTACT METAMORPHISM IN THE OSLO REGION

Metals	Minerals
Fe	Andradite, hedenbergite, oxidic and sulfidic iron ore
Zn, Cu, Pb	Sphalerite ZnS, chalcopyrite $CuFeS_2$, galena PbS
Mn	Andradite, hedenbergite, rhodonite $(Mn,Fe,Ca)SiO_3$
Bi, Ag	Bismuthinite Bi_2S_3, galena, sphalerite
Mo, W	Molybdenite MoS_2, scheelite $CaWO_4$
Co, As, Sb	Cobaltite CoAsS, arsenopyrite FeAsS, bismuthinite
Be, Ce	Helvite $(Mn,Fe,Zn)_4Be_3Si_3O_{12}\cdot S$, vesuvianite, allanite $(Ca,Fe)_2(Ce,Al,Fe)_3Si_3O_{12}OH$
Metalloids	
Si	Silicates in skarn, quartz
F, Cl, S	Fluorite CaF_2, scapolite, * sulfidic ore
B, P, Ti	Axinite, † apatite, ‡ sphene §

* Scapolite is a mixed crystal of many components, chiefly $3NaAlSi_3O_8\cdot NaCl$ and $3CaAl_2Si_2O_8\cdot CaCO_3$, with some (OH) replacing Cl (and CO_3).

† Axinite is $Ca_2(Mn,Fe)Al_2BSi_4O_{15}OH$, but sometimes considerably richer in Ca; may contain appreciable Na_2O.

‡ Apatite is $Ca_5(PO_4)_3F$, but may contain some Cl and OH replacing F.

§ Sphene, $CaTiSiO_4(O,OH,F)$.

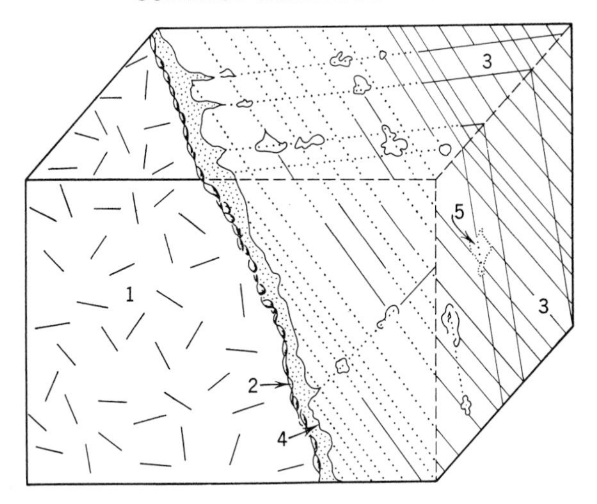

FIG. IV-16. Skarn rocks and pneumatolytic ore at Aranzazu, Mexico. (1) Granodiorite; (2) garnetized border of granodiorite (exaggerated); (3) limestone; (4) garnet rock at immediate contact carrying some ore; (5) bodies of andradite-wollastonite-copper ore localized along intersections of fissures and bedding planes. (After A. Knopf, *Ore Deposits as Related to Structural Features,* edited by Newhouse, Princeton University Press, 1942.)

The commonest pneumatolytic minerals are listed in Table IV-2. Other interesting minerals are those of the humite group:

Nordbergite	$Mg_2SiO_4 \cdot Mg(OH,F)_2$
Chondrodite	$2Mg_2SiO_4 \cdot Mg(OH,F)_2$
Humite	$3Mg_2SiO_4 \cdot Mg(OH,F)_2$
Clinohumite	$4Mg_2SiO_4 \cdot Mg(OH,F)_2$

Danburite, $CaB_2Si_2O_8$, has been found in dolomite contacts.

By pneumatolysis of shale, tourmaline and various aluminosilicates with fluoride may develop:

Topaz	$Al_2SiO_4(OH,F)_2$
Micas:	
Muscovite	$KAl_3Si_3O_{10}(OH,F)_2$
Lepidolite	$K(Li,Al)_3(Si,Al)_4O_{10}(OH,F)_2$
Zinnwaldite	$K(Li,Fe,Al)_3(Si,Al)_4O_{10}(OH,F)_2$
Phlogopite	$KMg_3AlSi_3O_{10}(OH,F)_2$

Sometimes intense boron pneumatolysis may produce datolite, axinite, and rare minerals like kotoite, $Mg_3B_2O_6$, fluoborite, $Mg_3BO_3(F,OH)_3$, ludwigite, $(Mg,Fe'')_2Fe'''BO_5$, and others.

The conclusion is that limestone is especially susceptible to pneumatolytic contact metamorphism, with formation of skarn rocks and often with useful ore deposits. The minerals often develop in perfect crystals, and the deposits belong to the best-known mineral occurrences in the world (Franklin Furnace, Clifton-Morenci, Auerbach, Berggieshübel, Banat, Concepción del Oro, etc.). Silicate rocks are not usually as intensively altered, but by introduction of fluorine and lithium, and the formation of greisen, feldspars may change into topaz, zinnwaldite, or other micas. Tin is sometimes introduced, forming cassiterite, SnO_2. Other introduced minerals in argillites are, for instance, molybdenite, apatite, or beryl.

To sum up the important points: Metamorphism in contacts of deep-seated rocks is very common, and the products of metamorphism (disregarding the pneumatolytic action) will vary regularly in accordance with the chemical composition of the pre-existing contact rock. Another factor of decisive importance for the mineral development is the temperature. It varies with the nature of the intruding rock and with the distance from the contact. Thus we may distinguish between an inner and an outer contact zone. They grade into each other by imperceptible transitions, but the mineral associations in the typical inner contact zone—and this is the only zone we have considered so far—are markedly different from the associations in the outer contact zones.

These problems will not be discussed further here. The question of the general relationships between the minerals and mineral associations, on the one hand, and temperature and pressure, on the other, is the real core of the study of metamorphic rocks. It will be discussed on a broader basis in connection with the facies principle and the general process of regional metamorphism. Although typical contact metamorphism may appear to be well defined and geologically seems to stand out as an isolated natural phenomenon, a closer study reveals that it is complex and variegated and that by gradual transitions it passes into other kinds of metamorphism. Its proper place in geology cannot be appreciated fully if it is treated as an isolated phenomenon. It should be considered in connection with, and as a part of, the general system of rock metamorphism and metasomatism.

4 · STRUCTURE OF METAMORPHIC ROCKS

Such phenomena as cleavage, foliation, drag folding, etc., can be readily observed in the field and should be used as a tool by every working geologist. . . . They have been utilized with remarkable results not only in pure scientific geology, but also in economic geology where both the geologist's reputation and large sums of money depend upon the correct elucidation of the structure.

Gilbert Wilson, *Proc. Geol. Assoc.,* 1946.

The regional metamorphism affects large rock volumes. Its typical locale is at some depth in active orogenes. Both temperature and pressure are elevated, and shearing forces make the rocks creep and flow keeping them in constant agitation, and in a steady state of differential movement. Such processes in rocks cannot be directly observed in nature, nor can they be adequately investigated by experiments. For this reason relict structures become of special importance in the study of metamorphism. They show us the sequence of events. The study of transitional rock series from unmetamorphic to completely altered rocks is the basis on which the principles of metamorphic geology rest.

Relict Structures

Mineral relics often indicate the temperature and pressure that are obtained in the pre-existing rock. They appear whenever a rock has been successively subjected to various physical conditions. If a mineral, stable in the earlier facies, is also stable in the later facies, it will, unless stress action sets in, be preserved in its original form and present a *stable* relic. This is a very common case, for example, quartz in various rocks, feldspar in amphibolites. But when a mineral, or a definite association of minerals, becomes unstable, it may yet escape alteration and appear as an *unstable* relic.

The relics are *proterogenic,* that is, representative of an earlier, premetamorphic rock, or of an earlier stage of the metamorphism. *Hysterogenic* products are of later date, and are formed in consequence of changed conditions after the formation of the chief metamorphic minerals.

Discrimination between the stable and unstable relics throws light on many common and important petrographical phenomena. We may here mention uralitization (alteration of pyroxene to hornblende) as an example. It has been assumed that the pyroxenes will be converted into amphiboles by metamorphism. According to the facies conception, this statement must be much restricted. It is true that uralitization may take place when pyroxene-bearing rocks are brought under the conditions of the amphibolite facies (which is the name proposed later for a certain facies). But even in the rocks of amphibolite facies diopside may be a stable constituent in those rocks in which femic lime is present in excess over the hornblende ratio. In such a rock, diopside is seemingly in a process of alteration into hornblende; yet the diopside is not an unstable mineral. It is only unstable in the presence of excess of magnesia.

Let us consider a gabbro under the conditions of the amphibolite facies. Then two cases are possible. Either the gabbro has a composition at which the diopside is stable in the amphibolite facies (a rather rare instance), or there are, in the rock, ferrous oxide and magnesia in excess of the hornblende ratio. In the former case the

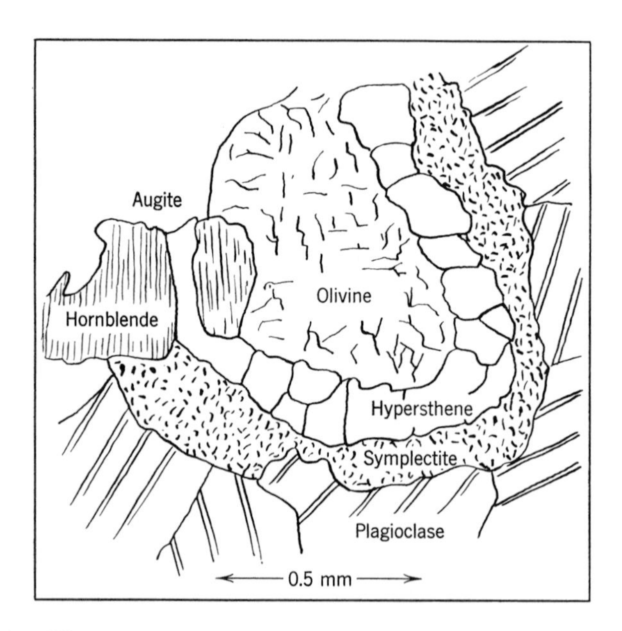

FIG. IV-17. Olivine with a corona of hypersthene and spinel. (Symplectite in the drawing is a microscopical intergrowth of spinel and hypersthene.) Olivine in the present rock is unstable in contact with alumina-rich hydrothermal solutions and reacts metasomatically according to the following scheme:

$$(Mg,Fe)_2SiO_4 + Al_2O_3 = (Mg,Fe)SiO_3 + MgAl_2O_4$$
$$3 \text{ olivine} + 2 \text{ alumina} = 2 \text{ hypersthene} + 3 \text{ spinel}$$

From an olivine gabbro, Seiland, northern Norway.

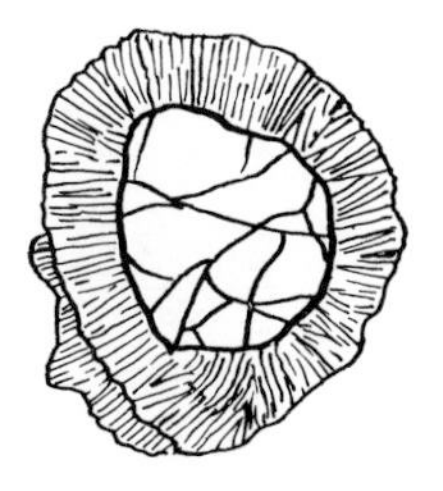

FIGS. IV-18 and IV-19. *Left:* Uralite inset, with hornblende cleavage and augite contours showing (010), (110), (100). *Right:* Garnet porphyroblast, with a reaction rim of hornblende and chlorite. Regardless of the chemical composition of the reaction rim, it is referred to as kelyphite.

original diopside of the gabbro will preserve itself as a stable relic; in the latter case it will be uralitized. If, nevertheless, we find diopside in such a rock, it is an unstable relic. Such gabbroid rocks are not rare. In the Fennoscandian Archean large areas of gabbro have been completely uralitized or had already crystallized primarily in the same facies.

A common phenomenon, fairly illustrative of the tendency towards equilibria, is the formation of armors around such minerals, which have become unstable in their association but have not been brought beyond their fields of existence in general (the *armored relics*). Thereby the associations of minerals in actual contact with one another become really stable. If, however, the constituent minerals of a rock containing armored relics are named without noting this phenomenon, it may be taken as an unstable association.

Many phenomena at the contacts of two minerals, such as coronas and reaction rims, are such armors (comprehensively described by J. J. Sederholm, 1916). See Figs. IV-17, 18, 19.

Structure relics are perhaps of still more importance, directly indicating the nature of the pre-existing rock and the mechanism of the metamorphic deformation. The interpretation of relics was compared by Sederholm to the reading of palimpsests, parchments used for the second time after original writing was nearly erased. Every trace of original structure is important in attempting to reconstruct the history of the rock and in analyzing the causes of its metamorphism.

In sedimentary rocks the most important structure is *bedding* (stratification, layering) which originally was approximately horizontal. In metamorphic rocks deformed by folding, faulting, or other dislocations, the sum of all deformations can be referred to the original horizontal plane, and the deformations can be analyzed (Fig. IV-20).

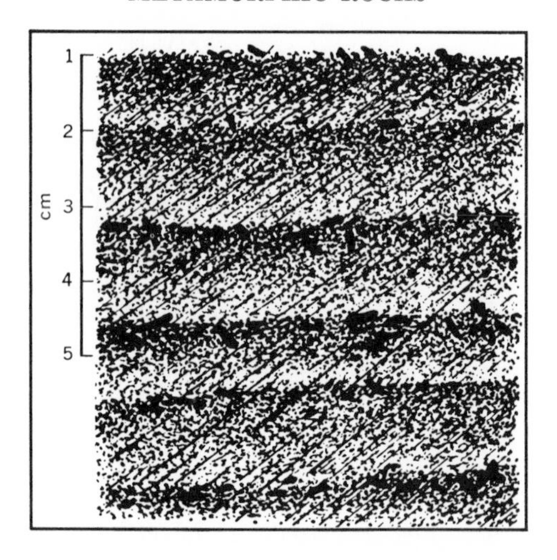

Fig. IV-20. Relict banding in varved staurolite-mica schist. The staurolite porphyroblasts preferably develop in the clayey darker bands of the varves. A secondary transversal schistosity has not been able to obliterate the primary banding. (From Suistamo, east Finland, after Eskola, 1939.)

Helicitic structures are seen in recrystallized rocks in which the direction of the original layering is reflected in lines of inclusions that follow the original rock structure and are preserved within new crystals that have grown astride the layering. The helicitic structures often reveal that certain porphyroblasts (garnets, for example) have rotated during recrystallization. (Fig. IV-21.)

In metamorphites derived from coarse-grained sediments—sand or

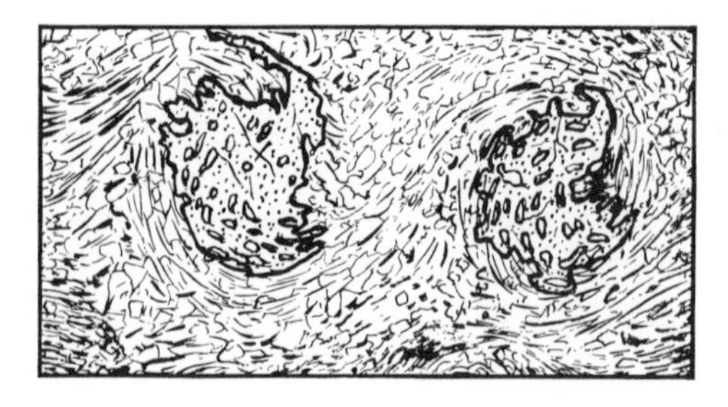

Fig. IV-21. Garnetiferous phyllite. Crystalloblasts of garnet about 2 mm long, with helicitic inclusions arranged in curves, showing that the garnets rotated during their growth. Groundmass essentially of quartz and muscovite with some biotite, apatite, and magnetite. Near Harlem, Dutchess County, New York.

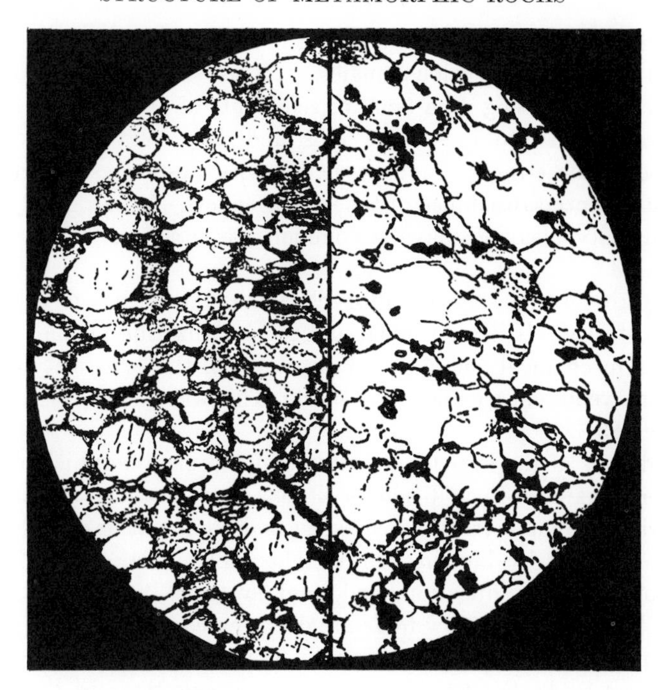

Fig. IV-22. Blastopsammitic structure in a Silurian graywacke (×35). The left half shows the rock but slightly recrystallized with easily recognizable clastic quartz grains, kaolinite metastasis, and graphite dust. The right half shows the same rock after further recrystallization; the quartz grains are deformed, but the sandstone structure is still perceptible; kaolinite, recrystallized muscovite, graphite, and patches of limonite are evenly distributed. (From Saxony, Germany, after K. H. Scheumann, 1925.)

conglomerate—the outlines of the coarser grains or pebbles are often clearly seen. The resultant relict fabric is called *blastopsammitic* or *blastopsephitic*, respectively. See Fig. IV-22.

Fabric relics may also be inherited from igneous rocks. Blastoporphyritic structures are particularly persistent, but should be carefully distinguished from porphyroblastic structures of purely metamorphic origin (see page 286).

Fissility and Schistosity

One of the earliest secondary structures to develop in sediments of low metamorphic grade is the *slaty cleavage* (also referred to as *flow cleavage* or *fissility*), which grades into *schistosity* (different from fracture cleavage, strain-slip cleavage, etc.). It is developed normal to the direction of greatest shortening of the rock mass, and cuts the original bedding at various angles. (It is shown in Fig. IV-20.) Tectonic forces acting on a book of sediments of heterogeneous layers will throw them into a series of folds, and slaty cleavage develops in response to the stresses imposed on the rock system as a whole due to the differential resistance of the several layers. Consequently, folding and slaty cleavage have a common parentage, as illustrated in Fig. IV-23.

In the rock series slate-phyllite-schist the slaty cleavage will grade into schistosity. It is a chemical and recrystallization phenomenon as well as a mechanical one, and the directions of the schistosity become the main avenues of chemical transport. This will be discussed further on page 285.

The details of the structural features of the metamorphic rocks are very complicated. In addition to simple slaty cleavage grading into schistosity, we have fracture cleavage and combinations of the two; we have folding by bending, shear folds, flow folds, drag folds and their combinations, lineation, and joint systems.

The classic areas for flow folds are deep-seated metamorphic regions like Canada and Scandinavia, or regions underlain by easily flowing salt deposits. As the temperature-pressure conditions in the earth's crust gradually change from deeper to higher levels, so the resulting types of folding must grade into each other and defy, therefore, a schematical classification. But the geographical extent of fold systems within the same geological units has been followed continuously for thousands of miles with more or less constant trend. For example in the Appalachian Mountains the fold system trends NE–SW as a whole, but several bends are obvious. Again such bends are well known in the western Alps or in the Carpathian Mountain system.

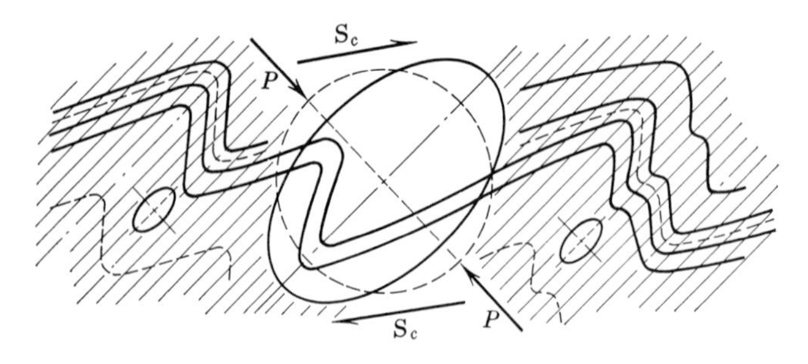

Fig. IV-23. The general relationship between deformation folding and slatey cleavage caused by pressure *PP* or the couple *Sc Sc*. (After G. Wilson, *Proc. Geol. Assoc.*, 1948.)

TABLE IV-3

STRUCTURAL ELEMENTS IN METAMORPHIC SEDIMENTARY ROCKS

(After E. Cloos, *Maryland Geol. Survey*, Vol. 13, 1937)

1. Planar structures
   - *a.* Beddings, stratification
   - *b.* Axial planes of folds
   - *c.* Cleavage
     - Flow cleavages
     - Fracture cleavages
   - *d.* Fractures and joints
     - Cross joints
     - Longitudinal joints
     - Oblique joints
2. Linear structures
   - *a.* Axes of folding
   - *b.* Intersection of bedding and flow cleavage
   - *c.* Intersection of bedding, flow cleavage, and fracture cleavage
   - *d.* Stretching parallel to the axis of folding (*b*)
     - Stretched conglomerate, tourmaline, hornblende, volcanic blebs
   - *e.* Slippage on bedding planes normal to the axis (*b*)
     - Striation on bedding plane, volcanic blebs
   - *f.* Slippage on cleavage planes

Table IV-3 summarizes the structural elements encountered in sedimentary rocks. More detailed information on these interesting features is given in many excellent textbooks on structural geology.

In some ways such studies may be considered introductory to the more abstruse subject of petrofabrics (rock fabric, structural petrology, "Gefügekunde"), which is rapidly becoming recognized as one important tool in the investigation of complex areas. In this branch of geology a formidable terminology has developed; and the results, which for many areas have proved to be exceedingly valuable, yielding information otherwise not available, are still in some cases debatable. A complete presentation of the methods and theories is far beyond the scope of this book. For special studies of both structural geology and structural petrology, a list of references is given at the end (page 365).

Mechanism of Deformation of Rocks

Rocks may support, without permanent deformation, shear stresses of short duration, and yet yield continually when subject to smaller stresses for a long period of time. This behavior is explained in part by applying the theory of dimensional analysis as was done in Part I, Section 4, to the crust of the earth as a whole. For the deformation of rocks, the following factors are important: (1) confining pressure, (2) shear stress, (3) temperature, (4) time, and (5) the presence of solutions. The behavior of rocks under stress may be divided into three parts: (1) *Elastic deformation,* practically instantaneously achieved and as rapidly reversible on removal of the external forces acting on the rock. (2) *Flow,* which is that part of the deformation in excess of the elastic deformation but does not result in a notable loss of cohesion. It is characterized by change of shape as a function of time at constant stress. (3) *Rupture,* that part of the deformation characterized by loss of cohesion. Frequently, flow grades into rupture, with a progressive loss of cohesion, until complete separation occurs.

Flow may be classified under two heads: (1) viscous flow, which is simple and has been well defined by physicists; (2) plastic flow, which is complicated and covers a multitude of different behaviors, some of which are imperfectly understood.

In compact, crystalline rock no viscous flow has been detected. D. Griggs (1940) classified the mechanism of the flow phenomena in crystalline aggregates as follows:

1. Cataclasis—deformation characterized by mechanical granulation.
2. Gliding—intragranular twin and translation-gliding dominant, with secondary intergranular adjustment consequent to the change of grain shape resulting from intergranular movement.
3. Recrystallization—deformation by molecular rearrangement through the medium of solutions, local melting or solid diffusion, intergranular motion dominant.

Under the influence of shearing stress a crystal will deform plastically, but, as distinct from viscous flow in a liquid, gradually larger stresses are required to maintain a constant rate of deformation. This phenomenon is called *hardening.* During metamorphic recrystallization rocks are able to yield under very small stresses of long duration. If the stresses developed within the crystal exceed the elastic limit, permanent deformation (plastic strain) ensues, and the accompanying differential movement of the particles manifests as plastic flow. Physical properties like *brittleness* or *ductility* are not absolute, but are determined by a variety of factors (temperature, pressure, duration of deformation, etc.). Pure quartz is brittle and fails to yield plastically even at very high pressures, but quartz-sand moistened by a weak solution of Na_2CO_3 yields easily at pressures from 300 to 3000 atm.

Creep may be defined as the relaxation that takes place in strained crystals below the yield value; *recovery* is the relaxation in hardened crystals below the yield value. Variation in temperature and confining pressure has little effect on the critical stress necessary for plastic flow (*fundamental strength*, see Fig. IV-24), but flow may be greatly increased by even moderate degrees of isomorphous replacement of one type of ion by another in the space lattice.

Let us consider the kinds of movement listed above.

1. *Cataclasis* is easily understood. Plastic strain cannot proceed indefinitely in a crystal; when the *ultimate strength* of the material is reached, rupture will result. It may be either greater in magnitude or less than the elastic limit. In this way large rock units are moved in relation to each other.

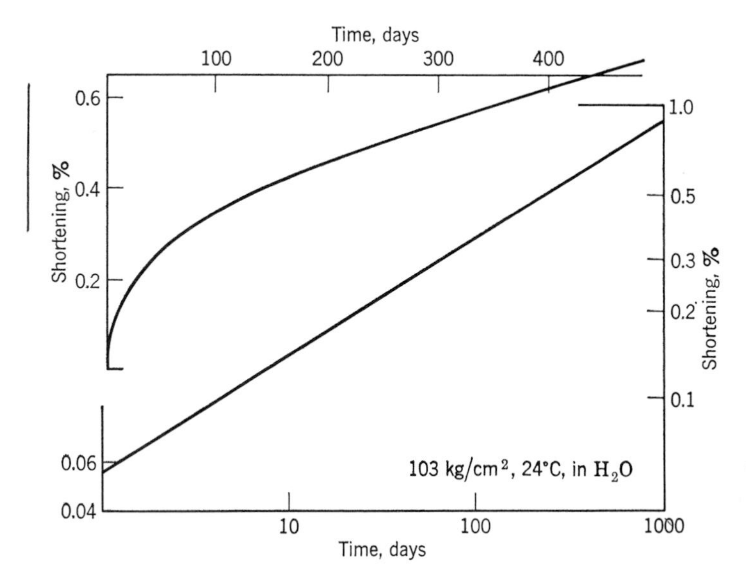

FIG. IV-24. Creep of alabaster at low stress, immersed in water. No rupture occurs. Upper curve: plotted on normal coordinates. Lower curve: same data plotted on double logarithmic coordinates. The mechanism of flow is here supposed to be solution and recrystallization solution at points of greatest stress, and deposition at points of lesser stress. Dry specimens exposed to higher stresses will creep and rupture; at lower stresses the creep rate decreases markedly and becomes zero at 92 kg/cm^2. This corresponds to the "fundamental strength" of the alabaster under the given conditions. (After D. Griggs, 1940.)

2. *Gliding* is very complex. Shearing stress acting on a rock is not homogeneous but will concentrate on surfaces predetermined by reversible imperfections in the crystal lattices. The overstrained crystal will then *slip,* that is, a whole layer of ions parallel to so-called glide planes or glide lines in the space lattice will be displaced relative to the rest of the crystal. Two contrasted types of gliding, twin gliding and translation gliding, can be distinguished. (See Fig. IV-25, *A* and *B*.)

Again stresses that strain the crystals make the energy barriers in the lattice unsymmetrical. Atoms, therefore, move in one preferred direction, resulting in a transport along the greatest stress component. The mechanism is a sort of directed self-diffusion and results in creep in this direction. One factor in gliding is therefore slip, another is creep, probably effected by atomistic jumps initiated by the strain energy of the crystal.

3. *Recrystallization* and differential transport of matter along grain boundaries is a change-of-phase transfer mechanism. At the stressed contact surfaces of the minerals, solution will take place, as already explained, and matter will recrystallize on unstrained surfaces. The transfer may be described symbolically as solid → fluid → solid, or solid → solution → solid. The process requires activation, diffusion,

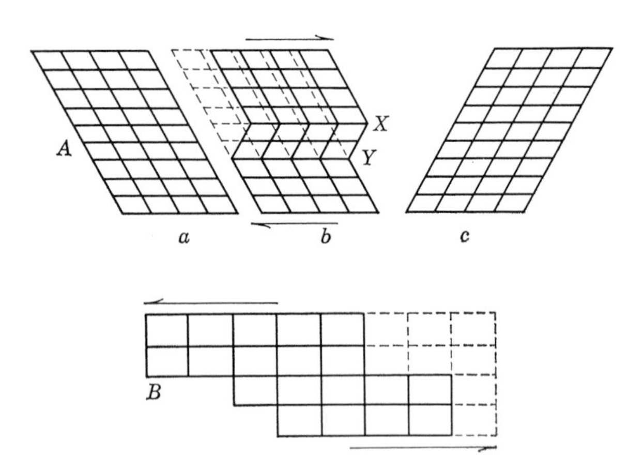

Fig. IV-25. (*A*) Twin gliding in a simple point system. Points are represented by intersections of full lines. Arrows indicate direction and sense of gliding in a horizontal glide plane. (*a*) Before twinning. (*b*) After development of a twin lamella *XY*. (Broken lines indicate position of points before twinning.) (*c*) After complete twinning. (Further deformation by twin gliding is impossible.) (*B*) Pure translation in a simple point system.

and consolidation. New crystals grow at the expense of old intensively strained crystals. This is probably the largest factor in creep of rocks.

5 · THE RECRYSTALLIZATION

τὸ ὑγρα μεικτα μάλιστα τῶν σωμάτων.

Aristotle.*

Experiments have shown that a mineral will not recrystallize in the absence of solvents unless it is heated to at least a critical temperature which is characteristic of the mineral. However, the critical temperature of recrystallization decreases if the mineral has suffered a previous plastic deformation, the recrystallization in this case deriving its energy from the strain forces that become released either by annealing or by contact with fluids.

In the metamorphic theory developed by Becke, van Hise, Grubenmann, and Niggli, it was assumed that the recrystallization of the rock minerals was effected by watery solutions that circulated in the pore spaces dissolving and precipitating the chemical constituents of the rock. In this manner a rather small quantity of liquid is presumably sufficient to effect recrystallization of large rock bodies.

This kind of metamorphism is easily demonstrated experimentally. A simple example is afforded by the salts sodium iodate and potassium chloride, which show a reaction point at about 38°C, as also indicated by the temperature curves of their solubility products:

$$NaIO_3 \cdot H_2O + KCl \xrightarrow{38^\circ} KIO_3 + NaCl + H_2O$$

In a moist mixture of solid sodium iodate and potassium chloride, potassium iodate and sodium chloride were seen to develop in the temperature range of 40° to 50°C, but the reaction was not complete after 24 hours. At 37°C the reaction had already reversed itself.

Circulating Solutions

The function of circulating solutions in silicate-rock recrystallization is limited to low-grade rocks and sheared metamorphites in general, and is particularly important in cementation and diagenesis of sediments.

* An early Latin translation, *Corpora non agunt nisi fluida,* became well known and was taken to indicate that no reaction could take place in the solid state. However, the original Greek dictum simply states that fluids react, or mix, more readily than solids.

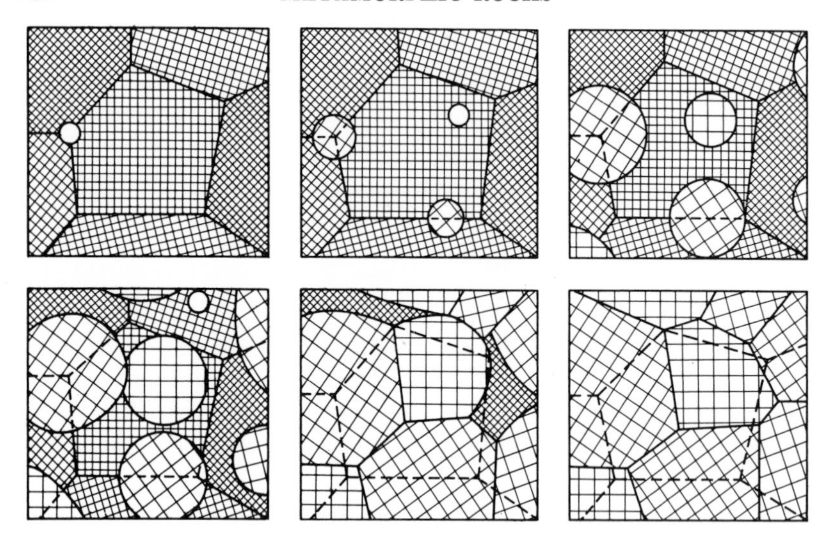

FIG. IV-26. The process of recrystallization. (After G. Sachs, from Hedvall, *Die Reaktionsfähigkeit fester Stoffe,* 1937.)

The solutions behave like ground water, the laws of the movement of which have been thoroughly analyzed by King Hubbert (1940). It is characteristic that, as the rocks become more consolidated and the channelways in the rocks more constricted, the movement of the solutions becomes gradually slower. By burial and orogenic pressure the rocks soon become tightly compacted, all fissures and cavities close completely, and the only "channelways" left are grain boundaries and fine fissures in the mosaic structure of crystals of molecular—or even of atomistic—dimensions. (See Fig. IV-27.) Long before this state is reached, the permeability becomes practically zero, that is, the resistance to the flow of the solutions is so great that all mechanical movements stop, even if steep pressure gradients obtain.

The old and rather obvious idea that aqueous solutions streaming through rocks should dissolve matter at one place, carry it along, and deposit it at another place does not apply to metamorphic rocks in general. Only in cleaved rocks and in the loosely packed sediments and their low-metamorphic equivalents is the permeability good enough to permit circulation. In these rocks water acts as a carrier of dissolved matter, and under certain conditions low-temperature minerals, quartz, calcite, chlorite, zeolites, and others, are deposited freely in open spaces and between grains. Under other conditions again the solutions dissolve certain minerals, resulting in a leaching that may be highly selective (usually lime, iron, and manganese are removed easily, whereas alkalies and silica remain), leaving behind either an open skeleton structure or replacing the dissolved compounds by simultaneous precipitation of new minerals (silica, for example, is often precipitated whereas the other constituents are removed).

Small changes of this kind are observed in many rocks, but typically they affect all rocks through hot-spring action in areas of thermal activity, and in other rather special situations.

Thus Fenner (1910) has described the processes that become operative immediately upon the extrusion of lava on a lake bed. Shrinkage and cracking of the cooling basaltic glass make the lava permeable to heated vapor and waters from the lake. The hot waters circulating in the cooling rocks—Fenner has calculated that under the conditions described a maximum temperature of 180°C may have been reached—had several functions. They served as a medium in which chemical reactions took place; they dissolved the volcanic glass and thus produced wider channels for circulation; they added material for the formation of new minerals which replaced the old ones volume by volume.

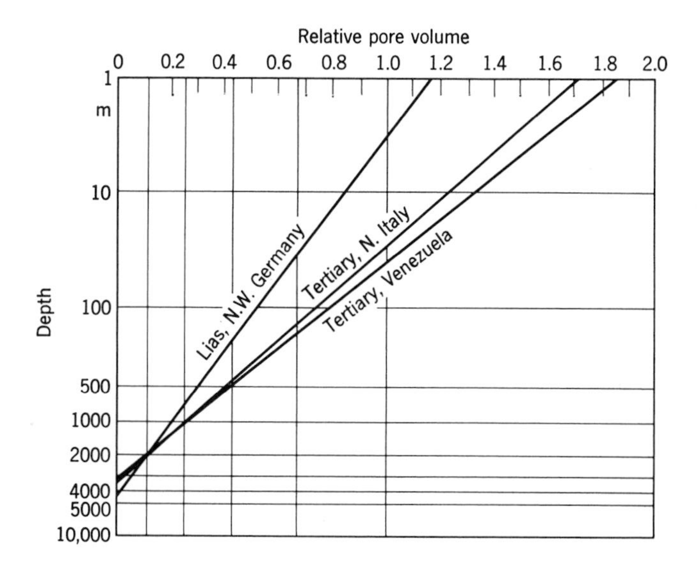

Fig. IV-27. There is a linear relation between the relative pore space and the logarithm of the depth. However, at great depths the relation is not exact, and so the depth at which the pore volume actually becomes zero is probably greater than that shown in the figure. But whatever the depth, the pore space must eventually be so restricted that no water can stream through the rock.

Changes now in progress have been studied in the thermal areas of Iceland, of the Yellowstone National Park, of Hawaii, and elsewhere. The effect on basalt is usually that water, carbon dioxide, and some soda are introduced into the basalt, with complete recrystallization of all primary minerals:

$$\text{Plagioclase} + \text{pyroxene} \rightarrow \text{zeolite} + \text{chlorite}$$

Diffusion and Ichors

Strongly compacted rocks become impermeable to circulating solutions (Fig. IV-27); *diffusional movements* may then take place. If a rock whose minerals are in mutual equilibrium under hydrostatic pressure is affected by directed pressure (shearing stresses), then free energy gradients are created parallel to the pressure gradients. The high-pressure regions will then function as sources, and the low-pressure regions will function as sinks for diffusive flows of matter in an effort to iron out the free energy differences. These fluxes also generally carry heat from high- to low-pressure sites. It is important to note that, in principle, all atoms and molecules will have a tendency to escape from the loci of high pressure, but their *tendency to escape* is proportional to their molecular volumes (see equation 15 in Appendix), thus compositional gradients are created.

In this way the striped and banded structures of crystalline schists and gneisses are qualitatively explained. The mineral molecules of large specific volume (feldspars, especially potash feldspar) first respond to the pattern of shear, and start migrating down the pressure gradients to become concentrated in pressure shadows and zones of slip; thermal energy, i.e., heat, will flow along the same avenues and further raise the mobility of the molecules; the result at elevated temperature being a gneissic rock with the mobile granitic material forming streaks and veins in a more rigid frame of darker colored rocks.

In rock bodies exposed to directed pressure the minerals will take up the pressure and the fluid ("pore fluid") will escape from places of high pressure. Rising pressure in such rocks always causes an increase of the solubility (at higher temperature a decrease of the melting point) at the contact points that are mostly strained. Clapeyron's equation (Appendix, equation 19) describes the relation between the pressure and a temperature of reaction in a closed system:

$$\frac{dT}{dP} = \frac{T}{\Delta H}(V' - V)$$

Applied to the solution (or melting) of a stressed mineral, V denotes the molecular volume of the stressed mineral and V' the volume of the solution. But because the solution disappears from the spot of the reaction, its volume may be neglected and we arrive at Poynting's law:

$$\frac{dT}{dP} = -\frac{TV}{\Delta H}$$

which shows that the strained parts of the minerals will go into solution, whereas the same minerals will be precipitated from the solution at places of no strain.

Thus shearing stress in rocks greatly facilitates the recrystallization processes. If we remember that in all actual cases the differential pressure is accompanied by: tectonic movement, squeezing, and the opening up of channelways along which heat and fluxes of matter are conducted; that it agitates the reacting particles, reduces their grain size, thus increasing the total area of reacting surfaces and renewing surfaces of contact; then it becomes evident that the directed pressure has an important function as a catalyst, materially aids recrystallization and deformation, and may bring about a complete metamorphism (and metasomatism) of many rock bodies.

The strain pattern of metamorphic rocks can now be studied with fairly high precision.

All metamorphic rocks of Dutchess County, New York (briefly to be described on page 343), are sheared. At the lower stages of metamorphism, no vestiges of moving solutions on the shear planes have been found (except quartz veins). Farther to the east, however, the sheared rocks show the effect of "hydrothermal" solutions, the main thoroughfares of which have been shear planes, enhancing thereby the schistosity and imparting to the rocks a marked foliated character. These shear planes represent the chief avenues by which solutions permeated the sediments, metasomatically transforming them into schists, and finally vanished among the low-grade argillaceous slates in the west.

The precise mechanism that permits flowing movements along the shear planes is still obscure. It may be a process of lateral secretion due to mechanical pressure gradients (see page 298). The general effect is that well-developed shear planes, often conspicuously studded with crystalloblasts (for example, garnet, kyanite, staurolite not present in the host rock), forming thin subparallel mineral layers, intersect the host rock as if it were injected by a multitude of fine veins. To be sure, in certain instances the material of the shear zones should more properly be considered as solid but highly comminuted material squeezed in among the schistosity planes in a dough-like state, together with the pore solution. The explanation is not clear; we are still in the dark as to how solid matter can be forced through narrow openings and move long distances. But observations from practically all metamorphic areas studied with this problem in mind indicate the reality of the process. *In sheared rocks migration is led chiefly along the shear planes and is minimal across the schistosity or foliation.*

In high-metamorphic gneiss areas, layers, bands, and zones of augengneisses conformably following the foliation for long distances corresponded to the shear-plane veins just described and testify to the chemical transport along preferred zones.

The chemical composition of the fluxes is to a large extent governed by the volume relations of the constituent molecules: in high-grade metamorphic rocks it approaches a granitic composition, and when generated in large amounts it may have a strong "granitizing" effect on the surrounding rocks. This fluid was called *ichor* by Sederholm (see page 353).

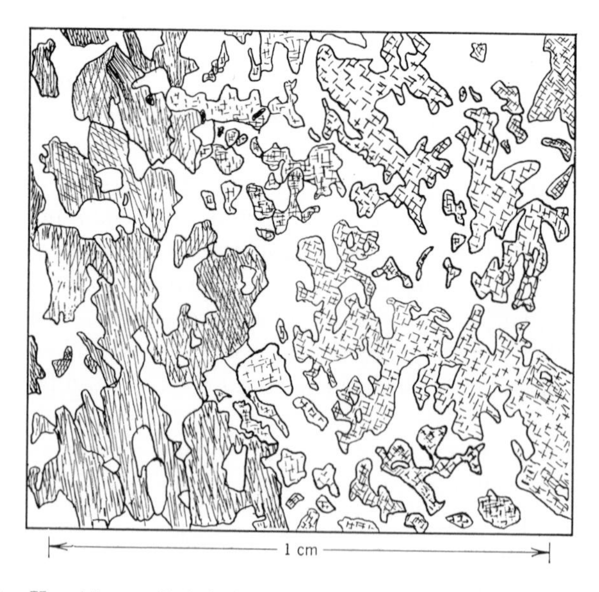

Fig. IV-28. Xenoblasts. Left half shows hornblende, right half pyroxene growing in a plagioclase groundmass. (Transition amphibolite—pyroxene skarn, Kristiansand, southern Norway.)

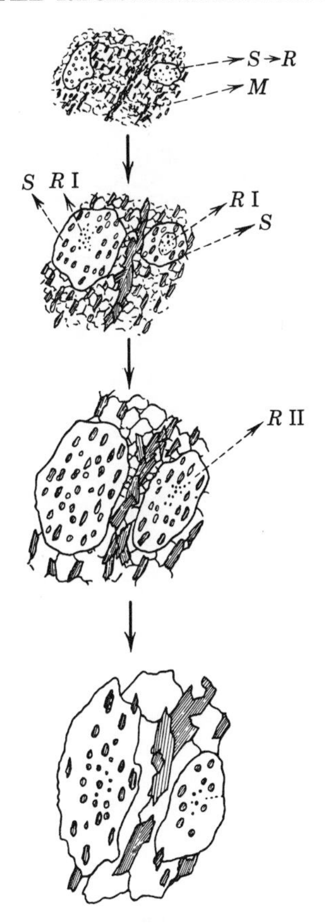

FIG. IV-29. Schematic illustration of the mechanism of the growth of two porphyroblasts of quartz in biotite schist. *Top picture:* $S \rightarrow R$ = original porphyroblast of quartz; M = matrix. *Next picture:* The original porphyroblast has become a relict core = R I, in a younger shell of new quartz (S), showing helicitic structure. *Third picture:* The large quartz porphyroblasts show helicitic structure (biotite inclusions indicating the trend of rock schistosity) and palimpsest structure (R II = vestiges of the original porphyroblast). *Bottom picture:* Recrystallization has erased all traces of original porphyroblast which still shows helicitic structure. (From biotite schist, Yanai, Japan; after S. Iwao, *Jap. J. Geol. Geogr.*, 1938.)

Crystalloblastic Fabric

The Greek word βλαστος (growth, bud, sprout) has been used in describing various kinds of crystals growing in a solid environment. A *crystalloblast* is a crystal that has grown during the metamorphism of a rock. Recrystallized rocks exhibit generally a crystalloblastic

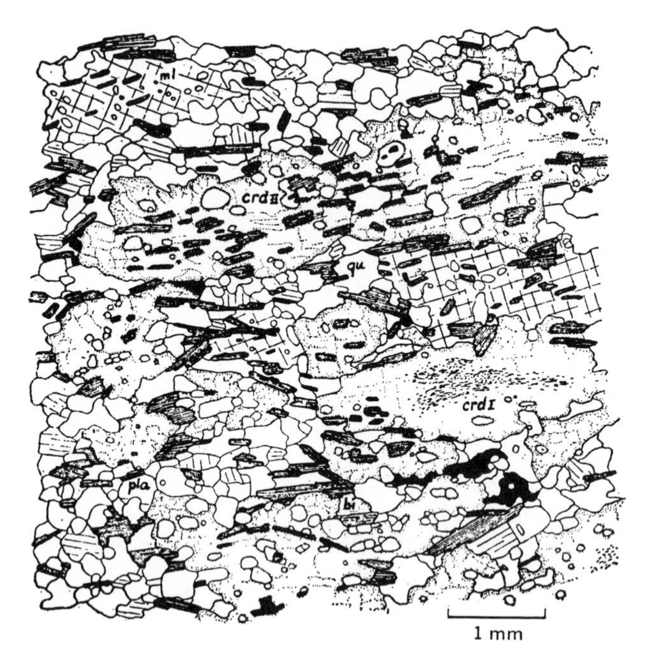

FIG. IV-30. Young xenoblasts (metablasts) of cordierite (crd II) and microline (cross-hatched) growing at the expense of the original quartz (qu) and biotite (bi). The original bedding of the rock is preserved as a helicitic structure in the metablasts. (Cordierite-kinzigite, Merimasku, Finland; after Anna Hietanen, *Bull. Geol. Soc. Am.*, 1947.)

fabric, being attributed to simultaneous growth of all the component crystals, as distinct from the regular sequence of crystallization in magmatic rocks. The majority of the minerals in metamorphic rocks are irregular in outline, *xenoblasts;* but some minerals are frequently bounded by their own crystal faces, *idioblasts.* Larger crystals are often packed with small inclusions of other minerals exhibiting the so-called sieve structure (*poikilitic* or *diablastic* structure).

Crystals of prismatic or tabular habit (amphiboles, micas, chlorites) often develop a preferred orientation, with their greatest dimensions in subparallel position. This is partly the result of rotation and deformation of crystals, and partly due to growth of new crystals, with appropriate orientation. The process results in the so-called crystallization schistosity (page 292).

Granoblastic refers to a non-directed rock fabric, with minerals forming grains without any preferred shape or dimensional orientation. *Lepidoblastic, nematoblastic,* and *fibroblastic* refer to rocks of scaly, rod-like, and fibrous minerals, respectively. *Porphyroblasts* in a meta-

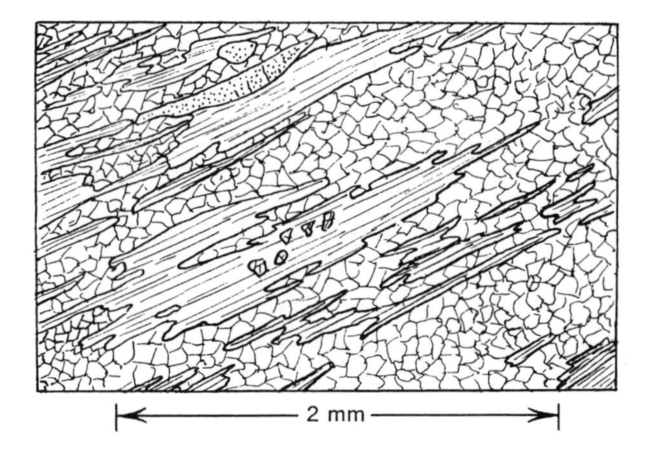

FIG. IV-31. Lepidoblastic quartz-muscovite schist. Upper left, two tourmaline grains.

morphic rock show in their structural relations analogies to inset crystals in a porphyric igneous rock. Minerals that have developed as new crystalloblasts independent of a primary seed crystal are sometime called *holoblasts*.

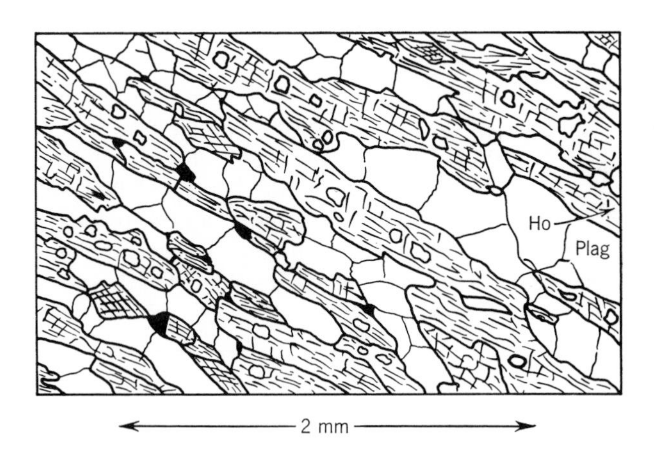

FIG. IV-32. Nematoblastic amphibolite (hornblende and plagioclase and some ore grains).

Idioblastic Series and Force of Crystallization

The metamorphic minerals may be arranged in an idioblastic series (crystalloblastic series) in order of decreasing force of crystallization. (Table IV-4.) Crystals of any of the listed minerals tend to assume

TABLE IV-4

Idioblastic Series of Metamorphic Minerals

(After F. Becke, 1903)

Sphene, rutile, magnetite, hematite, ilmenite, garnet, tourmaline, staurolite, kyanite
Epidote, zoisite
Pyroxene, hornblende
Ferromagnesite, dolomite, albite
Muscovite, biotite, chlorite
Calcite
Quartz, plagioclase
Orthoclase, microcline

idioblastic outlines at surfaces of contact with simultaneously developed crystals of all minerals of lower positions in the series.

When a crystal is in chemical equilibrium with a surrounding solution it will not grow, and the total molal free energy of the crystal equals the free energy of its components in solution, $F_{xl} = F_{\text{sol}}$. If a pressure is exerted on the crystalline phase, its free energy will increase (equation 15, Appendix), the result being that the crystal dissolves as demanded by Poynting's law.

If the surrounding solution becomes supersaturated, the free energy of the crystalline components in the solution increases, and we obtain the relation $F_{xl} < F_{\text{supersat.sol}}$. Consequently the crystal will grow and it will continue to grow even under an excess pressure (stress), that is until the free energy of the stressed crystal becomes equal to $F_{\text{supersat sol}}$. For this reason every crystal is able to execute a pressure of growth when in contact with a supersaturated medium. Equation 15 gives us the quantitative relation between change in free energy with pressure

$$\left(\frac{\partial F}{\partial P}\right)_T = V$$

We can define the linear force of crystallization as the excess pressure which a given crystal can endure (or just grow against) at a given degree of supersaturation (Ramberg, 1947):

$$\text{Force of crystallization} = \left(\frac{\partial P}{\partial F}\right)_T = \frac{1}{V} = \frac{d}{M}$$

where d is the specific gravity of the crystal and M the molecular weight.

The force of crystallization as here defined is seen to be inversely proportional to the fictive molal volume of the crystal. This concept harmonizes very well with the empirical Becke series of Table IV-4, which shows that minerals of close-packed lattices (typical neosilicates) rank highest in the idioblastic series. In decreasing order are the inosilicates, the phyllosilicates, and, last, the tectosilicates with open framework lattices. Discrepancies are to be expected, and in the Becke idioblastic series albite would seem to have a rather high position. In many low-grade albite schists (examples from New Zealand, Fennoscandia, eastern New York State, etc.) albite actually occupies a much lower position. Its normal position appears to be with quartz and the other feldspars.

The force of crystallization cannot, however, be identified directly with the power of minerals to grow in the form of idioblasts. The development of crystal faces during the growth of a mineral in a solid medium depends on factors different from the simple ability to execute a certain pressure of growth against the solid surroundings. Ramberg lists the following factors:

1. The force of crystallization.
2. The degree of supersaturation.
3. The concentration and mobility in the enclosing rock of the chemical constituents of the porphyroblast.
4. The energy of activation of a germ of the porphyroblast.
5. The effect of the surrounding minerals and of the size of the porphyroblast on its escaping tendencies (vapor, tension, activity, etc.).
6. The plasticity of the surrounding minerals.

Furthermore the form energy must be considered. Such a mineral as actinolite often shows idioblastic prism faces but poorly developed terminal faces; minerals such as mica, chlorite, and chloritoid, show a higher degree of form energy on their cleavage planes than on other faces. Again the different faces may show a different ability to absorb foreign particles. In a growing crystal of andalusite, the (110)-faces are able to shove away foreign particles much more effectively than the (001)-faces. Chiastolite is andalusite that is pigmented by graphite dust from the original bituminous argillaceous schists, and in cross section often exhibits a cruciform figure, showing that the dust was enclosed selectively by the edges of the prism faces. In andalusite, chloritoid, and less frequently in staurolite, cordierite, and in some other minerals, the pigmented parts habitually exhibit an hourglass structure.

Preferred Orientation of Crystals

In most crystalline schists there is a tendency in the constituent minerals to assume parallel or partially parallel crystallographic orientation. The shape and spatial arrangement of minerals such as mica, hornblende, or augite show a definite relation to the foliation in the schist or gneiss, that is, both foliation and fissility * of a metamorphic rock is directly related to the preferred position assumed by the so-called schist-forming minerals, mica, hornblende, chlorite, etc.

The different types of crystal structure react differently to pressure and stress in rocks. Amphibole crystals, more than pyroxene, show a tendency to orient themselves parallel to the direction of stress during metamorphism, thus inducing in the rock an easily detectable linear structure indicative of the direction of the mechanical forces that affected the rock. A good example of the difference of behavior between amphibole and pryoxene is seen in Fig. IV-28: amphibole but not pyroxene has reacted to the shearing forces which influenced the rock. Micas and other sheet silicates try to orient themselves with the *c*-axes parallel to the pressure gradient.

It has been maintained that crystals grow (1) with the axis of best thermal conductivity along the thermal gradient and (2) with the axis of largest compressibility along the pressure gradient. But actually this is not always so.

It is typical of many metamorphic rocks that mica or chlorite grows on the schistosity planes, forming thin, planar, or wavy surfaces, greatly adding to the schistose character of the rock. All sheet silicates show a perfect cleavage parallel to the planes of the nets, each net being kept firmly together by strong silicon-oxygen bonds, whereas the different nets are loosely stacked on top of each other by weak cation-oxygen bonds, and therefore susceptible of easy slips.

Differential movements in rocks often follow planes rich in mica or chlorite. Compared with other rock-making minerals, the sheet-silicates show a low hardness. They serve, therefore, as "lubricants" and are conspicuously present in thin veneers on thrust walls and slickensides in tectonically affected rocks.

Students of structural petrology distinguish between preferred orientation of inequidimensional grains according to their external crystal

* Knopf and Ingerson (1938) use *foliation* to denote a parallel arrangement of minerals. A foliated rock will split more or less easily into thin slabs parallel to the foliation plane. Where the rock splits into thin layers, it is called a schist. Where the layers are thicker and exhibit wavy surfaces, it is called gneiss. A slate is an extremely fine foliate. Fissility in rocks is analogous to cleavage in minerals.

form (*Regelung nach Korngestalt; Formregelung*), and preferred orientation of equidimensional grains according to their space-lattice structure (*Regelung nach Kornbau; Gitterregelung*).

Minerals that rank low in the idioblastic series, such as quartz, calcite, and feldspar, commonly exhibit equidimensional grains as xenoblasts without typical external form, and are thus unable to show *Formregelung*. A special universal-stage technique is in such cases necessary to demonstrate in detail the preferred orientation of the mineral grains according to their lattice structure.

The possible means by which preferred orientation may evolve are various and often complex. A great many textures can be interpreted in a variety of ways, depending upon the dominant belief of the observer. The chief factors involved are listed in Table IV-5.

TABLE IV-5

MECHANISMS RESPONSIBLE FOR PREFERRED ORIENTATION OF MINERALS IN METAMORPHIC ROCKS

(After F. J. Turner, 1948)

(*a*) Synchronous with rock deformation:
- (1) Rotation or other differential movement of existing crystals.
- (2) Plastic deformation (that is, by translation or twin gliding) or rupture of existing crystals.
- (3) Growth of new crystals under the influence of the deforming forces.
- (4) Selective complementary growth and suppression of existing crystals during recrystallization under the influence of the deforming forces.

(*b*) Subsequent to deformation or in undeformed rocks:
- (5) Growth of new crystals in a mechanically anisotropic rock.
- (6) Selective complementary growth and suppression of existing crystals during recrystallization in a mechanically anisotropic medium.

Recrystallization in Petrogenesis and Geochemistry

This phenomenon can hardly be overrated. We have seen that thermal energy in a crystal manifests in vibrational disorder → rotational disorder → substitutional disorder (which is a generalized type of diffusion), and, finally, in recrystallization. The temperature of recrystallization is lowered by stress because strain energy then is supplied and less thermal energy is required to drive recrystallization.

It is known that the activation energy of recrystallization in metals is about twice the activation energy required to make the metal atoms diffuse through their own solid structure. The reason for this, evidently, is that more bonds must be broken to transport an atom across a crystal boundary than merely to pass it along in the same structure. Buerger has further discussed this fact and pointed out that it implies that, whenever the temperature is sufficiently high to cause spontaneous growth of the crystals, it is already maintaining a very high level of diffusion. In this condition some atoms may be expected to be rather freely migrating through the remainder of the structure of the crystal. Thus, whenever the rock is in a condition to recrystallize, it is also something of a blotter for available atoms, thanks to temperature. It is, therefore, evident that *wholesale diffusion must play an important role in the transport of chemical material in metamorphism.*

Lindgren long ago pointed out that replacement occurs on approximately a volume-by-volume basis. Although the field evidence for this has been obvious, the mechanism for accomplishing it has been obscure. Diffusion suggests the mechanism. There is a tendency on the part of crystals to have their volumes determined by their largest atoms. Thus the volumes of the rock minerals are dominated by their oxygen atoms, and the volumes of the sulfides are dominated by the packing of the sulfur atoms. Thus diffusion supplies a mechanism for approximately maintaining volume during replacement.

In the study of metamorphism, it is customary to consider that some oxide, such as silica or alumina, has remained constant while other oxides have varied in the process. Evidently a closer approximation would be that the *oxygen* content has remained nearly constant while the wandering interstitial metals themselves have varied with the change. This shows the importance of calculating the content of the standard cell of metamorphic rocks.

6 · METAMORPHIC DIFFUSION AND METAMORPHIC DIFFERENTIATION

> *. . . whatever felspathisation once meant, I maintain that it does not mean an impregnation with felspar derived from a magmatic source. In felspathisation, material for the formation of felspar may be contributed both by the country rock and by the introduced solutions, magmatic or otherwise.*
>
> H. H. Read, *Meditations on Granite,* 1944.

During metamorphism chemical reactions take place among the constituents of the rock. The mechanism of the reactions is the same in metasomatism and in normal metamorphism, the difference being that metasomatism requires a long-distance migration of the chemical elements whereas normal recrystallization, simply corresponding to a relocation of the material already present in the rock, takes place over small distances.

Metamorphic differentiation is restricted to denote differentiation effected by diffusion processes, and does not include differentiation by flowing liquids or bodily movement of solid particles. The only potential that will create diffusion is the differences in the chemical activity of the several rock-making elements at different places in space. Like other chemical processes, the diffusion currents are driven by chemical forces which can be expressed in any one of the following terms: chemical activity, chemical potential, free energy, fugacity, or vapor tension. These terms are not identical but are related by simple equations.

Every chemical species present in a rock will try to reach a place where its activity is low. This fundamental law has been especially emphasized by Ramberg in several publications: *"Dispersion of minerals at places where the activities are great, migration of the dispersed atoms towards places where the activity is less and consolidation there, constitute the fundamental processes in metasomatism and metamorphic differentiation."*

Activity gradients in the rocks are created by five main factors: differences in pressure at constant levels, differences in temperature, differences in the chemical composition of the phases (minerals), differences in the size of the minerals, and differences in the surrounding milieu.

The mechanism of the migration of the different chemical species—ions, atoms, molecules—is highly debatable. It is probably extremely complex and involves diffusion in the intergranular films and mosaic fissures, as well as volume diffusion in the different phases (minerals, pore solution). All students of the subject believe that mechanical agents, that is, shearing stress or general differential movements in the rocks materially promote diffusion and reactions and reduce the temperature necessary for them.

Without postulating any "pervading solution," "pore solution," or "intergranular film," we may provisionally and noncommittally refer to the medium in which the chemical reactions take place as the *disperse phase.*

We shall revert to this question presently, but first we shall describe some of the results of metamorphic differentiation. There are four governing principles: the concretion principle, the secretion principle, the principle of enrichment in the most stable constituents, and the solution principle.

Concretions

The starting point of a concretion is a crystalloblast. In rocks of high metamorphic grade, single porphyroblasts may sometimes attain a very large size. (Garnet porphyroblasts from Sundfjord, Norway, weigh up to 2000 pounds; plagioclases from recrystallized anorthosite (both in Norway and South Russia) attain 5 feet in length; cordierites from Orijärvi, Finland, and garnets from the Adirondacks, New York, may all attain a length of 30 cm.) Turner has computed that small crystalloblasts of manganiferous garnet (25 per cent MnO) in a low-grade quartz-albite-epidote-muscovite schist of Otago, New Zealand, contain an amount of manganese equivalent to that present in a sphere of schist of about 1.4 mm radius. Thus diffusion through distances 1 to 2 mm could account for the growth of these porphyroblasts. In the case of larger and more spaciously distributed crystalloblasts the diffusion distances of certain chemical species are of a much higher order of magnitude. (See legend to Fig IV-36.)

Porphyroblasts are in themselves not regarded as typical concretions, but crystal aggregates, either monomineralic or polymineralic,

FIG. IV-33. Microline-quartz concretions in biotite gneiss. The upper third of the picture shows part of a concretion pegmatite. Next to it are individual concretions arranged parallel to the pegmatite and to the foliation of the gneiss. There is no genetical difference between the pegmatite body and the "augen" in the gneiss. (Polished specimen shown one-half natural size, from Kristiansand, southern Norway.)

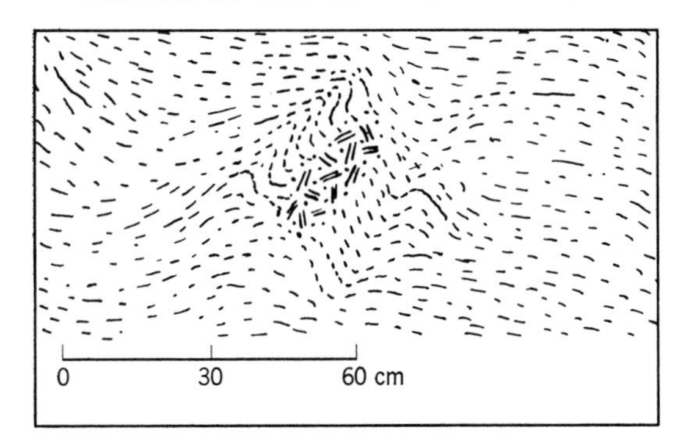

FIG. IV-34. A small concretion pegmatite in biotite gneiss. (Kristiansand, southern Norway.)

which have grown in solid rock during metamorphism represent true concretions. They may attain considerable dimensions. To be sure, they may constitute actual rock masses, and may then be called *petroblasts.*

Concretions may have different composition. Calcareous concretions in sedimentary and low-metamorphic rocks are well known. Quartz concretions also often show concretionary growth in low-metamorphic rock. During metamorphism of higher grade (in gneisses), quartz-feldspar concretions exhibiting pegmatitic structures are often encountered.

Concretion pegmatites are usually small bodies with dimensions of 1 cm to 1 m, and they are almost always comfortably surrounded by gneiss exhibiting a selvage of femic minerals (usually biotite). Fig. IV-33.

Obviously quartz and feldspar exhibit a smaller partial tension (activity) in the petroblastically growing pegmatite body than in the groundmass of the gneiss. The conditions causing this difference in stability must be sought either in the difference in grain size between pegmatite and gneiss, or in the difference in chemical milieu of pegmatite and the gneissic groundmass. It is probable that a grain of potash feldspar surrounded by quartz and albitic plagioclase (in pegmatite) shows less partial tension than when surrounded by hornblende and/or biotite (in gneiss). Thus there are partial activity gradients driving the diffusion.

All concretions are formed by diffusion along activity gradients and growth of the stable concretion under a certain mechanical excess pressure (see "Force of Crystallization," page 290).

Secretions

Cavities, open cracks, or fissures in a rock are prerequisite for secretions (or excretions) which grow by filling the cavity, starting from the inside walls and slowly growing towards the center. This is the opposite of the growth of concretions that proceeds from the center of the spheroid and outward.

The growth of secretions is easily explained by dispersion, diffusion, and consolidation of individual chemical species in the solid rock. By shear and tear (in orogenic areas) of solid rocks, cracks are opened, and mechanical pressure gradients are immediately set up which, in turn, create corresponding gradients in the chemical activities. Consolidated particles in the rocks will disperse, migrate towards the crack, and then reconsolidate in new minerals. In this way every molecular seam that opens will automatically be caulked and filled with minerals until the mechanical pressure in the rock becomes evenly distributed.

Secretion pegmatites represent the most conspicuous examples of this process; they are usually discordant elliptical or dike-like bodies, extending sometimes for hundreds of meters. A great many pegmatites of this type in southern Norway were investigated by O. Andersen (1932), who demonstrated that they preferably lie in gabbroid rocks (amphibolites) in which cracks form more easily than in granitic gneisses. One may now ask: Why were not these bodies formed by a magma (or solution) that flowed into the open cracks and congealed there? The answer is at hand. It is impossible to find any channels through which the magma streamed. The pegmatites are ductless bodies. No flowing solution was able to penetrate the country rock (see page 282) except along shear zones. Some of the shear-zone material may be squeezed into the cracks, but the origin of the shear-zone material is probably also by diffusion from the strained rocks into the semi-open shear planes. Consequently, the pegmatites were fed by diffusion. Additional evidence is seen in the structure of these pegmatites; zoning is common, usually with a core of quartz. The minerals of the marginal zone are, in the great majority of cases, oriented normal to the walls; this is particularly conspicuous in flakes of black biotite, standing erect like bristles. The outer zone is often relatively fine-grained and contains graphic granite. The inner zones are coarse-grained, with single crystals of quartz and feldspar that in many occurrences may attain 10 m. Rare minerals regularly present in certain occurrences are topaz, lithium mica, beryl, all often in large crystals; furthermore, there are tourmaline, spes-

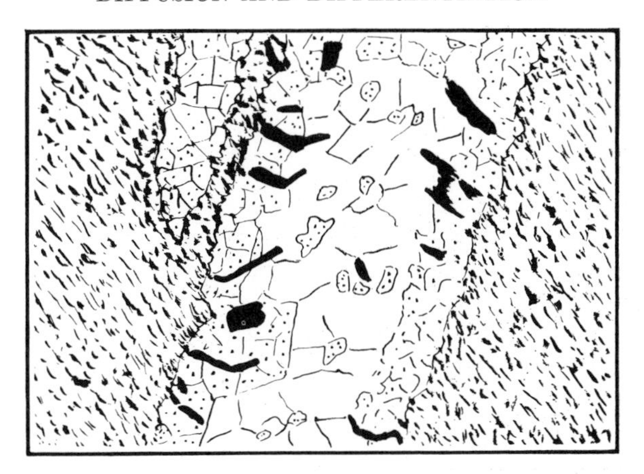

Fig. IV-35. Secretion pegmatite in biotite gneiss. The pegmatite has a zonal structure. The borders consist of plagioclase and quartz (dotted) penetrated by flakes of biotite (black) oriented normal to the walls. The core consists of pure microcline perthite (white) with some odd grains of quartz. The pegmatite dike is 4 cm wide. (Nelaug, southern Norway.)

sartite-garnet, and various minerals containing rare earths. It is evident that the granitizing elements, potassium, sodium, silicon, and some rarer elements, lithium, fluorine, beryllium, etc., must have migrated over very long distances in order to construct these large pegmatite bodies. Other types of pegmatite are discussed on pages 124 and 355.

Enrichments in Stable Constituents

During progressive metamorphism, or if two chemically incompatible rocks are brought together, various new minerals will become stable. It may then happen that one such new mineral—let us call it mineral A—is particularly insoluble in the disperse phase of the rock. As a consequence, A will immediately crystallize out, and the concentration of A in the disperse phase will drop. In poorly consolidated rocks circulating solutions may bring in fresh supplies, and mineral A may continue to grow by extracting the necessary elements from the incoming solutions. Likewise in sheared rocks fresh supply may circulate along the shear planes, and minerals in the shear-plane veins may grow by precipitation from the shear-plane solutions. Crystalloblasts outside the shear planes grow by direct diffusion through the solid rock. In any case, certain elements necessary for the growth of mineral A are extracted from the disperse phase, and either directly delivered to A, or *via* the shear-plane solutions delivered to A. Mineral A grows

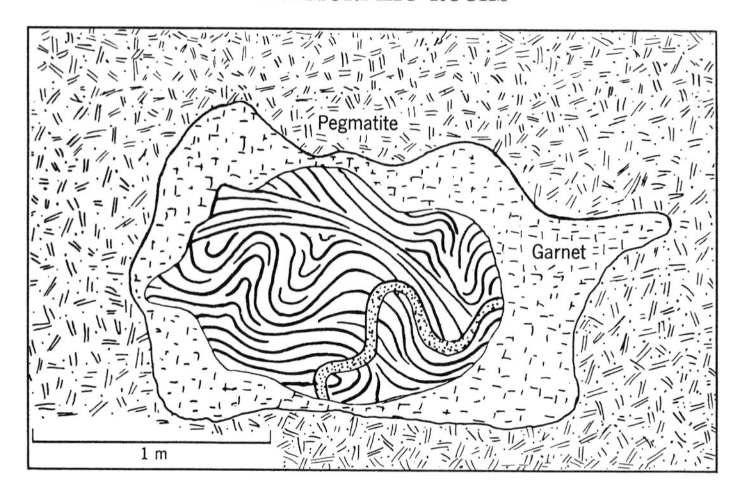

FIG. IV-36. Enrichment of garnet around a folded and contorted inclusion of crystalline limestone in granite. The garnet shell represents an enormous concentration of iron; it contains 25.9 per cent, the granite 0.33 per cent total iron oxides. The garnet shell corresponds roughly to a mass of 10 cubic m or 37.5 tons, containing about 9.5 tons of iron oxides. It takes, therefore, 3000 tons of granite, corresponding to a sphere 12.5 m in diameter, to supply the necessary iron. No feeders extend into the garnet mass, and the surrounding granite is so massive as to be hardly permeable for moving solutions. Diffusion distance of at least 6 m must therefore be assumed. (Kristiansand, southern Norway.)

and multiplies, introducing as often as not certain elements into the rock that were not originally present in large amounts. The introduced elements are often derived from neighboring rock masses.

In skarn rocks the effect of local or zonal enrichment in the most stable constituent is frequently seen. Likewise the process of pneumatolytical concentration of minerals at limestone contacts apparently operates on this principle. (See Fig IV-16.) For geochemical migration and metasomatic exchange in rocks the principle of enrichment in the most stable constituents is of great importance. It often operates in combination with the solution principle.

Solution Principle

The precipitation of the most stable mineral usually implies a simultaneous solution of certain other minerals. The rate of solution depends upon the surrounding conditions. Obviously the circulating fluids of permeable rocks are much more effective than the stationary "pore fluids" of dense rocks in dissolving minerals rapidly. If solution is effected by diffusion, the rate depends upon the activity gradients which, and this is important, are governed not only by concentration gradients but also by temperature and pressure gradients, etc.

Large rock masses are changed chemically by the removal of certain soluble constituents. For example, all basic igneous rock, and also amphibolites of similar composition will always lose CaO if they recrystallize in a low-temperature facies. Quite generally, a change from high temperature amphibolite into low-temperature greenschist implies a change in composition in that CaO is removed. (Actinolite greenstone $\rightarrow$ greenstone + CaO$\uparrow$, and greenstone $\rightarrow$ chlorite-albite

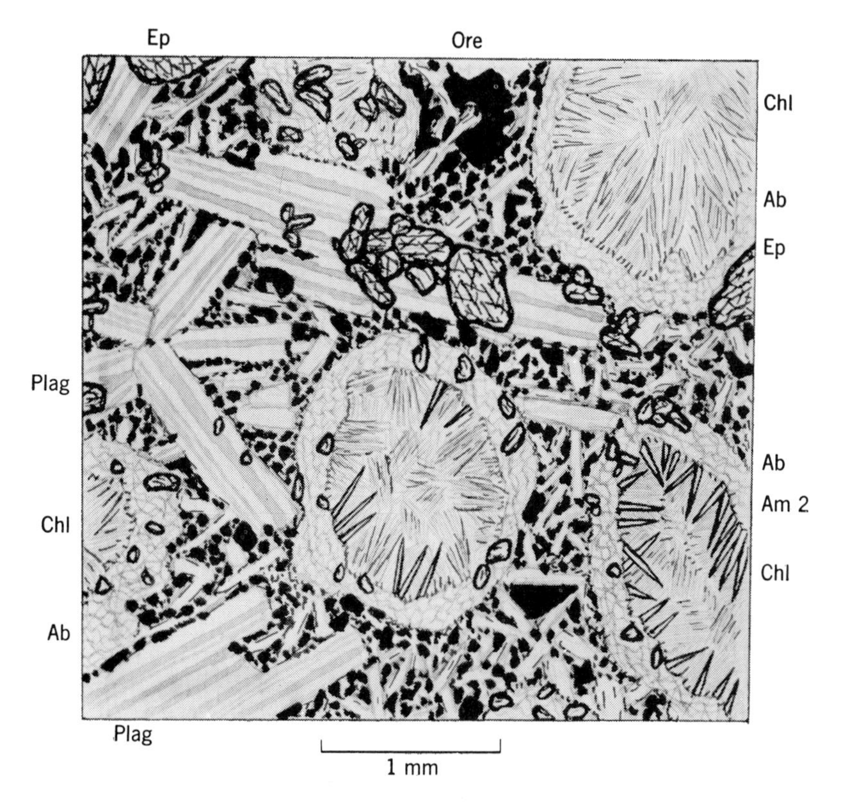

Fig. IV-37. Metamorphic basaltic lava of the Oslo region. Plagioclase is altered to albite and epidote, and nothing is left of the original dark constituents. Numerous vesicles are filled with green amphibole, epidote, chlorite, and freshly crystallized albite. Ab = aggregates of albite, Am 2 = secondary amphibole, Ep = epidote, Chl = chlorite. (After Sæther, 1945.)

fels + CaO↑; at both steps CaO is removed in solution.) We might look at this phenomenon as follows: In the low-metamorphic facies the only stable lime-rich mineral is $Ca(OH)_2$ (unless sufficient CO_2 is present); it never forms, however, but is always carried away in solution. *Consequently, low-temperature metamorphism of basic igneous rocks is almost always accompanied by appreciable loss of CaO.*

Low-grade contact metamorphism has the same effect; examples are basalts of the outer contact zone at Oslo. (See Fig. IV-37 and Table IV-6.)

In the same manner the expulsion of carbon dioxide in medium-grade and high-grade metamorphism of limestone and dolomite in the presence of silica may be regarded as a removal of the most soluble component. Whether released in the gaseous state or in aqueous solutions, the carbon dioxide represents the most soluble component in the pore fluid under these physico-chemical conditions. At other places, under different physico-chemical conditions, carbon dioxide may become less soluble and may reappear in calcite crystalloblast in greenschists and in the calcite-quartz veins that are abundantly present in many low-grade schists.

TABLE IV-6

SCHEME OF THE REACTIONS TAKING PLACE IN BASALTIC ROCKS OF THE OUTER-CONTACT METAMORPHIC ZONE OF THE OSLO REGION

(After Sæther, 1945)

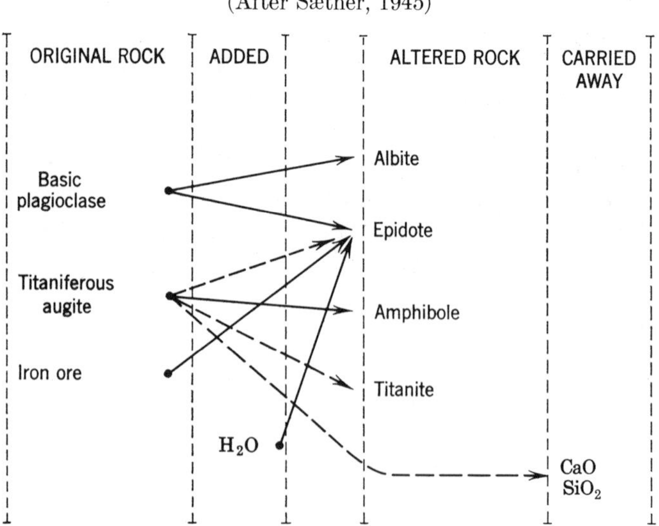

Metamorphic Differentiation in a Gravity Field

This is treated mathematically in the Appendix, equations 25–28. A summary in a non-mathematical language is given below:

1. In a *homogeneous* phase (mixed crystal, magma) chemical species with fictive volume smaller than that of the average surrounding phase will be gradually concentrated at greater depths. Species with larger fictive volumes will be concentrated at higher levels. See Fig. III-59.
2. In *heterogeneous* systems (rocks), species (minerals) with small fictive volume are commonly discontinuously enriched downward, and species with larger volume are discontinuously enriched toward the top, the several species being arranged in horizontal layers.
3. It can be shown that interionic bonding forces (chemical affinities) are strong enough to disturb this gravitative arrangement, in that heavy elements with strong affinity to light phases will, if the light phases are more abundant, concentrate in these light phases at the top, and vice versa for less abundant light elements with strong attraction to heavy phases. Thus practically *all* uranium has concentrated in oxygen at the very top layer of the crust of the earth.

Metamorphic differentiation caused by gravitation is in a different class from other kinds of metamorphic differentiation. Inconspicuous, perhaps, in the limited sphere of direct observation of man, its demesne is, nevertheless, extensive in space and time and its results are magnificent. It rules the endogenic geochemical cycle, leads the heat economy of the earth, and, indeed, is responsible for planetary structures.

7 · REGIONAL METAMORPHISM AND MINERAL FACIES

Becke, in his criticism of the facies classification . . . has cited examples of non-equilibrium rocks. It would not be difficult to add further examples to the list. Any mineral assemblage in which one mineral is in process of arising from another must be regarded as being in unstable equilibrium.

Tilley, Facies of metamorphic rocks, *Geol. Mag.*, 1924.

Materials and Processes

Regional metamorphism and mountain building are interrelated and represent the logical consequences of prolonged sedimentation in a slowly subsiding geosynclinal basin. As the sediments accumulate, they gradually become exposed to increasing temperature and pressure, they become sheared and deformed, and a general recrystallization ensues. The temperature increase is mainly due to two things (1) the heat flow from the interior of the earth, and (2) radioactive heat generated in the sediments themselves.

1. Obviously heat from the interior of the earth is introduced regionally and locally, partly associated with magmas, partly in the form of "emanations" following certain main avenues, determined by a variety of factors. Taken over a large volume, and over a long period of time, the amount of heat thus introduced will correspond to about 40 cal/cm^2/year, which is the general flow of heat of the earth (page 7). The heat capacity of most rocks is 0.2 cal/g or about 0.5 cal/cm^3. Consequently the heat required to raise the temperature by one degree of a pile of sediments 25 km thick is $0.5 \times 25 \times 10^5 = 1.25 \times 10^6$ cal/cm^2. The general heat flow of 40 cal/cm^2/year is therefore able to effect a rise in temperature of $40/1.25 \times 10^6$ 30°C per million years.

2. The amount of radiogenic heat can be expressed in a similar way. From the data given on page 7 we derive that the radioactive heat generated in granite will raise the temperature of the granite by 33°C per million years; and the heat generated in gabbro will raise its temperature by 10°C in a million years.

Since there is more radioactivity in the geosynclinal sediments than in the average rock of the crust, the geosynclinal regions must be hotter than the encasing crustal rocks.

As soon as the metamorphic processes start to work, many of the sedimentary minerals become unstable, and the sediments lose volatile constituents (water, carbon dioxide, etc., that are adsorptively present on mineral surfaces or inside the mineral lattices).

In geological discussions the fact that sediments at the very incipience of metamorphism regularly change their chemical composition has often been neglected. However, these changes are not to be neglected, and H_2O, CO_2, etc., are just as important as many other of the rock-making constituents, and serve to characterize the rocks in which they occur. Isochemical regional metamorphism *senso strictu,* therefore, does not exist. Some chemical changes always happen, but in many rocks they are small and genetically unimportant. It has become customary, and is expedient in an attempt to survey and classify the various petrographic types, to distinguish between the normal metamorphism (with but small changes in the total chemical composition) and the metasomatic metamorphism (with essential changes in the chemical composition). But from a genetical point of view this distinction is illogical and obscures the broader relations.

Regional metamorphic rocks always have suffered a change in composition, for transportation and transfer of chemical matter and heat are not only concomitant with, but actually essential parts of, the earth processes constituting metamorphism. By and large, we find that rocks of low-grade metamorphism have changed the least, chemically, and rocks of higher metamorphic grade have changed more. A distinction according to the degree of chemical alteration would be roughly a distinction between degrees of metamorphism.

The normal geosynclinal sediments are sand, clay, and limestone. From mixtures of these sediments the usual types of regional metamorphic rocks develop under the influence of incoming material and increasing temperature and pressure, the derivatives of the continuous clay (shale)-limestone series being the most interesting petrographically and most sensitive to temperature-pressure changes.

Any sedimentary unit will recrystallize according to the rules of the several mineral facies, as systematically surveyed in this chapter under *ACF* Diagrams, the complete sequence of events being a progressive change of the sediment by deformation, recrystallization, and alteration in the successive stages:

Greenschist facies → epidote amphibolite facies → amphibolite facies → granulite facies

The conditions of the granulite facies correspond to the highest possible regional metamorphism. At still higher temperatures, differential melting of the rock system sets in, that is, differential anatexis with formation of palingenic magmas.

Table IV-7 summarizes the metamorphic series of rocks that develop from the several types of common sediments and usually converge toward a granitic composition regardless of the nature of the original material. The chemical changes in the progressive series are characterized by increase in alkalies and silica, and decrease in magnesium, calcium and iron.

By the regional metamorphism a stationary temperature gradient is established in the mountain masses that controls the geographical distribution of the several mineral facies.

By and large, the outer parts of the geosynclinal region are less affected, and in the ideal case these marginal parts contain unmetamorphosed sediments, clay, sand, and limestone, etc., which gradually change into metamorphic rocks of successively higher facies as they extend into the central parts.

TABLE IV-7

METAMORPHIC SERIES

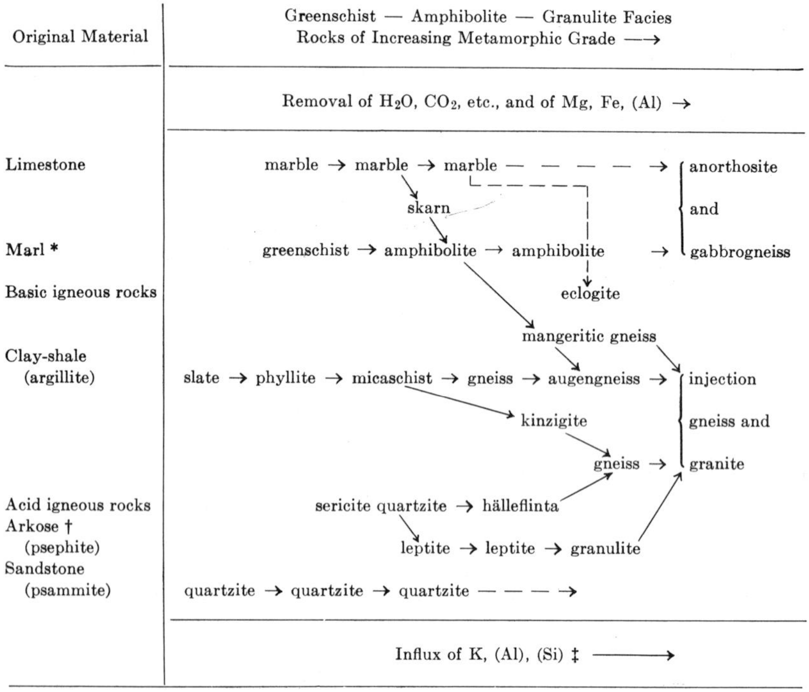

Original Material	Greenschist — Amphibolite — Granulite Facies Rocks of Increasing Metamorphic Grade ⟶
	Removal of H_2O, CO_2, etc., and of Mg, Fe, (Al) →
Limestone	marble → marble → marble — — — — → anorthosite
	skarn — and
Marl *	greenschist → amphibolite → amphibolite → gabbrogneiss
Basic igneous rocks	eclogite
	mangeritic gneiss
Clay-shale (argillite)	slate → phyllite → micaschist → gneiss → augengneiss → injection
	kinzigite — gneiss and
	gneiss → granite
Acid igneous rocks	sericite quartzite → hälleflinta
Arkose † (psephite)	leptite → leptite → granulite
Sandstone (psammite)	quartzite → quartzite → quartzite — — — →
	Influx of K, (Al), (Si) ‡ ⟶

* Original basic igneous rocks (gabbros, basalts) show a composition related to that of marl and yield analogous metamorphic products. Not listed are ultrabasites (peridotites, etc.), which by metamorphism become serpentine, chlorite or talc schist, soapstone, hornblende schist, pyroxene or olivine masses.

† Original acid igneous rocks (granite, diorite, rhyolite) show a composition related to that of arkose and yield analogous products. Leptite is primary, fine-grained rock, usually showing tufaceous or blastporphyric relict structures; or it is derived from argillaceous sediments. Hälleflintas are dense rocks of conchoidal fracture, genetically related to leptites. Kinzigites, characterized by containing aluminum silicates, usually also rich in magnesia, are metasomatic gneisses, but probably argillites also enter into their constitution. Granulite used to be a neutral name designating medium- to fine-grained gneisses. In Germany and Fennoscandia, granulite means a garnet gneiss with but little mica and a rather characteristic structure (granulite facies).

‡ Contrary to previous opinion, the exchange of matter is often practically non-existent in the early stages of metamorphism (except for the "mobile" constituents). Not until the onset of incipient melting does K or Na migrate over longer distances. The other cations are even more sluggish (Mehnert, 1960).

This change, as demonstrated by the succession of mineral facies in the present surface rocks, corresponds to the temperature gradient within the original mountain mass. A rapid change in facies indicates a previous steep temperature gradient and vice versa. This relation of mineral facies and temperature gradient of the regional metamorphism puts us in a position to estimate the depth of erosion or, indeed, the original altitude of eroded mountain ranges.

It is of paramount importance to obtain better information about the temperature-pressure conditions of the recrystallization and thus to show the relation between the chemical and mineralogical composition of all varieties of rocks. A large scale attempt in this direction was the development of the facies classification of rocks.

Definitions and Principles of Mineral Facies

Regional metamorphism has affected rocks of the extensive orogens of the earth, for example, the old Paleozoic Caledonian mountain ranges, the young Paleozoic Hercynian areas, and the Tertiary Alpine mountain ranges. Above all, it has influenced the pre-Cambrian rocks, which are composed predominantly of crystalline schists.

According to Rosenbusch, the crystalline schists may be divided into orthogneisses formed from pre-existing igneous rocks, and paragneisses formed from pre-existing sediments. He also showed that the two groups could be distinguished by chemical criteria.

Michel-Lévy (1888) distinguished three main "étages" in the formation of the crystalline schists, and Becke and Grubenmann (1910) demonstrated that the same original material may produce radically different metamorphic products according to the effective temperature and pressure during the metamorphism. Grubenmann distinguished three different depth zones: Epizone, Mesozone, and Katazone.

By and large, it may be said that in the eroded mountain ranges, rocks of the katazone are encountered in the central parts; toward the marginal parts are found rocks of the mesozone and the epizone.

The minerals that make up the essential rock constituents in any one zone, and are in mutual chemical equilibrium, have been called *typomorphic* by Becke. Grubenmann regarded the following minerals as typomorphic for the epizone: chloritoid, garnet, antigorite, hornblende, chlorite, epidote, zoisite, albite, glaucophane, sericite, biotite.

For the mesozone: kyanite, staurolite, almandite, anthophyllite, grünerite, hornblende, epidote, zoisite, soda-rich plagioclase, muscovite, biotite, alkali hornblende.

For the katazone: sillimanite, almandite, rhombic pyroxene, olivine, monoclinic pyroxene, omphacite, pyrope, cordierite, spinel, anorthite, albite, jadeite, potash feldspar, biotite, acmite.

Grubenmann and Niggli (1924) adhered formally to the idea that the zones represent the normal sequence of layers in the crust of the earth. But, since from a physico-chemical point of view, the zonal distinctions correspond to a division according to temperature and pressure, and since metamorphism in the crust is typically associated with abnormal conditions, they emphasize that in many places the natural sequence may be reversed.

Long before the depth zones of Grubenmann and Niggli had been proposed, G. Barrow (1893) had worked out a classification principle of regional metamorphic rocks of the Highlands of Scotland according to the so-called index minerals. In his work we find the germ of that principle in metamorphic geology which later British authors (E. B. Bailey, H. H. Read, C. E. Tilley), supported by the work of V. M. Goldschmidt on the hornfelses of the Oslo region, used to contrive a well-defined system. Finally, through the supplementary investigations of Eskola, it was perfected and shaped into a tool of highest value: the system of the mineral facies of rocks.

Let us recall the discussion of the mineralogical composition of the hornfelses as previously presented. Now we may regard all hornfelses as belonging to one mineral facies, the facies of the rocks of the inner contact aureoles (the hornfels facies). The mineral development is determined here by the chemical composition of the original sediments, that is, if the chemical composition of the original sediment is known, the minerals that will develop in the corresponding hornfels can be calculated. Although nature always introduces elements of uncertainty, and equilibrium conditions may not have been attained in all rocks, a good general picture of the hornfels minerals is thus obtained that statistically corresponds closely to the actual conditions.

By referring all varieties of rocks to definite facies, that is, to definite temperature-pressure regions "so similar that a definite chemical composition has resulted in the same set of minerals," the rock kingdom is broken up into a series of natural divisions.

A mineral facies, in the sense of Eskola, "comprises all the rocks that have originated under temperature and pressure conditions so similar that a definite chemical composition has resulted in the same set of minerals, quite regardless of their mode of crystallization, whether from magma or aqueous solution or gas, and whether by direct crystallization from solution . . . or by gradual change of earlier minerals . . ." (P. Eskola, 1921).

Ramberg (1952) says: "Rocks formed or recrystallized within a certain *P-T* field, limited by the stability of certain critical minerals of defined composition, belong to the same mineral facies."

In a way the mineral facies correspond to the zones of metamorphism, in the sense used by Barrow. Rocks within the same zone may be called isofacial, or isograde, as proposed by Tilley, who, furthermore, proposed the term isograd for a line of similar temperature-pressure values.

In going from an area of unmetamorphic rocks into an area of progressively more highly metamorphic rocks, each zone of progressive metamorphism is defined by an index mineral—the first appearance of which marks the outer limit of the zone in question. The sequence of index minerals in progressively metamorphosed pelitic rocks is: chlorite, biotite, almandite garnet, staurolite, kvanite, and sillimanite.

A line can be drawn on the map indicating where biotite first appears. This line is the biotite isograd. The less metamorphosed argillites on one side of this line lack biotite, whereas the more metamorphosed rocks on the other side contain biotite. An isograd can be drawn for each mineral. Actually the isograds are surfaces, and the lines we draw on the map are the intersections of these surfaces with the surface of the earth.

However, an isograd may be related to *rate* of reaction, indicating the attainment of a sufficient rate such that a mineral will appear—some garnets may grow at room temperature, but the rate of the growth is too low for the garnet to appear. One may here use the word *isoblast* (Yoder, 1952).

Goldschmidt applied the name *Facies-Gruppe* to denote the relationship of several geologically related facies. For the same thing Vogt later used the term *facies series.*

Minor departures from the standard mineral assemblages within a facies due to small differences in temperature and pressure, but not due to chemical differences, may find their expression in terms of subfacies.

Critical minerals occur only in one facies and not in any other (pyrope, jadeite).

Typomorphic minerals occur stably in several facies.

False equilibria may occur if certain minerals form metastably outside their true range of stability under permission of Ostwald's rule. It appears, for example, that andalusite often presents itself with a false passport as a stably growing crystal in that it incontestably replaces older minerals, whereas sillimanite actually represents the stable phase.

ACF Diagrams

If equilibrium is attained during metamorphism, a definite relation between the constituent minerals and the "fixed" chemical components of the rocks should exist. The quantitative factors governing the mineral development are, however, so complicated that some explanation is necessary. As shown by chemical analyses of rocks, no less than twelve oxides must be regarded as essential for the mineral development. According to the mineralogical phase rule, a maximum number of twelve minerals in equilibrium with each other can thus occur in these rocks.

(1) CO_2, when present, forms a carbonate, either calcite or dolomite.

(2) P_2O_5 forms apatite.

(3) TiO_2 forms sphene. It may also go into ilmenite, or instead of creating any new mineral it enters as a substitutive oxide into biotite, garnet, and other minerals.

(4) Fe_2O_3 forms hemate, or goes into magnetite, or replaces Al_2O_3 in various minerals.

(5) SiO_2 governs the degree of silicification of the minerals. In a first approximation we shall limit our attention to rocks oversaturated with silica, in which the minerals always exhibit the highest possible degree of silicification.

(6) Na_2O enters usually into the albite molecule, which in most facies forms solid solution with anorthite (plagioclase). In such cases the composition of the plagioclase is fixed by the ratio $CaO:Na_2O$ in the original rock, allowing, however, for "femic" CaO and for small amounts of CaO consumed by CO_2 and P_2O_5 (see 1 and 2) and for possible amounts of Na_2O present in hornblendes.

(7) H_2O determines whether or not hydrous mineral phases will develop. Most rocks are "open" to H_2O, i.e., the rock "system" may exchange H_2O with the environment. H_2O is therefore a "mobile" component and requires special consideration. (See equation 31 in Appendix.)

The oxides 1 to 6 play a passive role in the mineral development, for, within the usual range of variation, these oxides are of no consequence in the formation of any other minerals than those already considered. More active and complicated are K_2O, Al_2O_3, FeO, MgO, CaO.

It is customary in petrographic calculation to unite FeO and MgO. In the following graphical representations we shall do this, although it must be emphasized that FeO and MgO cannot substitute for each other in all minerals of the facies of low-grade metamorphism.

Now we have reduced the "active" oxides in metamorphic rocks to K_2O and the following three components: Al_2O_3, (FeO + MgO), CaO. These three components control to a large extent the mineral development in many metamorphic rocks. On them Eskola has therefore based his triangular *ACF* diagrams, which are of great importance in the facies classification of rocks. The following are the rules for calculating the position of a chemical rock analysis in an *ACF* diagram:

The chemical analysis is first recalculated into cation percentages. Correct for calcite and apatite by subtracting $5/3 \cdot P + C$ from Ca. If one wants to correct for the accessories, ilmenite, sphene, and magnetite, the percentages of these minerals are first estimated by micrometric analysis as *i*, *s*, and *m*, respectively. Then from the percentage of $Fe^{\cdot\cdot}$ subtract $\frac{1}{2}i + \frac{1}{3}m$, likewise subtract $\frac{1}{3}s$ from Ca and $\frac{2}{3}m$ from $Fe^{\cdot\cdot\cdot}$. *A*, *C*, and *F* can now be reckoned as follows:

$A = Al + Fe^{\cdot\cdot\cdot} - (Na + K)$; $C = Ca$; $F = Mg + Fe^{\cdot\cdot} + Mn$. They are finally recalculated so that $A + C + F = 100$. Note that *A* is given by the total aluminium and ferric iron less the quantity that would be required to combine with total alkali as feldspar.

The *ACF* triangles used in this book have one feature different from those of Eskola. In the *A*-corner, the atom Al is here taken as unit, whereas Eskola used Al_2O_3. Thus the position, for instance, of the mineral zoisite (epidote) $Ca_2Al_3Si_3O_{12}(OH)$ is here presented by a plot dividing the *C*-*A* join in the proportion 2:3. In Eskola's diagram the proportion would be 2:1½ because 3Al is represented by $1\frac{1}{2}Al_2O_3$ (see Fig. IV-38).

Descriptions of Mineral Facies from Low to High

Rocks of the greenschist facies recrystallized at low temperatures and often under high shearing stress. Chlorite schist, epidote-albite schists, actinolite schists are typical, they are all green, hence the name. Serpentinites, talc schists, phyllites, and muscovite (sericite) schists are common. Plagioclase is not stable in these rocks but breaks up into epidote and albite. (Pages 313 ff.)

Rocks of the epidote amphibolite facies recrystallized in a somewhat higher temperature range. Amphibolites with epidote and either albite or oligoclase are typical. Mica schists (garnet-), biotite schists, staurolite schists, kyanite schists are common, also cordierite-antophyllite (gedrite) schists, chloritoid schists occur. Only sodic plagioclase is stable in these rocks. (Pages 317 ff.)

Rocks of the amphibolite facies recrystallized around 500°C. Amphibolites of hornblende and plagioclase are typical, they often carry quartz and/or biotite, or garnet. Sillimanite-muscovite schists (gneisses) are common. (Pages 320 ff.)

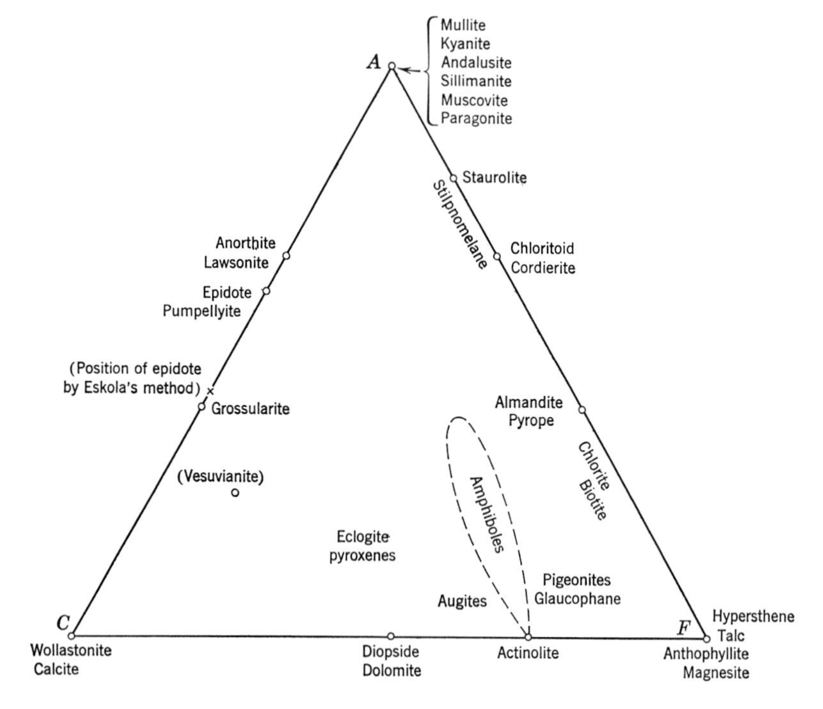

Fig. IV-38. Plots of all minerals of systematic importance in the *ACF* classification.

Rocks of the granulite facies have their greatest extension in old pre-Cambrian areas, but are also found in younger deeply eroded mountain chains. Hypersthene (-garnet)-plagioclase gneisses, usually with quartz, but occasionally with olivine or spinel are typical. Amphibolites with hypersthene or diopside are found, also sillimanite gneisses. The temperature range is probably from about 500°C to about 700°C (?). (Pages 322 ff.)

Zeolitic Facies. These rocks were newly defined by Fyfe et al. in 1958 as transitional between diagenesis and regional metamorphism. Fyfe's definition is based on Coomb's (1954) descriptions from Southland, New Zealand, of andesitic volcanic sands buried to depths of 6–9 km, corresponding to load pressures of 2000–3000 atm and temperatures at 200–300°C. The sand is completely converted to assemblages rich in zeolites, for example, laumontite, albite, and quartz (adularia may develop and pumpellyite may replace laumontite).

Zeolite-facies rocks with laumontite as the chief zeolite are also abundant in the northwestern Coast Range, California.

Important relations are:

$$\text{calcite} + \text{kaolinite} + \text{quartz} + \text{water} = \text{laumontite} + CO_2$$

$$CaCO_3 + Al_2Si_2O_5(OH)_4 + 2SiO_2 + 2H_2O = CaAl_2Si_4O_{12}\cdot 4H_2O + CO_2$$

and additional SiO_2 and H_2O give:

$$\text{calcite} + \text{kaolinite} + \text{quartz} + \text{water} = \text{heulandite} + CO_2$$

At constant P and T it is possible to obtain, for a system isochemical with respect to the inert components, mineral assemblages characteristic both of the zeolite facies and of the greenschist facies depending on the relative values of the chemical potentials of water and carbon dioxide. (E-An Zen, 1961.)

Greenschist Facies. Metamorphic rocks of the lowest grade, recrystallized in the range of 100 to 250°C (or perhaps less), and often under rather high shearing stress, are grouped together in the mineral facies of the greenschists.

The common minerals stable in this facies are quartz, albite, potash feldspar, muscovite, kaolinite, chlorite, serpentine, talc, epidote (zoisite), iron-rich actinolite, and the carbonates calcite, dolomite, and magnesite. Accessories are tourmaline, rutile, spessartite, and others.

The *ACF* diagram, shown in Fig. IV-39, demonstrates that epidote is a common constituent in most of the rocks. However, the chemical adjustment is not good. The low temperature and, generally, the near-surface conditions have hindered the attainment of internal equilibrium; moreover, the equilibrium conditions are difficult to define, and their interpretations exceptionally complicated.

A general feature in low-grade metamorphism is that the sediments tend to lose water and carbon dioxide. If these constituents find easy egress to the surface of the earth, they are delivered into the hydrosphere and atmosphere and disappear from the rocks. However, in many situations part of the water and carbon dioxide will be retained, and the partial vapor pressure of them will increase in the rock during the metamorphism.

The phases crystallizing under high pressure of water or carbon dioxide will be different from those appearing if these gases are absent.

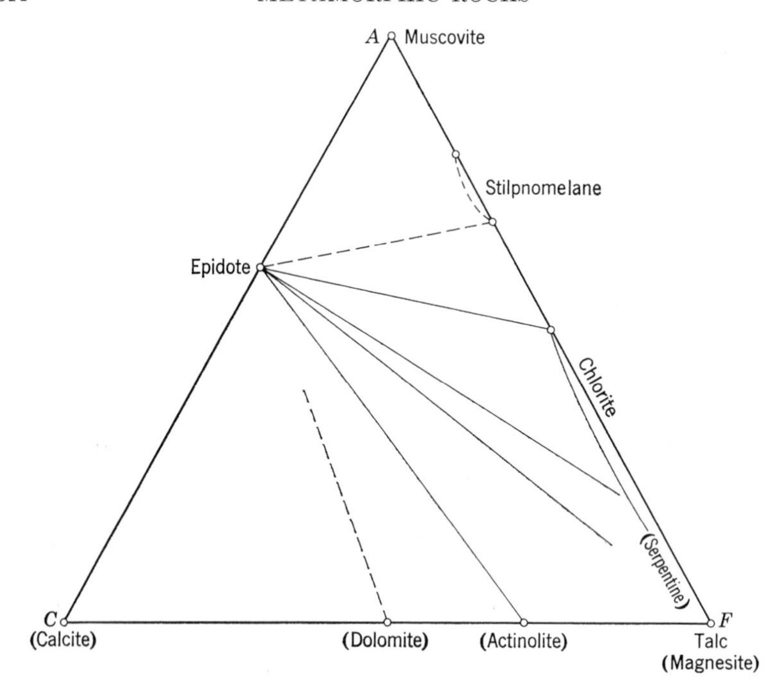

FIG. IV-39. *ACF* diagram of the mineral associations in the greenschist facies.

If the carbon dioxide pressure is insufficient, neither calcite nor dolomite will be able to grow. But field observations indicate that circulating carbonate-bearing solutions are almost always present in this type of rock, that is, in lime-rich rocks almost always calcite (or dolomite) are able to form. To be sure, no lime-rich silicate close to the *C*-corner is stable in this facies. Thus no choice is left to a lime-rich rock; either calcite has to develop or the rock has to change its composition. If much carbon dioxide is present, magnesite will develop instead of talc in the *F*-corner. The carbonates, therefore, grow together with the silicates as "supernumerary" phases, the amount of which is determined by the carbonate content of the rock-making solutions. Talc and magnesite, therefore, are frequently associated; so are talc and calcite (or dolomite), although according to Fig. IV-39 they should be separated by actinolite.

The *F*-corner of the diagram represents both Mg and Fe. In many rock-making minerals FeO and MgO substitute freely for each other, and it is customary in petrographic calculations (and for many rocks satisfactory) to regard FeO and MgO as one component. But because of the low temperature in the greenschist facies, this substitution does not take place in all its minerals.

We have the equation:

$$\begin{array}{l} 5Mg_3Si_4O_{10}(OH)_2 + 6CaCO_3 + 4SiO_2 \rightarrow \\ \quad \text{talc} \quad + \text{calcite} + \text{quartz} \rightarrow \\ \quad \quad \text{greenschist facies} \rightarrow \\ \quad \quad \quad 3Ca_2Mg_5Si_8O_{22}(OH)_2 + 6CO_2 + 2H_2O \quad \text{(IV-1)} \\ \quad \quad \quad \quad \text{tremolite} \quad + \quad \text{gas} \\ \quad \quad \quad \quad \quad \text{higher facies} \end{array}$$

Under the conditions of the greenschist facies the association calcite + talc may be regarded as the low-temperature "modification" of the pure Mg-tremolite. However, in tremolite, part of magnesia may be replaced by iron (actinolite), whereas talc is unable to take iron into solid solution. Let us consider actinolite as a solid solution of two components, $Ca_2Mg_5Si_8O_{22}(OH)_2$ and $Ca_2Fe_5Si_8O_{22}(OH)_2$. At low temperatures the magnesian component breaks up into calcite and talc, but the ferrous component remains stable. Schematically we have:

$$\begin{array}{lcl} \text{Calcite + talc + quartz} & \nwarrow & \\ & & \begin{cases} xCa_2Mg_5Si_8O_{22}(OH)_2 \\ yCa_2Fe_5Si_8O_{22}(OH)_2 \end{cases} \quad \text{(IV-2)} \\ \text{Fe-rich actinolite} & \swarrow & \\ \quad \text{greenschist facies} & \leftarrow & \text{higher facies} \end{array}$$

indicating that in rocks of the greenschist facies the association calcite, talc, actinolite does not necessarily mean lack of chemical adjustment.

Analogous difficulties are encountered in interpreting assemblages with chloritoid, which is an iron-alumina silicate into which no, or very little, magnesia can enter. Consequently chloritoid is unable to form in magnesia-rich rocks. The general position of chloritoid is uncertain, however. It is possible that it requires special conditions for its development (perhaps higher temperature than in the usual greenschist facies, perhaps special pressure conditions). It may actually define a subfacies of somewhat higher grade.

Stilpnomelane is a major constituent of many low-grade schists in southern New Zealand, in the western Alps and probably has a wider general distribution in rocks of the chlorite zone than has usually been recognized (Turner).

It should be remembered that Fig. IV-39 pictures only the mineral facies of highest degree of silification (stable in contact with quartz). At the *F*-corner serpentine is a stable phase if silica is insufficient for the formation of talc. Serpentine forms a continuous series of solid solutions with chlorite; consequently, the extension of the chlorite series towards the *F*-corner is determined by the amount of silica.

The potassium content represents another complication. The reaction

$$\underset{\text{orthoclase}}{KAlSi_3O_8} + \underset{\text{kaolinite}}{Al_4Si_4O_{10}(OH)_8} \rightarrow \underset{\text{muscovite}}{KAl_3Si_3O_{10}(OH)_2} + \underset{\text{quartz}}{4SiO_2} + \text{Al-hydroxide} \quad \text{(IV-3)}$$

appears to take place at very low temperatures (probably under intensive diagenesis). Muscovite, therefore, is the stable mineral in the *A*-corner; but, if potash is deficient, kaolinite remains stable. It should be clearly recognized, therefore, that in rocks of the greenschist facies alternative equilibrium assemblages, depending upon the presence of absence of additional components of the system, are possible, and that where circulating aqueous solutions have been active in metamorphism, the composition of the schists may have been materially affected by removal of lime or other bases. As emphasized by Turner, a basaltic lava may be converted, with but slight change in chemical composition, to the assemblage albite-epidote-actinolite-chlorite, which represents a state of equilibrium in a closed system. But under the rules of metamorphic differentiation (emigration of the most soluble compound, page 300) the bulk composition of the rock then tends to change so that it can be expressed as a simpler assemblage such as albite-epidote-chlorite or even albite-chlorite. If carbon dioxide is introduced as an additional component, the stable parageneses is albite-epidote-chlorite-calcite.

In conclusion, it may be stated that the systematic investigation of rocks of this facies is very complicated. Iron and magnesia cannot be regarded as one component but must be treated separately; the different equilibria are influenced by rock pressure, H_2O-pressure, CO_2-pressure, the Mg/Mn/Fe ratio, the Al/Si ratio, and other factors. Consequently, Fig. IV-39 gives no easy clue to the various mineral associations. It must be read *cum grano salis*, and for each rock we want to investigate we must look for all possible additional information in the field and in the laboratory.

Epidote Amphibolite Facies. With increasing metamorphism biotite becomes a stable mineral. Biotite is common and easy to see in rocks, it is, therefore, an excellent index mineral, and in many areas of progressive metamorphism the biotite isograd has been accurately mapped. It seems practical to choose this isograd as the boundary between the greenschist facies and the next higher epidote amphibolite facies.

Biotite will form from chlorite (antigorite):

$$\underset{\text{antigorite}}{Mg_3Si_2O_5(OH)_4} + \underset{\text{orthoclase}}{KAlSi_3O_8} = \underset{\text{biotite}}{KMg_3AlSi_3O_{10}(OH)_2} + \underset{\text{quartz}}{2SiO_2} + \underset{\text{water}}{H_2O}$$

and (IV-4)

$$\underset{\text{antigorite}}{5Mg_3Si_2O_5(OH)_4} + \underset{\text{muscovite}}{3KAl_3Si_3O_{10}(OH)_2} = \underset{\text{biotite}}{3KMg_3AlSi_3O_{10}(OH)_2} + \underset{\text{amesite}}{3Mg_2Al_2SiO_5(OH)_4} + \underset{\text{quartz}}{7SiO_2} + \underset{\text{water}}{4H_2O} \quad \text{(IV-5)}$$

The temperature at which these reactions go to the right is not easy to ascertain, but it may lie around 250°C. This is, therefore, the lower boundary of the epidote amphibolite facies.

Another interesting reaction is

$$\underset{\text{chlorite}}{(Mg,Fe,Mn)_9Al_3Al_3Si_5O_{20}(OH)_{16}} + \underset{\text{quartz}}{4SiO_2} = \underset{\text{garnet}}{3[(Mg,Fe,Mn)_3Al_2Si_3O_{12}]} + \underset{\text{water}}{8H_2O} \quad \text{(IV-6)}$$

that is, chlorite of certain compositions goes directly into garnet with release of water. This reaction is particularly sensitive to the Mn/Fe/Mg ratio. With high manganese content it proceeds to the right at lower temperatures than shown in equations IV-4 and IV-5; that is, in areas where manganiferous garnets develop, the garnet isograd is on the low-temperature side of the biotite isograd. In most areas, however, the garnet is a common almandite, and reaction IV-6 takes place at higher temperatures than IV-4 and IV-5. With very high magnesium content reaction IV-6 does not take place unless at very high pressure (and temperature), that is, in the eclogite facies.

In the study of garnet isograds, the chemical composition of the several garnets is, therefore, of the utmost importance.

It is interesting that anthophyllite is the stable mineral of the *F*-corner. Synthetic studies have demonstrated that talc is stable to 750 to 800°C (see Fig. IV-8). The fact that in this facies it is replaced by anthophyllite would seem to indicate that the vapor pressure in the "pore fluids" usually is less than the rock pressure.

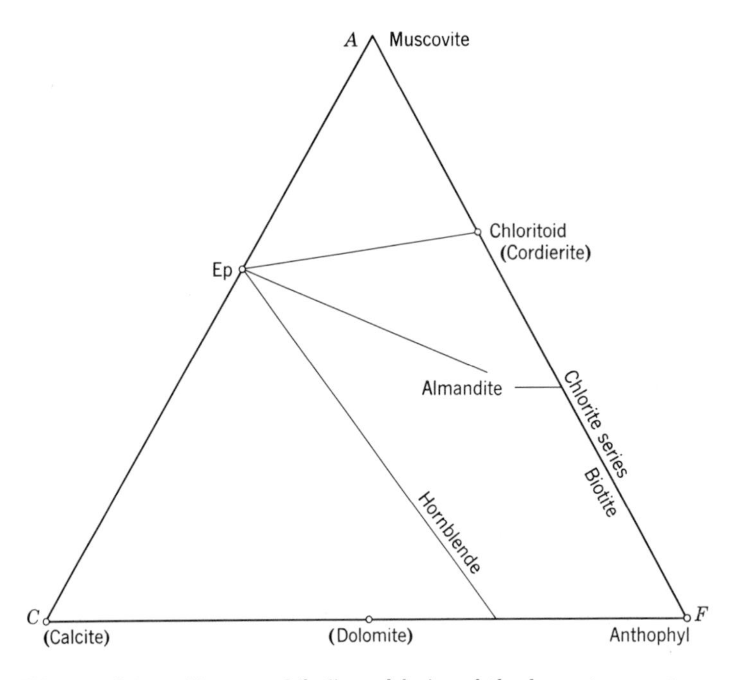

Fig. IV-40. Albite-epidote amphibolite subfacies of the lower temperature range of the epidote-amphibolite facies.

The most important mineral reactions in the epidote-amphibolite facies are those involving plagioclase and epidote. The thermochemical relationships of these minerals have been discussed previously (Fig. IV-12). The point of interest is that the composition of plagioclase in contact with epidote is a temperature (and pressure) indicator. In the low-grade parts of the epidote-amphibolite facies, albite is the only stable plagioclase; in the higher parts oligoclase of 20–30An becomes stable.

Thus the equilibrium relation between plagioclase and epidote as illustrated in Fig. IV-12 makes it logical to establish two subfacies: From lower to higher metamorphic grade:

Albite-epidote amphibolite subfacies, which is transitional into the greenschist facies. The distinguishing features are that in rocks of suitable compositions biotite develops. With deficient K_2O either manganiferous almandite or chlorite substitutes for biotite. Actinolite may be slightly more magnesian than in the greenschists. In certain

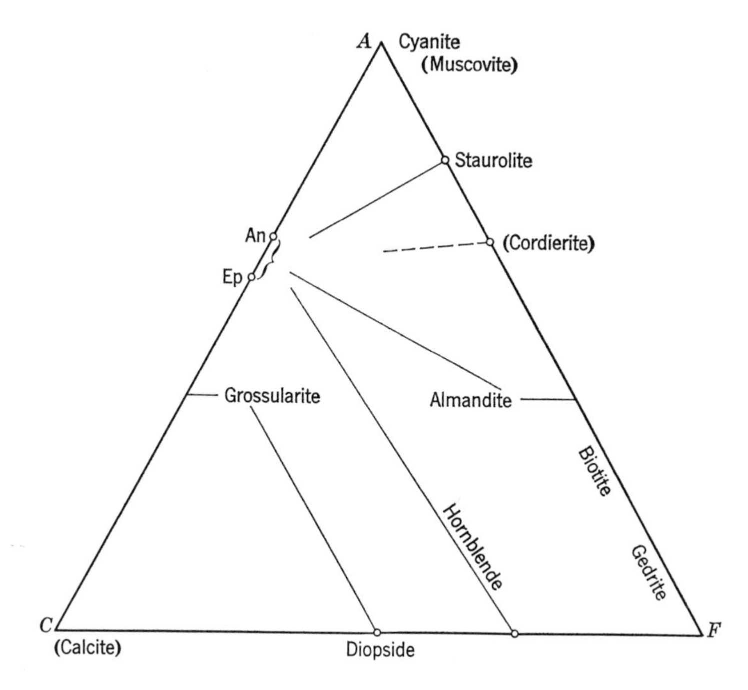

FIG. IV-41. The higher part of the range embraced by the epidote-amphibolite facies.

rocks of high iron content, chloritoid may develop. A parallel subfacies is probably the actinolite greenstone facies of Tilley (1924) and the greenstones of Th. Vogt (1927).

Oligoclase-epidote emphibolite subfacies will develop above temperature t_2 (ca. 400°C?) in Fig. IV-12. Here belong the *Alpine amphibolite facies* of Angel and Scharizer (1932), characterized by the paragenesis hornblende-oligoclase-epidote, and the *prasinite facies* (Woyno, 1912; Weg, 1931), based upon the assemblage oligoclase-epidote-barroisite.

It requires high pressure (stress), but is otherwise similar to the Alpine amphibolite facies. Barroisite is a blue-green sodic amphibole related to the glaucophane series.

In the higher part of the range embraced by the epidote-amphibolite facies, the anorthite content of the plagioclases attains 30 to 40 per cent.

In this facies, iron and magnesium are frequently unable to replace each other. The iron-rich mineral chloritoid can take but very little magnesia in solid solution and is, therefore, unable to form in iron-poor rocks. Cordierite may develop instead but requires a somewhat higher temperature for its formation. Similarly, staurolite is an iron-mineral, and cordierite substitutes for it in magnesium-rich rocks.

A serious problem in determining the equilibrium conditions for all these rocks is the presence of unstable relics and hysterogenic products.

Amphibolite Facies. In the classical investigations of Eskola on the relations between the chemical and mineralogical composition in

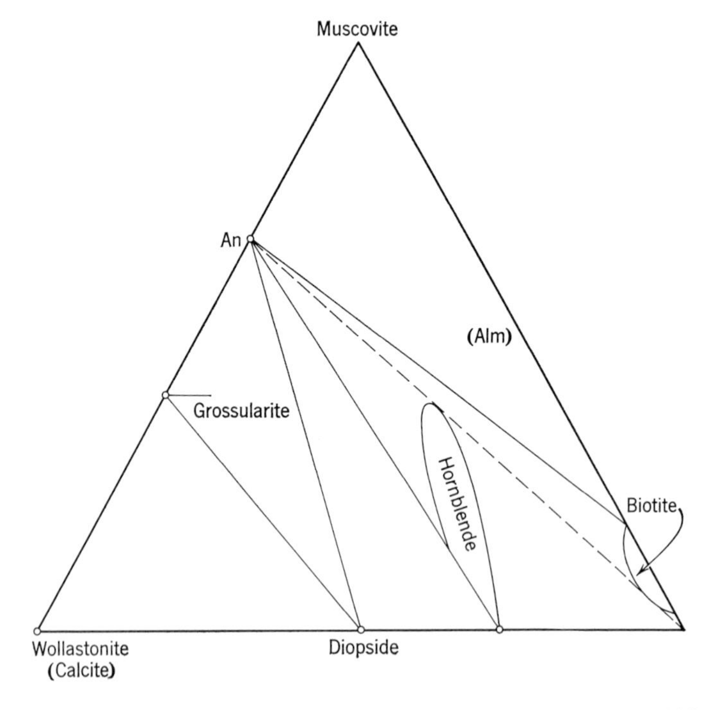

FIG. IV-42. Amphibolite facies. *ACF* diagram for rocks with excess SiO_2 and K_2O.

the metamorphic rocks of the Orijärvi region (*Bull. comm. géol. Finlande*, 1915) the first principles of the mineral facies and the laws of mineral assemblages in the amphibolite facies were laid down. In this facies amphibole always occurs if the rock composition permits. Anorthite is stable (epidote is unstable except in lower temperature range of the facies), and the association hornblende-plagioclase is characteristic.

The facies has a wide distribution in pre-Cambrian gneisses and in the deeper parts of the orogenic mountain ranges. The epidote-amphibolite facies is also represented in these areas, but it is more typical for the more elevated parts of the folded mountains. In the pre-Cambrian not only supracrustal formations (leptites, crystalline limestones, etc.) but also the synkinematic intrusive-like rocks (hornblende gabbros, diorites, granites) belong to the amphibolite facies. The formation of the minerals of the amphibolite facies of the intrusive complexes is perhaps metamorphic in origin, but it is also possible that the intrusions were effected by low-temperature "emanations" or solutions, and that the equilibrium relations of the amphibolite facies governed the primary crystallization of these rocks.

The mineral associations in rocks of sufficient silica are shown in Fig. IV-42. In rocks deficient in silica, olivine, spinel, and corundum may occur.

The typical mineral associations of the amphibolite facies are governed strictly by the bulk chemical composition of the rock. It simplifies matters that during this kind of metamorphism, water always appears to have been present in sufficient quantities to produce the typomorphic hydroxyl-bearing minerals. For instance, the formation of biotite depends only on the proportions $K_2O/(Mg,Fe)O/Al_2O_3$. In rocks containing sufficient K_2O to bind all $(Mg,Fe)O$ and Al_2O_3 in mica, the amount of mica is accurately determined by the amount and the proportions of these oxides. An excess of K_2O then goes into potash feldspar. For such rocks the *ACF* diagram reflects quantitatively the proportions of the several chief minerals other than potash feldspar and quartz (and contents of Ab in plagioclase).

In rocks showing insufficient K_2O relative to Al_2O_3 for the formation of mica, one or two of the following minerals occur: andalusite, cordierite, anthophyllite (or cummingtonite). The amount of these minerals in relation to mica is determined by the amount of K_2O, and cannot be pictured quantitatively in the *ACF* diagram. See Fig. IV-43.

The maximum temperature attained in this facies appears to be around 500°C. Wollastonite is present in the upper temperature range

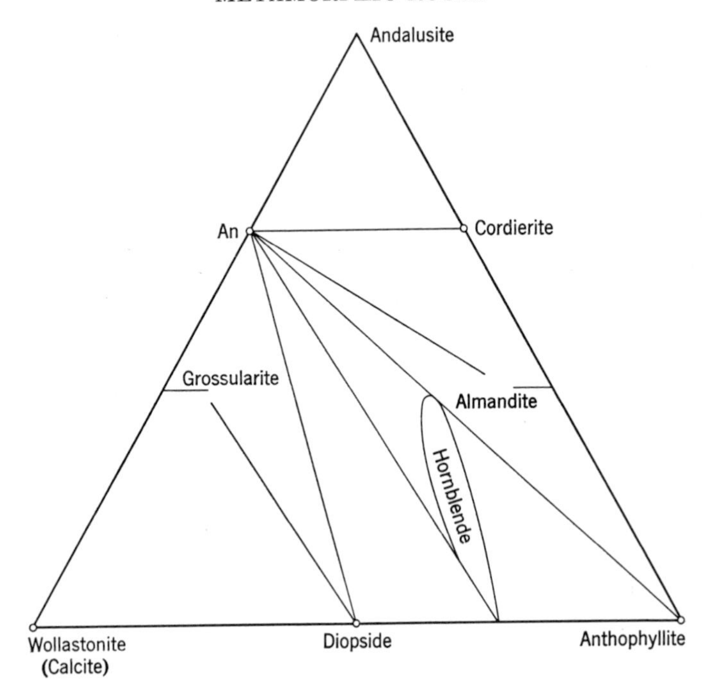

FIG. IV-43. Amphibolite facies. *ACF* diagram for rocks with excess SiO_2 and deficient in K_2O.

of the facies, but in the larger parts of the facies the assemblage quartz + calcite is stable.

It is not always obvious where to put the division lines between the several facies. It is often a matter of convention or of expediency; Table IV-8 summarizes the proposals of various authors.

Granulite Facies. The granulite facies is comprised of rocks which have been affected by the highest pressure-temperature conditions of regional metamorphism and granitization. At high pressures, rocks of granitic composition will be melted between 600 and 700°C. (See Fig. III-41.) The temperature range of the granulite facies is, therefore, from about 500°C (corresponding to the transition amphibolite facies–granulite facies) to about 650°C (corresponding to the transition into anatectic "magmatic" rocks). A modern treatment of the granulite facies has been given by H. Ramberg (*J. Geol.,* 1949).

Rocks of granulite facies have their greatest extension in old pre-Cambrian areas (Uganda, Peninsular India, Norway, Greenland) but are also found in younger, eroded mountain chains. In the Caledonian

TABLE IV-8

CLASSIFICATION OF SOME MINERAL FACIES

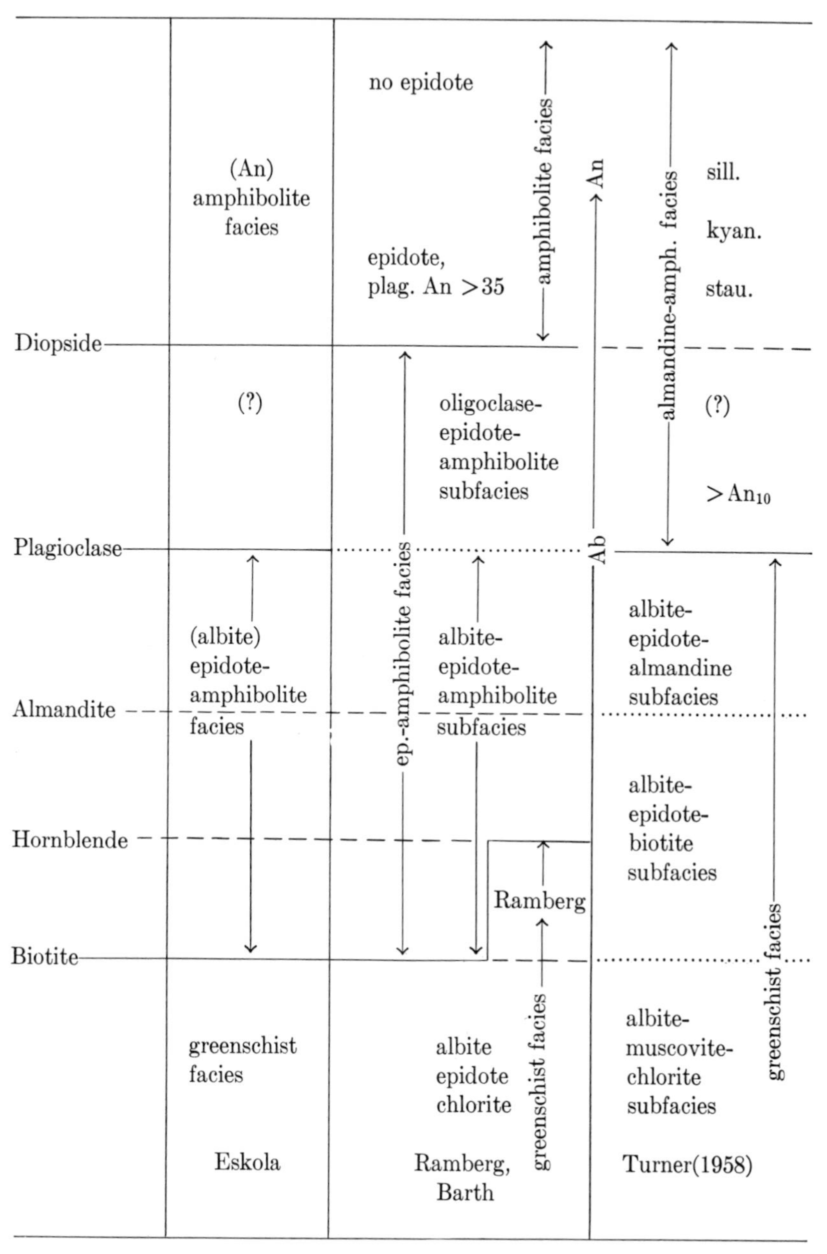

chain in Norway, for example, the granulite facies is found in two situations: (1) in the roots of the chain in the west coast district and (2) in the form of overthrust rocks of charnockitic-mangeritic relation in the central mountain area.

In this facies the basic rocks are rich in hornblende, whereas ortho-rhombic pyroxene is the most common ferromagnesian mineral in the acid members. Diopsidic pyroxene, biotite, and garnet are commonly found in basic as well as in acid members. Actinolitic hornblendes are unstable in the highest parts of the amphibolite facies and in the granulite facies; they split up into hypersthene and diopsidic pyroxene:

$$\underset{\text{actinolite}}{Ca_2(Mg,Fe)_5Si_8O_{22}(OH)_2} \rightleftharpoons \underset{\text{diopside}}{2Ca(Mg,Fe)Si_2O_6} + \underset{\text{hypersthene}}{3(Mg,Fe)SiO_3} + \underset{\text{quartz}}{SiO_2} + H_2O \quad \text{(IV-7)}$$

The hornblendes of the granulite facies are rich, therefore, in aluminum (aluminous "common" hornblendes). See Fig. IV-44.

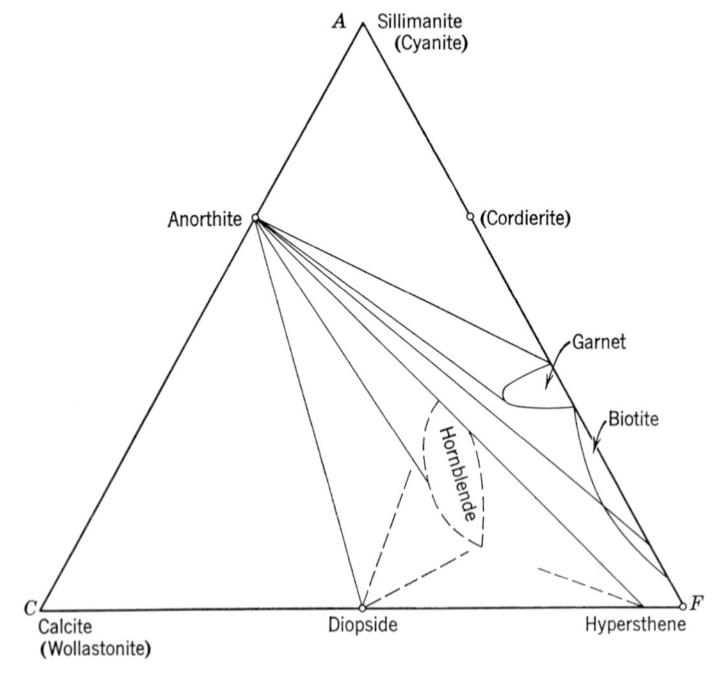

FIG. IV-44. Mineral associations in the granulite facies for rocks with excess SiO_2.

The following reactions are characteristic:

$$\underset{\text{hornblende}}{NaCa_2(Mg,Fe)_4Al_3Si_6O_{22}(OH)_2} + \underset{\text{garnet}}{(Mg,Fe)_3Al_2Si_3O_{12}} + \underset{\text{quartz}}{5SiO_2} \overset{\text{amphibolite facies}}{\rightleftharpoons}$$

$$\overset{\text{granulite facies}}{\underset{\text{hypersthene}}{7(Mg,Fe)SiO_3} + \underset{\text{plagioclase}}{NaCa_2Al_5Si_7O_{24}} + H_2O} \quad \text{(IV-8)}$$

It is noteworthy that hornblende will be stable at higher pressure-temperature conditions in SiO_2-deficient rocks than in those which are quartz-bearing. In other words, the reaction

$$\underset{\text{hornblende}}{NaCa_2(Mg,Fe)_4Al_3Si_6O_{22}(OH)_2} + \underset{\text{garnet}}{(Mg,Fe)_3Al_2Si_3O_{12}} + \underset{\text{hypersthene}}{3(Mg,Fe)SiO_3}$$

$$\leftrightharpoons \underset{\text{olivine}}{5(Mg,Fe)_2SiO_4} + \underset{\text{plagioclase}}{NaCa_2Al_5Si_7O_{24}} + H_2O \quad \text{(IV-9)}$$

takes place at higher pressure-temperature conditions than does reaction IV-8. Hornblende, rhombic pyroxene, anorthite or olivine, garnet, and/or spinel can, therefore, coexist in equilibrium in unsaturated rocks in the pressure-temperature interval between reactions IV-8 and IV-9. It should be remembered, however, that even in silica-rich rocks equations IV-8 and IV-9 are not restricted to a given curve in the pressure-temperature field, but are rather related to an interval because the minerals taking part in the reactions are mixed crystals. It is also probable that the reaction point for the pure magnesium members is situated so low that enstatite is replaced by anthophyllite (or cummingtonite).

Sanidinite Facies. Comprising rocks of extreme thermal metamorphism, this facies embraces some of the lime silicates described in the chapter on reactions between carbonates and silicates. The *ACF* diagram of this facies is Fig. IV-45.

Eclogite Facies. Rocks of the eclogite facies are very different from all other rocks and do not belong to the normal products of either contact metamorphism or regional metamorphism. The supposed pressure-temperature conditions of formation are shown in Fig. IV-49.

As distinct from all other facies, plagioclase is not present. To be sure, feldspar has never been observed paragenetically associated with the typical eclogite minerals, omphacite and eclogite garnet (pyrope). And these latter minerals are not known from any other rocks but are critical for the eclogite facies. Typomorphic minerals are diopside, enstatite, olivine, kyanite, rutile, and, in rare cases, diamond. Calcite is also stable.

Omphacite and eclogite garnet exhibit considerable chemical variation in their *ACF* relations. Equivalent amounts of aluminum and sodium, which are subtracted from the projection values, play the same role in jadeite ($NaAlSi_2O_6$) as in albite of the normal facies, but are here able to substitute for calcium and magnesium in the

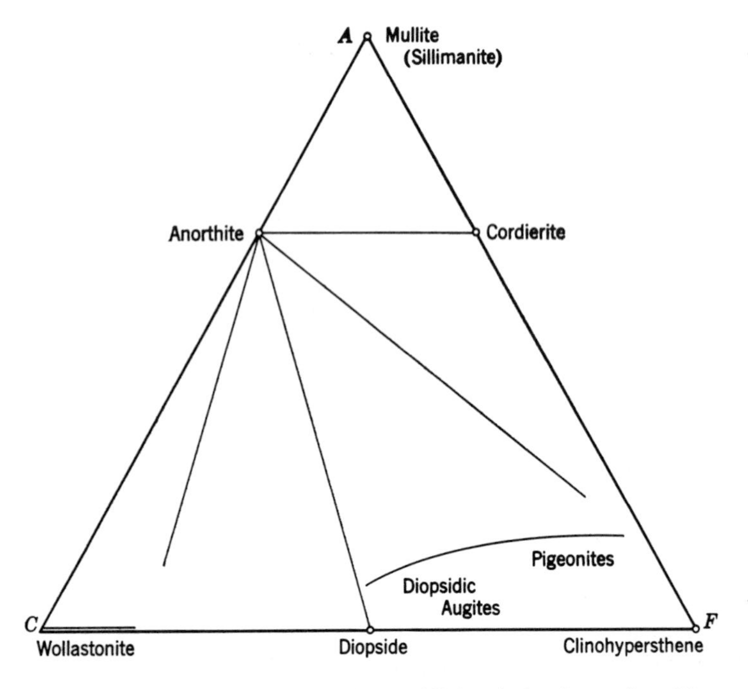

FIG. IV-45. Mineral associations in the sanidinite facies for rocks with excess SiO_2.

pyroxene lattice. The *ACF* triangle is not well adapted, therefore, to express the rules of association in this facies. (See Fig. IV-46.)

The typical eclogites contain but two minerals: garnet and pyroxene, for both of them exhibit a high degree of solid solubility for other substances. With increase or decrease in the Fe,Mg-content beyond the limits of solid solutions in these two minerals, hypersthene or kyanite, respectively, will enter as an additional mineral. See fields 1 and 2 of Fig. IV-46.

Some eclogites occur in association with crystalline limestones; it is possible, therefore, that compositions poor in magnesium and iron, corresponding to fields 3 or 4 might occur. But no good examples are on record.

The eclogite pyroxenes show great variations in the Ca/Mg/Fe ratio and in the amount of silica; jadeite, omphacite, and chloromelanite are the chief molecules. The eclogite pyroxenes as well as all other eclogite silicates are very poor in titania that enters into rutile or

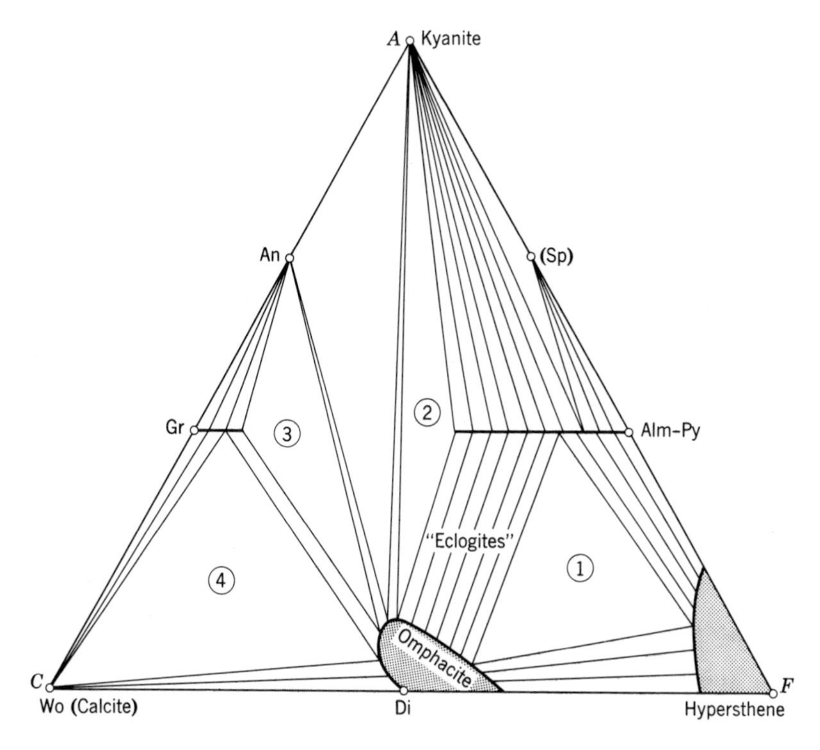

FIG. IV-46. *ACF* diagram of the mineral associations of the eclogite facies.

ilmenite. Olivine occurs in eclogites low in silica. The high density of all eclogite minerals is noteworthy, and reflected in the density of the rock itself. The specific gravity of eclogites ranges from 3.35 to 3.6, whereas other rocks of gabbroic compositions show 2.9 to 3.1. The density relations of jadeite and its chemical equivalent mixture of nepheline and albite have been discussed on page 242. Likewise the formation of rutile is accompanied by augmentation of the specific gravity; by change of facies from amphibolite to eclogite, the following reaction takes place:

$$3CaSiTiO_5 + Ca_2Mg_5Si_8O_{22}(OH)_2 = 5CaMgSi_2O_6 + 3TiO_2 + (H_2SiO_3)$$

sphene + tremolite = diopside + rutile

Density: 3.5 3.0 3.26 4.24

Natural tremolitic hornblendes are low in silica, and H_2SiO_3 is not actually produced. The volume of the product is, therefore, 5.7 per cent less than that of the sphene-tremolite association.

Another transformation worth considering is that from graphite to diamond. Graphite is stable at ordinary pressure; and only above ca. 20,000 atm (depending upon the temperature), corresponding to a depth of 70–80 km, is diamond stable.

It is possible that many silicates possess high-pressure modifications that are as yet unknown to science. A graphic illustration is the recent discovery of coesite, the high pressure modification of SiO_2.

Eskola has suggested that at great depths anorthite may lose silica and form a pyroxene-like mineral of compact structure (this is analogous to albite losing silica and changing into jadeite):

$$CaAl_2Si_2O_8 - SiO_2 = CaAl_2SiO_6$$

traces of which have been found in solid solution in some of the eclogite pyroxene.

Again olivine has a high-pressure modification of spinel structure. See page 237.

Hornblende occurs in many eclogites as a hysterogenic product; it is often alkali-bearing. Another characteristic hysterogenic product is kelyphite: peripheral alteration of garnet into pyroxene or amphibole. Complete gradation is often traced from unaltered eclogite through eclogite-amphibolites containing relict garnet and omphacite, together with newly generated plagioclase and hornblende, to amphibolites of normal composition.

Eclogite may occur in the following ways:

1. Schlieren and lenses in garnet-bearing anorthosite.
2. Schlieren and lenses in garnetiferous charnockite, pyroxene granulite, or pyroxene gneiss.
3. Lenticular masses in granite gneisses and migmatites.
4. Schlieren and segregations in peridotite and dunite.
5. Sheet-like masses, associated with amphibolite, in paragneisses.
6. Nodules, associated with peridotite and other ultrabasic types, in kimberlite pipes of South Africa. They may represent products of magmatic recrystallization under deep-seated conditions (the so-called griquaite of R. Beck).
7. Irregular masses in glaucophane schists.

Backlund concluded from studies of the Caledonian mountain region that shearing stress of great magnitude and high temperature are the conditions essential for development of eclogites, and that extreme hydrostatic pressure is not required. On the other hand, it is possible that eclogites, and even rocks of unknown dense minerals (see page 251), incapable of surviving at surface conditions, have been formed at great depths, but during the slow cooling and unloading prior to exposure at the earth's surface they have undergone complete retrogressive metamorphism to assemblages of the amphibolite (or granulite) facies.

In this sense eclogite might be regarded as the high-pressure "modification" of gabbro (or amphibolite). Eskola has pointed to the probable existence of a continuous shell of eclogite under the sial crust.

The chemical composition of eclogites is restricted to gabbroic and ultrabasic rocks. It is possible that only rocks of high melting temperature remain wholly crystalline under the conditions of the eclogite facies.

Glaucophane Schist Facies. This facies is closely related to the eclogite facies and is characterized by soda amphibole (glaucophane, crossite), soda pyroxene (jadeite, aegerite), eclogite garnet, lawsonite, and pumpellyite. Other minerals are quartz, aragonite, sphene, muscovite, chlorite, stilpnomelane, and, locally, epidote and albite. Most of the soda is present in a mineral other than plagioclase; as in the eclogite facies, the *ACF* diagram does not, therefore, present adequately the various mineral associations. Fig. IV-47.

One of the rock types making up the California glaucophane schists is true eclogite, but greenschist assemblages such as quartz-muscovite-chlorite-pumpellyite are also common. It has been suggested therefore, that the glaucophane schist facies is essentially equivalent to the greenschist and albite-epidote amphibolite facies, and should not be retained as an independent facies. However, the stability of lawsonite—and this mineral occurs only in the glaucophane schist facies—with a molar volume approximately as small as that of anorthite in spite of its two molecules of water, and the presence of eclogite garnets and jadeite which also exhibit unusually small molar volumes, indicate that this facies belongs in a field of high pressure.

Coleman recently reported aragonite as a stable constituent, again supporting this view; for aragonite has a smaller volume than calcite. For the first time aragonite is here reported as a stable mineral of an association (metastably it is frequently found in a variety of ways), it is of interest, therefore, to investigate into its stability relations.

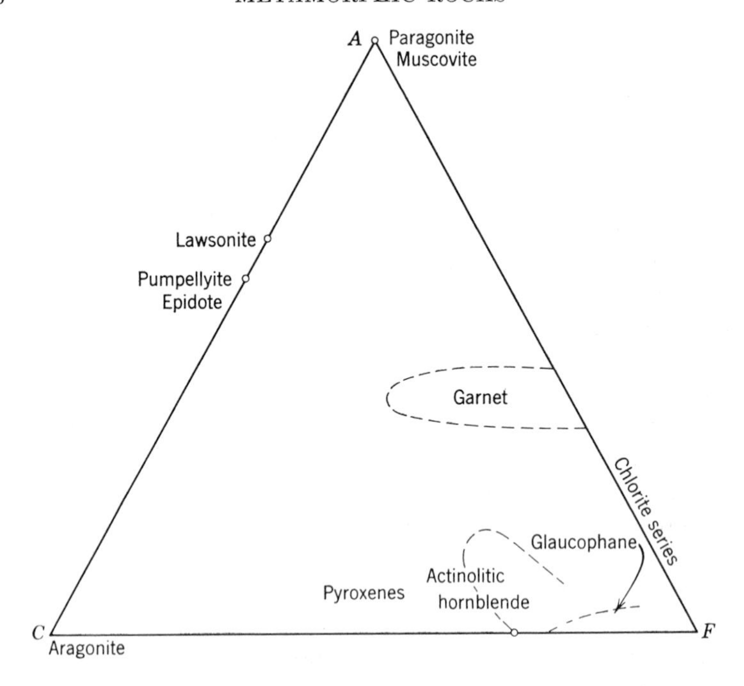

Fig. IV-47. *ACF* diagram of glaucophane schist facies.

Laboratory measurements of the heat effect of the transition aragonite → calcite have shown that the change in free energy at room temperature and pressure is

$$\Delta G = G_0' - G_0 = -190 \text{ cal/mole}$$

(the primed symbols refer to calcite, the unprimed to aragonite).

The specific volumes for calcite and aragonite are 37 and 34 cm^3/mole, respectively, giving

$$\Delta V = V_0' - V_0 = +3 \text{ cm}^3\text{/mole}$$

The drop in free energy shows that calcite is the stable phase, but the increase in volume indicates that aragonite may become stable at increased pressure.

From equation (15 in Appendix) we obtain the free energy at pressure p,

$$\text{for aragonite} \quad G_p = G_0 + pV_0$$
$$\text{for calcite} \quad G_p' = G_0' + pV_0' = G_0 - 190 + p(V_0 + 3)$$

Calcite and aragonite at equilibrium must have the same free energy; $G_p = G_p{}'$ if,

$$p = \frac{190}{3}\ \text{cal/cm}^3 = \frac{190 \cdot 42.5}{3}\ \text{kg cm/cm}^3 \backsim 2700\ \text{atm}$$

Consequently, above 2700 atm and 25°C, aragonite is the stable form.

The phase diagram of $CaCO_3$ has now been determined experimentally, and is given in Fig. IV-48. It is worth noting that there are two pressure phases of $CaCO_3$ that are stable in respect to calcite, but not to aragonite; furthermore that calcite II probably has a dis-

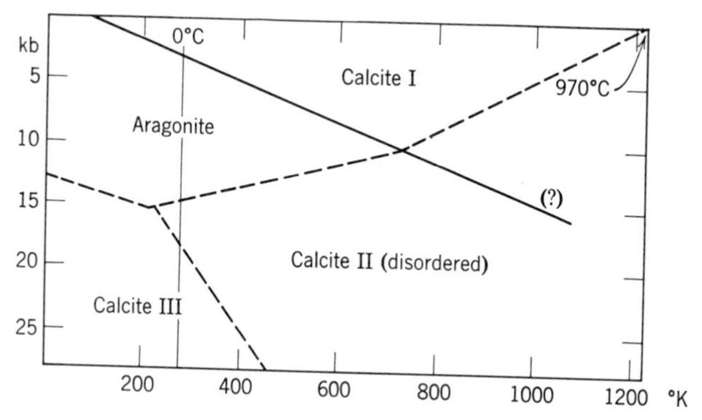

Fig. IV-48. Phase diagram of $CaCO_3$. Heavy line and stippled lines refer to stable and metastable phases, respectively. Aragonite is the stable high-pressure phase, but for calcite to convert into aragonite requires a reconstruction of the lattice and a high initial energy which is often lacking. Calcite, in the absence of transition to aragonite, is itself metastable. The transitions I → II and II → III are thermodynamically possible, since ΔV is negative for each with increasing pressure. These are structurally minor "displacive" transitions which take place readily in the metastable field. (J. M. Jamieson, 1957.)

ordered crystal lattice being the effect of not only elevated temperature but also elevated pressure.

The exact relations of the eclogite facies and the glaucophane schist facies are not well understood. Perhaps both assemblages in certain situations are able to form metastably at lower pressures outside their respective fields of stability.

Survey of the Facies

The parameters responsible for regional metamorphism may vary from rock to rock in relative intensity and importance; the mineral assemblages of the metamorphic rock will vary accordingly.

A graphic illustration of the several possibilities is furnished by a diagram using temperature and pressure as axes. Each special kind of metamorphism corresponds to a mineral facies; each mineral facies can, in its first approximation, be assigned to a specific pressure-temperature region. (See Fig. IV-49.)

The absolute values of pressure and temperature in Fig. IV-49 are subject to some conjecture. It is probably unlikely that any rock at the present surface has been more deeply buried than 20,000 m. This corresponds to a geostatic pressure between 5000 and 6000 atm. It is conceivable, however, that shearing stress has exceeded this value for shorter periods in orogenic regions. Provisionally, we may place the upper part of the eclogite facies at 6000 atm, but it is possible that some of the "tectonic" eclogites and some glaucophane schists may have reached higher pressure values.

The upper temperature limit of metamorphism is reached by melting (anatexis, pages 349 ff.). Rocks of granitic composition under sufficiently high water vapor pressure will melt completely as low as ~650°C.

Some water probably will be "squeezed" out as explained on page 130, so that sufficient water rarely will be present to effect an extreme depression of the melting interval. It is unlikely, however, that rocks of granitic composition would stay solid above 700°C. To be sure, this value may be taken as the upper limit of the normal re-

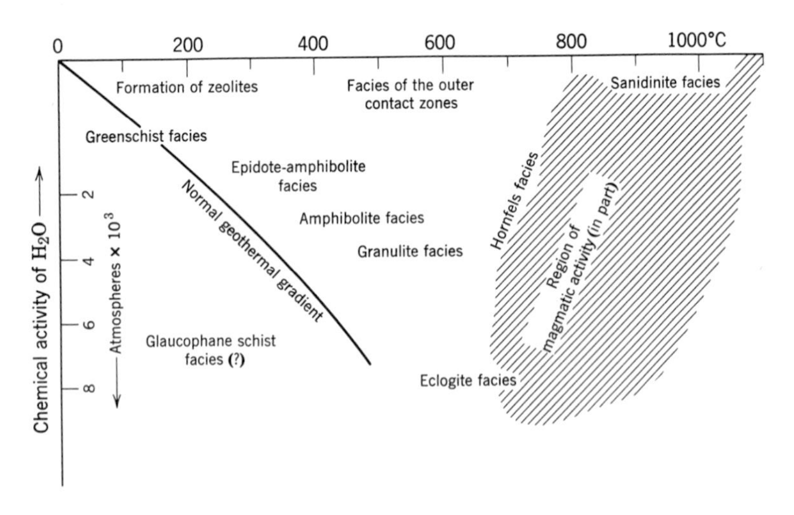

FIG. IV-49. Survey of the principal mineral facies of rocks in their relation to temperature and pressure.

gional metamorphism, which equals the upper limit of granulite facies. If any facies is to be placed at still higher temperatures, it means that the corresponding rocks are limited in their chemical composition to more refractory mixtures. This may apply to some eclogites (see page 325). Parts of the hornfels facies likewise may extend into higher temperature regions, and the sanidinite facies is probably completely above 700°C. Extreme cases are encountered where fragments and splinters of adjacent rock are picked up by, and become completely immersed in, a flowing lava. Basaltic lava picking up a granitic country rock will promptly melt it and incorporate it in the lava. But if the lava picks up, for instance, some aluminum-rich shale (which is highly refractive) it will be a long time before it softens. It will react with the lava, however, usually surrounding itself with a reaction rim, and if of dense structure the inside of it will just heat up and undergo recrystallization without much injection of magmatic "emanations." In such cases we actually encounter magmatic temperature: 1000 to 1200°C (= pyrometamorphism).

The lower limit of metamorphism may be put approximately at 100°C. Below this temperature diagenic processes are supposed to take over, that is, compaction and cementation of sediments. Zeolite formation also belongs to this stage.

The Artificial Facies of Yoder

A discussion of the mineral facies cannot be concluded without mentioning a most important series of experiments conducted by H. S. Yoder (1952) at the Geophysical Laboratory in Washington, giving compelling evidence that the relation between facies and P,T is not at all so certain and as absolute as has been thought.

Work in the system $MgO—Al_2O_3—SiO_2—H_2O$ gave as one result that at approximately 600°C and 1200 atm it is possible to have different mineral assemblages suggestive of every one of the now accepted metamorphic facies in stable equilibrium. These different mineral assemblages ("artificial facies") observed by Yoder are the result of differences in the water content, and are not related to variation in temperature and pressure.

As an example: The mineral clinochlore, corresponding to one of the most common rock-making chlorites, shows an upper limit of stability, either alone or in association with talc of 680°C, for example. This is, indeed, a super high temperature for any kind of metamorphism—almost a magmatic temperature. But according to the tenets of the mineral facies, chlorite is strictly limited to the greenschist facies, around 200°C.

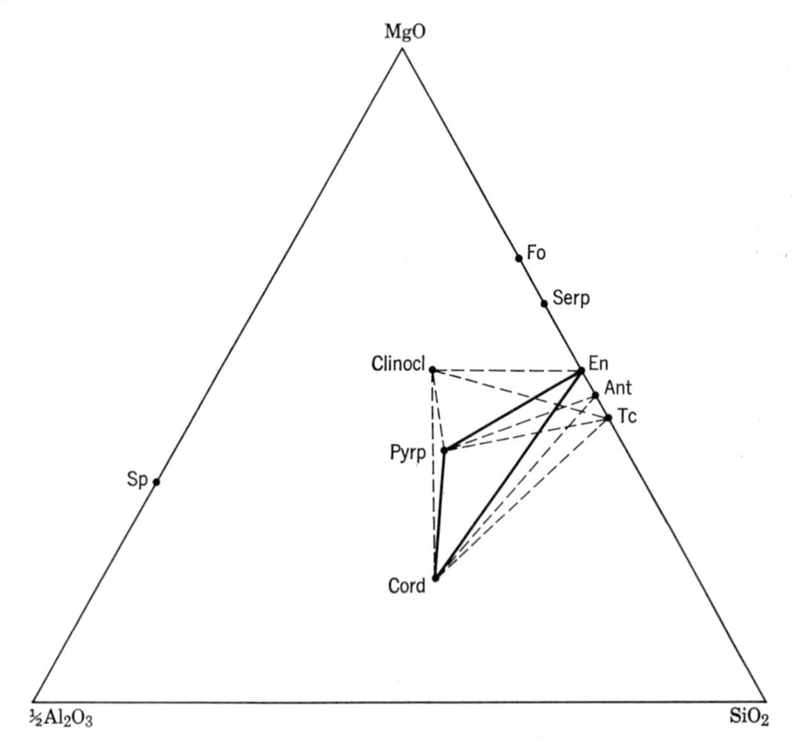

FIG. IV-50. A one-point projection of the $MgO—Al_2O_3—SiO_2—H_2O$ system showing some of the constituent minerals. Ant = anthophyllite $Mg_7Si_8O_{22}(OH)_2$, En = enstatite $MgSiO_3$, Fo = forsterite Mg_2SiO_4, Pyrp = pyrope $Mg_3Al_2Si_3O_{12}$, Serp = serpentine $Mg_6Si_4O_{11}(OH)_6 \cdot H_2O$, Sp = spinel Mg Al_2O_4, Tc = talc $Mg_3Si_4O_{10}(OH)_2$, Clinocl = clinochlore $Mg_5Al_2Si_3O_{10}(OH)_8$. The assemblage shown in full lines refers to the anhydrous phases, cord-pyrp-en; the hydrous phases are joined by dashed tie-lines.

It will take one example to illustrate the principle: Some of the minerals in the system $MgO—Al_2O_3—SiO_2—H_2O$ are shown graphically in Fig. IV-50, and the area around the cordierite-pyrope-enstatite association is selected for study at a constant temperature-pressure combination: 600°C and 1200 atm, but under varying concentrations of H_2O. The various mineral associations which then develop are shown graphically in Fig. IV-51.

Interpreted in terms of the mineral facies of rocks the associations are:

1. Enstatite-cordierite gneiss = granulite facies (~600°).
2. Anthophyllite-garnet gneiss = amphibolite facies (~500°).

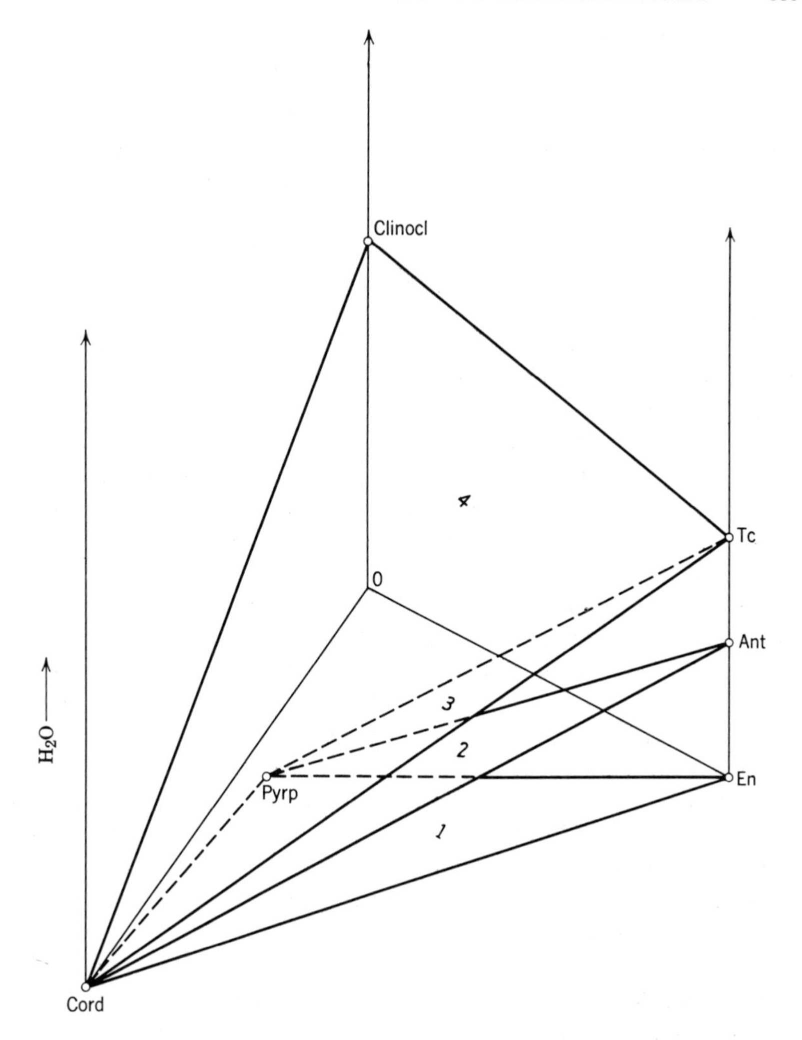

Fig. IV-51. A perspective projection of the system cordierite-pyrope-enstatite-water, constructed after observed and inferred data of Yoder (1952). Activity of water increases normal to the bottom face cord-O-en. (1) The water-free association is cordierite-pyrope-enstatite as shown on the bottom face of the illustration. With a small amount of water, first enstatite yields part of itself to hydration (replaced by anthophyllite); with increasing concentration of water, eventually all enstatite disappears and the association becomes (2) cordierite-pyrope-anthophyllite. With more water, talc starts to form at the cost of anthophyllite and the association eventually becomes (3) cordierite-pyrope-talc. With still more water, pyrope yields to hydration (replaced by chlorite) and the final association stable in the excess-water region is (4) cordierite-talc-chlorite.

3. Anthophyllite-talc schists = epidote-amphibolite facies ($\sim 400°$).
4. Talc-chlorite schists = greenschist facies ($\sim 200°$).

According to the facies geology, we go through a temperature range from 600° to 200°C, while the experiments prove that the temperature actually was constant; chemically we just go through a small range of composition from no water to 5–6 per cent of water. The mineral association corresponding to the granulite facies is water-free. With increasing water content, the minerals indicate facies of successively lower grade.

We must conclude, as Yoder has done, that the accepted mineral facies do not represent reliable temperature-pressure indicators. Evidently, the trouble is that the series of mineral facies chosen by Eskola are not isochemical: they differ markedly in their water content, and this difference is sufficient to cause great differences in the mineral associations. What was interpreted by Eskola and others as a change in temperature-pressure, is actually, in some rocks, the result of a change in the bulk composition.

Therefore, a sequence of progressive mineral facies established in one area, cannot be used as temperature indicators in another area. And again: The present composition of a metamorphic rock is not the same as that which prevailed before the metamorphism. During the metamorphism the rock was soaked in pore liquids and vapors which have now escaped, very much in the same manner as the "volatiles" escape from a magma, leaving behind a solidified rock quite different in composition from the magma out of which it crystallized.

These facts, in the correct apprehension of the facts, put the geologist in a position to understand and interpret many of the observations that had seemed mysterious.

Yoder mentions that the occurrence of eclogite with amphibolite, or even with low-grade rocks, may be understood. To be sure, in an area of isofacial rocks the field geologist may frequently observe abnormal rock variants indicating a change in mineral facies. An example is a chlorite schist occurring with pre-Cambrian rocks of the amphibolite facies near Tampere, Finland (Seitsaari, 1954). In this case the feldspar of the chlorite schist is a bytownite, proving, that the chlorite was not formed at the temperatures of the greenschist facies ($\sim 200°$) but actually at higher temperatures ($\sim 500°$ (?)) corresponding to the amphibolite facies.

In view of these results the questions arise: Are the established mineral facies "no good"? Is there no relation between temperature (pressure) and the several mineral facies of metamorphic rocks?

Although Yoder has proved that there is no absolute relation between temperature and facies, it seems that, to the field geologist, and to the laboratory man as well, the established facies classification still remains as the best system of classification of metamorphic rocks; and in a majority of cases the facies will indicate the temperature-pressure conditions under which the several rocks recrystallized.

The point is that, generally speaking, there is a regular relation between the chemical activity of H_2O and the facies.

The role of water in metamorphism is determined by at least four variable parameters: rock pressure, temperature, water pressure, and the amount of water present. However, these four variables are geologically related: during a normal progressive regional metamorphism, rock pressure and temperature are interdependent, and the amount of water and the pressure of water are related to the encasing sediments and to the degree of metamorphism in such a way that, generally speaking, the low-grade metamorphic facies are characterized by the presence of an excess of water, the medium-grade by some deficiency in water, and the high-grade metamorphic facies by virtual absence of water. In the usual digrammatic illustration of the mineral facies of rocks, temperature and pressure (depth) are taken as coordinates; in regional-metamorphic rocks we may now add a third, dependent coordinate: the activity of H_2O running parallel to, but in opposite direction of, the pressure axis.

Thus, Fig. IV-49 illustrates the usual case but does not take into account all the exceptions that may occur. For the mineral facies of a rock is not a one-valued function of temperature (and pressure). So complex are the geological processes that in addition to temperature and rock pressure a great number of additional variables enter into the picture: stress, rate of growth, position in the gravitational field, pressure of water, of CO_2, and of other volatiles, absolute amount of water, of CO_2, etc. Thus, at the present state of our knowledge, and for a long time to come, it will be impossible from the simple mineral facies of a rock to infer with certainty the temperature at which the metamorphism took place.

8 · PETROCHEMICAL CALCULATIONS

The chemical petrology of metamorphic rocks is rather complicated. Compared to igneous rocks, the variation in rock composition is much larger, and the complexities of the constituent minerals more confusing. Furthermore, a much greater number and variety of metamorphic minerals exist because the variations in P and T are also much larger than by igneous-rock formation. Construction of *ACF* diagrams and arranging the rocks into several mineral facies are the best means to keep order in the jumble and to obtain an approximate survey of the chemical variations and relations.

However, the *ACF* diagrams take care of only three of the many chemical components of the rocks and cannot, therefore, reflect the multitude of changes and trends in a rock series or display adequately the broader chemical relations.

The CIPW norm (page 65) was meant for igneous rocks, and to this day it has, indeed, been restricted to igneous rocks. The introduction of a similar method of recalculation in metamorphic petrology meets with many difficulties. Quite apart from the fact that the metamorphic minerals are more complicated, chemically, than are the igneous minerals, the question arises: Which mineral facies should be represented in the standard or normative set of minerals? For rocks which have arrived at equilibrium in the greenschist facies exhibit minerals and mineral associations that are very different from those in rocks of, say, amphibolite facies.

In regional metamorphism there are three main groups of rock:

1. Catarocks centering around the granulite facies.
2. Mesorocks centering around the amphibolite facies.
3. Epirocks centering around the greenschist facies.

It is necessary, therefore, to make norms corresponding to each of these major groups of rock.

The catanorm. Fortunately, catametamorphic rocks correspond in their mineral content so closely to that of igneous rocks that the CIPW norm can be used for both. Thus the catanorm, given in cation percentages, is analagous to the old CIPW norm.

The first step is to recalculate the chemical analyses into cation percentages. It may be repeated here that, by adopting the cation percentage method, the mineral equations become simplified (page 67).

The unit is the cation, and it is expedient to relate the symbols of the standard minerals to a quantity containing one cation. Then the sum of all cations in a mineral formula represents the number of mineral units. Consequently, $Na_2O \cdot Al_2O_3 \cdot 2SiO_2 = 6Ne$ (because the nepheline formula written in this way contains 6 cations), and $MgSiO_3 = 2En$ (because the formula contains 2 cations). Thus the mineral equations become, for instance,

$$3Fo + 1Q = 4En,$$

$$Mg_2SiO_4 + SiO_2 = 2MgSiO_3$$

It is important to note that the number of cations on both sides of the equations must be the same. Thus one may calculate, easily and rapidly, not only the catanorm but any other mineral association, including the mode: any combination of cations into mineral molecules can be made without affecting the sum, which remains 100.

We shall see that only in this way is it possible to calculate "norms" for metamorphic rocks.

The mesonorm has been constructed for rocks belonging to the mesozone. In regional metamorphic geology the zonal distinction corresponds to a division according to temperature (and pressure); rocks of the catazone transform with decreasing temperature into rocks of the mesozone and of the epizone. Rocks of the mesozone center around the amphibolite facies. Those of the epizone are near the greenschist facies of Eskola. In mesozonal rocks the chemical adjustment is usually rather good. Many minerals foreign to the catazone are present: hornblende, biotite, as well as garnets (almandine-pyrope-grossularite), sphene, and others. Clearly, the chemical composition of rocks made up of such minerals should not be rendered in terms of the vastly different minerals of the catanorm. Rather, one should have another set of normative minerals—the standard minerals of the mesonorm—which should be combined so as to reflect the actual mineral composition of mesozonal rocks.

It is difficult to normalize minerals in the mesozone because they form extensive and complicated series of mixed crystals. One example is biotite. Its composition has to be idealized, and the following norm formula is proposed:

$$\text{Bi} = \tfrac{1}{8}[\text{KAlMg}_3\text{Si}_2\text{O}_{10}(\text{OH})_2]$$

with Fe substituting for Mg.

Another difficult mineral is hornblende. For rocks with sufficient silica the following simple formula is adopted:

$$\text{Actinolite, Act} = \tfrac{1}{15}[\text{Ca}_2\text{Mg}_5\text{Si}_8\text{O}_{22}(\text{OH})_2]$$

with Fe substituting for Mg. For undersilicified rocks:

$$\text{Edenite, Ed} = \tfrac{1}{16}[\text{NaCa}_2\text{Mg}_5\text{-AlSi}_7\text{O}_{22}(\text{OH})_2],$$

with Fe substituting for Mg. For alkalic rocks:

$$\text{Riebeckite, Ri} = \tfrac{1}{15}[\text{Na}_2\text{Fe}^{2+}_3\text{Fe}^{3+}_2\text{Si}_8\text{O}_{22}(\text{OH})_2]$$

The accessory mineral typical of the mesozone is sphene:

$$\text{Sph} = \tfrac{1}{3}[\text{CaTiSiO}_5],$$

Spinel may occur

$$Sp = \frac{1}{3}[MgAl_2O_4]$$

Except for ilmenite (il) and acmite (ac), all other mesonormative minerals are in common with the catanorm.

The various steps in the calculation of the mesonorm are as follows:

1. Calcite is formed from CO_2 and an equal amount of Ca.
2. Apatite is formed from P and 5/3 times this amount of Ca.
3. Pyrite is formed from S and half this amount of Fe^{2+}.
4. Sphene is formed from Ti and equal amounts of Ca and Si.
5. The alkali feldspars are formed provisionally from K and Na combined in the right proportions with Al and Si to form Or and Ab.

6*a*. If there is an excess of Na over Al it is to be combined with an equal amount of Fe^{3+} to form riebeckite, $Na_2Fe_3{}^{2+}Fe_2{}^{3+}Si_8O_{22}(OH)_2$.

6*b*. If, as usually happens, there is an excess of Fe''' over Na, it is assigned to magnetite.

7. The remaining ferrous iron is combined with magnesium and called Mg', and the other remaining cations of aluminum and of calcium are called Al' and Ca' respectively. Thus the relations graphically shown in Figure IV-52 are obtained by calculating the values $Mg' + Al' + Ca' = 100$.
8. Anorthite is formed from Al'.
9. If there is an excess of Al' over Ca' it is calculated as corundum.

10*a*. From the remaining Ca' and Mg', hypersthene and/or actinolite, and/or diopside are formed.

10*b*. But if Hy can form, available Mg' is *first* used to convert the provisionally formed orthoclase into Bi:

$$5Or + 3Mg' = 8Bi.$$

11. If too much silica has been used, desilification is effected by
    a. converting Act into Ed:

$$5Ab + 15Act = 16Ed + 4Q.$$

b. If there still is not enough Si in the analysis further desilification is effected in the following order:

$$4Hy = 3Ol + Q$$

$$3Ol + 4C = 6Sp + Q$$

$$5Ab = 3Ne + 2Q$$

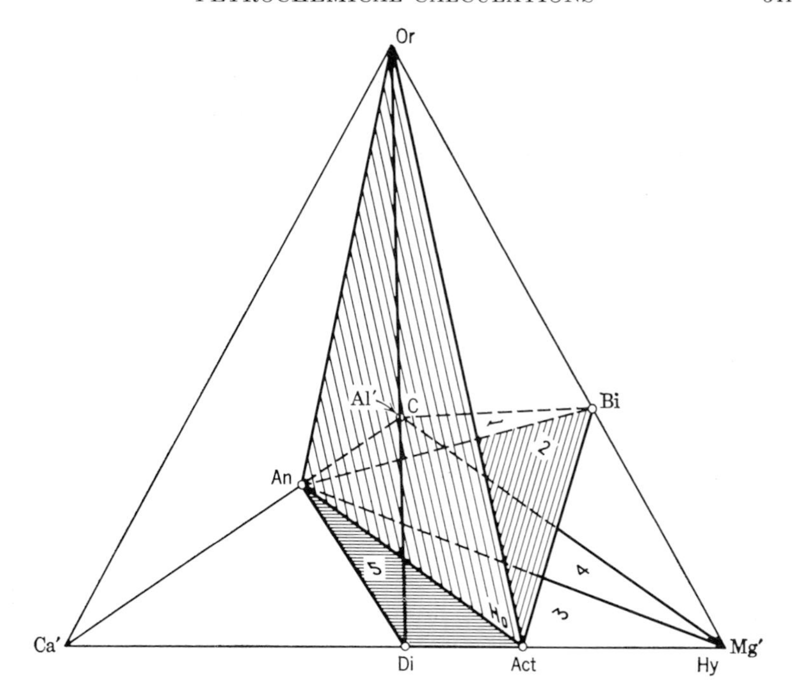

Fig. IV-52. Graphical illustration of the five chief mesonormative associations (Barth, *J. Geol.,* 1962).

(1) Or An Bi C	Gneisses and granites
(2) Or An Bi Ho	Hornblende gneisses
(3) An Bi Ho Hy	Amphibolites
(4) An Bi C Hy	Gabbroschists
(5) Or An Di Ho	Skarn-amphibolites

Amphibolite is chosen as an example illustrating the method of calculation. (See Table IV-9.) Among the metamorphic rocks, the amphibolites occupy a position rather similar to that of the basaltic-gabbroic rocks of the igneous suite. But, whereas basalts and gabbros have been extensively investigated and well-known chemical and mineralogical criteria exist to distinguish the various types, such as tholeiite, tephrite, basanite, etc., no corresponding characterization has been made of amphibolites.

The epinorm is designed to reflect the minerals and mineral associations in rocks of the epizone, gravitating around the greenschist facies. Zoisite, Zo = $Ca_2Al_3Si_3O_{12}(OH)$, is an important mineral substituting for anorthite. Chlorite (Chl) can be regarded as a mixed crystal between antigorite, $Mg_3Si_2O_5(OH)_4$, and amesite (daphnite),

TABLE IV-9

Method of Calculating the Mesonorm of Average Amphibolite from the Pre-Cambrian of S. Norway

Si	Ti	Al	Fe^{3+}	Fe^{2+}	Mn	Mg	Ca	Na	K	P	(H_2O)	
45.70	1.18	16.38	1.91	8.12	0.18	9.96	10.18	5.36	0.92	0.11	(2.56)	
							18			.11		Ap 0.29
1.18	1.18						1.18					Sph 3.54
2.76		0.92							0.92			Or 4.60 (provisional)
16.08		5.36						5.36				Ab 26.80 (provisional)
			1.91	0.95								Mt 2.86
at this stage the cations are added up in accordance with Step 7												
25.68		10.10			17.31		8.82					
10.10		10.10					5.05					An 25.25
15.08					9.43		3.77				(2.52)	Act 28.28 (provisional)
					2.76						(0.92)	Bi 7.36
5.12					5.12							Hy 10.24

Biotite is formed from the provisionally made Or(4.60) and the appropriate amount of Mg′ according to 10*b*. The remaining Mg′(5.12) is used for hypersthene. The norm now computed requires 4.62 Si in excess of the Si actually present. Consequently, desilification must be effected by 11*a*:

$$5.77\text{Ab} + 17.33\text{Act} = 18.48\text{Ed} + 4.62\text{Q}$$

The final norm thus becomes:

Ab	21.03	Bi	7.36	Sph	3.54
An	25.25	Act	10.95	Mt	2.86
		Ed	18.48	Ap	0.29
Plagioclase	46.28	Hy	10.24		
					6.69
			47.03		

The rather numerous analyses of metamorphic hyperites included in the average is reflected in the rather high amount of normative hypersthene.

$Mg_2Al_2SiO_5(OH)_4$, with Fe substituting for Mg. Excess Al_2O_3 is calculated as muscovite or as kaolinite if potash is deficient. In calcium-rich rocks, dolomite and calcite must develop, or else the rock has to change its composition.

By using the principles developed in the preceding pages there is no difficulty in computing the standard mineral assemblages of the epinorm.

However, the conditions of the greenschist facies: the low temperature, local changes in the H_2O pressure and CO_2 pressure, etc., are not favorable for the attainment of equilibrium. Many epizonal rocks therefore exhibit non-equilibrium associations and are difficult to decipher. Time will show whether or not the calculation of epinorms will be an aid to the understanding of such kinds of rock.

9 · EXAMPLES OF METAMORPHIC ROCK PROVINCES

Dutchess County in southeastern New York (Balk, 1936, Barth, 1936). Paleozoic sediments, forming a northeasterly trending geosyncline, have been profoundly modified by instrusion of various kinds of magmatic matter, and by mountain folding. Derivatives of the argillaceous sediments are petrographically the most important, and changes induced in them by orogenic processes are the most instructive. The metamorphism increases in intensity from northwest to southeast. The so-called Hudson River slate exhibits three different mineral facies, the "facies group" of the argillaceous rocks. They are, from northwest to southeast: (1) muscovite slate facies, (2) kyanite schist facies, (3) sillimanite gneiss facies.

(1) Muscovite slate facies is developed in most of the rolling country northwest of the Hudson Highlands underlain by the Hudson River slate, a black or olive-gray argillite interbedded with lenses of conglomerate, arkose, or graywacke containing chert. To the east, the slate grades imperceptibly into a lustrous phyllite, characterized by rusty cavities indicating oxidation of pyrite, and by countless veins of quartz or of quartz and calcite. In most exposures, the original bedding planes are distinguishable from secondary cleavage. The biotite isograd is here easy to map; west of it, muscovite is the only recrystallized mineral, but immediately to the east is biotite, and, within short distances, garnet, staurolite, and kyanite. The biotite isograd, therefore, marks the western border of the facies of next higher rank, kyanite schist facies.

(2) Kyanite schist facies. In it kyanite is the most characteristic mineral, although it is not necessarily found in all the rocks. Isofacial with kyanite-bearing rocks are typical biotite-garnet schists, staurolite schists, and certain other schists. Changes in the internal structure take place: isoclinal folding, fracture cleavage, and the development of shear planes. To the east, the schists grade imperceptibly into sillimanite gneiss facies.

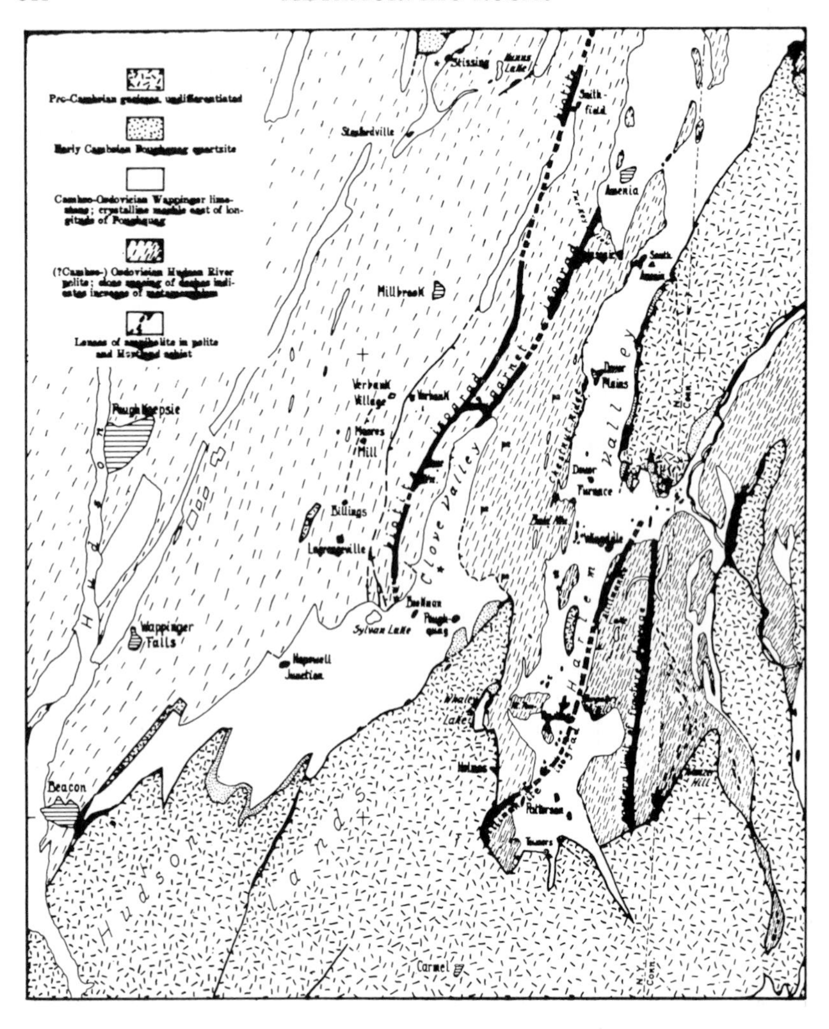

FIG. IV-53. Geological map of Dutchess County, New York, and vicinity. Various isograds are shown by heavy lines. (After R. Balk, 1936.)

(3) Sillimanite gneiss facies. The rocks on both sides of the sillimanite isograd are similar, the only difference (which it usually requires the microscope to reveal) being that, east of the isograd, kyanite is unstable and is replaced by sillimanite. However, structure also is slightly changed. Toward the east, the rocks are coarser and the schistosity less pronounced than in the schist facies. Again it is to be understood that not all rocks of this facies necessarily carry silli-

manite. Biotite-plagioclase gneisses are representative members of this facies and, farther southeast, biotite-perthite gneisses.

It should be noted that not only are the temperature and the pressure conditions important for a mineral to form but also the bulk chemical composition of the rocks. For instance, potash feldspar is certainly stable under the thermodynamical conditions of the kyanite

TABLE IV-10

Average Chemical Composition of Successive Members of the Rock Series Slate-Augen Gneiss

Number of Cations in the Standard Cells

	Slate	Schist	Plagioclase Gneiss	Augen Gneiss
Si	54.8	51.8	59.0	60.0
Ti	0.5	0.6	0.7	1.2
Al	16.9	21.5	15.0	14.7
Fe	5.8	5.7	5.4	3.6
Mg	3.6	3.4	2.2	1.1
Ca	0.1	1.2	1.8	2.6
Na	2.7	2.3	6.0	6.3
K	3.8	5.7	3.1	6.0
P	0.1	0.1	0.1	
H_2O	(9.8)	(7.9)	(1.9)	(1.0)

As previously explained, the standard cells represent approximately equal volumes. Thus by direct comparison it is seen that slate passes into gneiss by:

Adding	Subtracting
2.2 ions of K	2.5 ions of Mg
3.6 ions of Na	2.2 ions of Fe
2.5 ions of Ca	2.2 ions of Al
0.7 ion of Ti	0.1 ion of P
5.2 ions of Si	
Total: 14.2 cations, representing 34.2 valences	Total: 7.0 cations and 17.5 H-ions, representing 34.4 valences

schist facies, but it is chemically unstable (the excess Al_2O_3 of these rocks would react with it and transform it into biotite). Only after elimination of excess alumina, attained in a higher metamorphic grade, does potash feldspar become chemically stable.

The several rocks of the kyanite schist facies and the sillimanite gneiss facies are not simple products of a normal metamorphism. Chemical analyses show that metasomatic changes occurred, and that, generally speaking, the higher the metamorphism the greater the changes. These changes are shown in Table IV-10.

The conclusion is that a large-scale transport of material has taken place. It is reasonable to assume that the incoming ions were associated with intrusions and other orogenic events in the central and more intensely heated parts of the geosyncline. They will be called the "injected ions," and they represent, in a way, magmatic emanations (we shall use the word "emanation" in a general sense) which were injected into the slates and fixed in the slates through metasomatic reactions. The ions lost by the slate through these reactions will be called the "rejected ions."

In which direction did the rejected ions move? These are the ions that many petrographers want to move *away* from the magmatic source to make a "basic front." However, the theory of metasomatic exchange requires that the rejected ions move *towards* the magmatic source to make up for the loss. If they do, probably no trace of them will be left; they will disappear downward. It should be remembered, however, that water is rejected and most certainly travels away from the magma. It may possibly take along some of the rejected ions, that is, it will act as a "carrier" of some of the rejected magnesium, iron, and aluminum ions and will carry them away, although another fraction of the same ions, without company of water, will travel towards the magma.

Glen Clova, Scotland, described by Chinner (1960). In this area pelitic gneisses exhibit intimately interbedded layers of different composition and of widely varying oxidation ratio. Apparently no interchange of matter took place; the element migration was restricted to small-scale diffusion in connection with the recrystallization. Even the ferrous/ferric ratio did not equalize but remained different in the different layers; it governed to a large extent the mineral phases appearing.

The following reaction is important:

$$Fe^{2+} + \text{biotite} + \text{almandite} + O = \text{muscovite} + \text{iron oxide} + \text{quartz}$$

The confinement of rocks of varying oxidation ratio to well-defined sedimentary bands suggests that the oxygen content is premetamorphic; each band behaved as a unit "closed" to oxygen—the oxygen partial pressure in each unit being determined by the mineral assemblage and hence by the *original* oxygen content. Thus layers of a high oxidation stage will give rise to mineral associations different from those in layers of a lower oxidation stage, as shown in Table IV-11.

TABLE IV-11

RELATIONSHIP OF ASSEMBLAGES DEVELOPED IN PELITIC ROCKS OF VARYING OXIDATION RATIO WITHIN THE CHLORITE, BIOTITE, AND GARNET ZONES

(Chinner, 1960)

<table>
<tr><th></th><th>Barrow-Tilley Zones</th><th>Pelitic Metasediments of Low Oxidation Ratio</th><th>Pelitic Metasediments of High Oxidation Ratio</th></tr>
<tr><td rowspan="6">Increasing metamorphism ↓</td><td rowspan="2">Chlorite</td><td rowspan="2">Chlorite-sericite-ilmenite-magnetite</td><td>Chlorite-sericite-magnetite-hematite solid solutions</td></tr>
<tr><td rowspan="2">Garnet-chlorite-muscovite-magnetite-hematite solid solutions</td></tr>
<tr><td rowspan="2">Biotite</td><td rowspan="2">Biotite-chlorite-muscovite-ilmenite-magnetite</td></tr>
<tr><td rowspan="2">Biotite-garnet-chlorite-muscovite-magnetite-hematite solid solutions</td></tr>
<tr><td rowspan="2">Garnet (almandite)</td><td rowspan="2">Garnet-biotite-muscovite-ilmenite-magnetite</td></tr>
<tr><td>Biotite-garnet-muscovite-magnetite-hematite solid solutions</td></tr>
</table>

The metamorphism of the Stavanger area, southern Norway, has become well known through V. M. Goldschmidt's work (1920). The main point of petrologic interest is the progressive metamorphism of argillaceous sediments by granitic and trondhjemitic intrusions.

By low-grade metamorphism the sediment is changed into a quartz-muscovite-chlorite phyllite, into albite porphyroblast schist and, eventually, into a granite-like rock. Goldschmidt gives the following discussion of the transition sediment → albite porphyroblast schist:

In order to investigate the various possibilities of material transport he calculates the composition of various rock mixtures, the soda content of which would correspond to that of the albite porphyroblast schist. In this calculation the initial composition is taken as the quartz-muscovite-chlorite phyllite.

The possibilities are:

1. The material was introduced as trondhjemite; to 100 parts phyllite, 150 parts trondhjemite must be added.
2. The material was introduced as granite aplite; to 100 parts phyllite, 400 parts granite aplite must be added.
3. The material was introduced as albite; to 100 parts phyllite, 20 parts albite must be added.
4. The soda was introduced as albite; simultaneously silica and lime were introduced; water was removed. On 100 parts phyllite, 30 parts albite, 19.4 parts SiO_2, 3.9 parts Ca were added; 1.3 parts of water were subtracted.
5. The material was introduced and was fixed in the form of oxides of SiO_2, Na_2O, CaO; water was removed. To 100 parts phyllite, 26 parts SiO_2, 3.1 parts CaO, 2.8 parts Na_2O were added; 1.7 parts water were subtracted.

The results of these calculations prove that the introduction of soda did not occur as a simple addition of trondhjemite, granite, or albite; but, according to possibility 5, from solutions or vapors lime and soda were selectively absorbed, with simultaneous precipitation of silica.

The foregoing paragraphs sum up the main conclusions and the main results of the paper by Goldschmidt, which in 1921 was well-known and much discussed because it emphasized the importance of the functions exercised by the metasomatic solutions. The chemical composition of a metamorphic rock does not generally correspond to a mixture of the sediment plus the injected "magma." But from the magma certain oxides were selectively absorbed and fixed in the sediment, other oxides passed by without being fixed.

The volume relations were the chief reasons for Goldschmidt's rejecting the explanation of a simple mechanical mixing of the sediment and the intruded material. It is seen that the possibilities listed under 1 or 2 imply great quantities of foreign material to be introduced into the sediment. Goldschmidt's solution requires smaller quantities, although they are still appreciable. His solution can be represented by the following equation:

$$100 \text{ sediment} + 26.0\ SiO_2 + 3.1\ CaO + 2.8\ Na_2O - 1.7\ H_2O = \text{albite-porphyroblast schist}$$

This equation implies an increase of volume of more than 25 per cent. A metasomatism volume for volume is more satisfactory. By comparing the standard cells of sediment and schist we compare two units of approximately the same volume:

Sediment: $K_{5.1}Na_{2.1}Ca_{0.6}Mg_{2.4}Fe_{5.2}Al_{20.8}Ti_{0.6}Si_{51.0}C_{0.5}O_{160.0}H_{23.8}$

Schist: $K_{3.9}Na_{5.3}Ca_{2.8}Mg_{1.9}Fe_{4.0}Al_{16.1}Ti_{0.4}Si_{57.1}C_{1.0}O_{160.0}H_{9.8}$

Thus in order for the sediment to pass into schist without change in volume, the following ions must be added and subtracted:

	Adding	Subtracting
	3.2 Na-ions	1.2 K-ions
	2.2 Ca-ions	0.5 Mg-ions
	6.1 Si-ions	1.2 Fe-ions
	0.5 C-ions	4.7 Al-ions
		0.2 Ti-ions
Sum:	12.0 metal ions (34 valences)	7.8 metal ions and 14.0 H-ions (34 valences)

Rocks of the Sulitjelma area represent a series of facies covering the range from low greenschist facies to high amphibolite facies. They are summarized in Table IV-12.

10 · ANATEXIS AND GRANITIZATION

*Les roches stratifiées peuvent se changer en roches métamorphiques et, lorsque le métamorphisme est très énergique, elles passent même aux roches plutoniques les mieux caractérisées. Ainsi . . . dans les roches à base d'orthose, le gneiss passe insensiblement au granite. . . .**

Delesse, *Bull. soc. géol. France,* 1861.

There is a peculiar anomaly in the frequency distribution of igneous rocks.

Rocks of basaltic habit and composition (including pyroxene andesites) make up more than 98 per cent of all effusive rocks.

Granites and granodiorites make up at least 95 per cent of all deep-seated rocks, gabbros and other deep-seated equivalents of basalt taking less than 5 per cent.

* Stratified rocks may change into metamorphic rocks and, if the metamorphism is very strong, they may pass into rocks of plutonic character. Thus in rocks rich in potash feldspar, the gneiss passes insensibly into granite.

TABLE IV-12

Correlation between the Mineral Facies of the Intrusive Rocks and the Sedimentary Rocks of the Sulitjelma Area, Northern Norway

(Modified from T. Vogt, *Norges Geol. Undersk.*, 1927)

	Greenschist Facies——→		←——Epidote Amphibolite Facies				——→		←——Amphibolite Facies		
	Chlorite Muscovite ↑		Epidote Biotite Muscovite ↑		Garnet Epidote Biotite Muscovite ↑		Hornblende Garnet Epidote Biotite ↑		Hornblende Garnet Biotite ↑		
Sedimentary Series	Mica-chlorite schists	Calcite, muscovite→ →Epidote, H_2O	Epidote-biotite schists	Biotite, muscovite→ →Garnet, H_2O	Garnet-biotite schists	Calcite, Bi, Mu→ →Hornblende, H_2O	Hornblende-garnet-epidote schists	Epidote, (Mu)→ →Anorth, H_2O	Oligoclase-hornblende-garnet schists		
	↓ ↓ Quartz, albite ↑ ↑		↓ ↓ Quartz, albite ↑ ↑		↓ ↓ Quartz, albite ↑ ↑		↓ ↓ Quartz, albite ↑ ↑		↓ ↓ Quartz, plagioclase ↑ ↑		Plagioclase ↑
Intrusive Series	Chlorite-albite fels	Epidote, H_2O← ←Chlorite, CaO	Greenstones	Actinolite, H_2O← ←Chlorite, CaO	Actinolite greenstones	Hornblende, H_2O← ←Chlorite epidote actinolite	Epidote amphibolites	Anorthite, H_2O← ←Epidote	Amphibolites	Hypersthene An, H_2O← ←Hornblende	Gabbros
	↓ Chlorite		↓ Chlorite Epidote		↓ Chlorite Epidote Actinolite		↓ ...... Epidote Hornblende		↓ .... Hornblende (Diopside)		↓ .... Diopside Hypersthene

Gabbro phacoliths have been intruded into the sedimentary strata, often thinning out to sheet-like masses with concordant boundaries toward the surrounding schists. The metamorphism can be traced from high to low, starting with the truly magmatic gabbroic rocks traversing a number of transitional steps to the typical low-temperature greenschist-like gabbro derivatives. The metamorphism can also be traced from low to high; starting with practically non-metamorphic slates ending in highly metamorphic gneisses. Thus the two series have reached their present state along two different paths, the one from above, the other from below. The total range covered is from low greenschist facies to amphibolite facies.

The regional changes in the pressure-temperature diagram seem to have had a continuous course. The result is that all metamorphic facies between the extreme variations have been developed, and what may be called a complete series of facies is obtained. This is of great significance, and the main interest is attached to the investigation of the consecutive steps of facies, none of which is lacking.

Daly (1933) says: "The igneous rocks of the globe belong chiefly to two types: granite and basalt. To declare the meaning of the fact, that one of these dominant types is intrusive and the other extrusive, is to go a long way towards outlining petrogenesis in general."

Within orogenic regions the preponderance of granites and granodiorites is particularly striking and always demonstrable. This fact raises a major petrogenetic problem still essentially unsolved.

Fractional crystallization goes far towards explaining the volcanic rock suites and their associated intrusions, but fails to explain the batholithic suites of the orogenic zones. On the other hand, evidence of contamination is abundantly to be seen in the case of the batholithic rock suites, but fails to explain the genetical relationships among the volcanic rocks.

According to the theory of fractional crystallization, a residual magma of granitic or granodioritic composition should represent about 5 per cent of the original basaltic (or gabbroic) mother magma. We should expect, therefore, to find in the orogenic belts great amounts of primitive gabbros (or related rocks), but observations show this not to be the case. No mother gabbros are visible in the mountain chains. What, then, has produced the enormous granitic accumulations?

Our present knowledge leaves little reason to doubt that crustal rocks will begin to melt in the depth range 20–25 km. Magmas in the continents have certainly not existed since the beginning of the earth, but are generated in the depths of the geosynclines by differential melting. They are neomagmas, and the amount of liquid developed will depend upon the temperature, the chemistry of the encasing rocks, and the availability of water and other volatile substances. The sediments of the geosynclines are mainly arkoses, shales, and limestones; prior to metamorphism exaporites are also present and may materially lower the melting temperature. The melt is granitic or granodioritic in composition; it is unlikely that normal sediments can be a source of basaltic liquids. Thus a new magma is born (anatectic magma of Sederholm, neomagma of Goodspeed), and we have a mechanism by which granites and granodiorites are freed from their filial bondage to a basaltic mother magma.

Neomagmas may consist of completely metamorphosed material ranging from mobile grout-like consistency to a more homogeneous liquid, or even a melt which flows easily, has intrusive relationships to the country rock, and is sufficiently fluid to permit an abundance of crystals to form by fractional crystallization during flowage and later cooling stages.

Thus deeply eroded sections of old continents regularly exhibit an intimate intermingling of igneous, granite-like rocks with highly metamorphic crystalline schists and gneisses.

Continental rocks are weathered, transported, and deposited on the shelves; this is the external part of the geochemical cycle. By burying, downbuckling, folding, and orogenesis they again become incorporated into the consolidated rocks of the continents; this is the internal part of the geochemical cycle. A new cycle can start and the rocks are again exposed to the agencies of weathering. As explained in Part V most continental rocks are simply recycled modified sediments.

Metasomatic granitization, an old conception in geology, was introduced in consequence of the observation that granites sometimes have sharp contacts, sometimes show gradual transitions into the adjacent rocks, a fact which was interpreted differently by the neptunists and the plutonists of the old schools. The Norwegian geologist Keilhau considered continuous gradation as evidence of a metasomatism (a word which he used), and he introduced the term *granitification* (1836).

The French school of the last century studied metasomatism and feldspathization in relation to the fixation of mobilized material by solid rocks. From observations on granite contacts in the Pyrenees and elsewhere French geologists at the turn of the present century (Lacroix, Michel-Lévy, Barrois, Termier, and others) concluded that granitic magmas were accompanied by a halo of attenuated solutions capable of penetrating into the adjacent rocks as oil soaks into a piece of cloth. By this "a tache d'huile" mechanism the adjacent rocks were transformed, metasomatically, into rocks of granite-like composition, and thus prepared for the eventual assimilation by the intruding granitic magma.

The ideas were bitterly opposed by Rosenbusch and his German school, which has exercised a dominant influence on American thinking and has received further development and momentum through the teaching of Niggli in Zürich. Nevertheless, the "French" point of view was adopted by Fenner in his paper on the mode of formation of certain gneisses in the highlands of New Jersey (*J. Geol.,* 1914).

In Finland the early French ideas were nursed and critically examined by Sederholm and his Fennoscandian school, which, in more recent time, has shown a rich development. A theoretical fundament was established by Eskola, modified and expanded by Wegmann,

Backlund, Ramberg and others, and finally crowned by new ways of observation and verification in the field and in theory.

Simultaneously analogous ideas grew up in Germany (Scheumann, Erdmannsdörffer, etc.), in Great Britain (Holmes, Read, etc.), and in France (Perrin, Roubault, Roques, Lafitte and many others). Thus a continuous line of thought may be followed from the views of the French petrologists of the last century to the extreme views now advanced by some students.

Today nobody questions the phenomena of anatexis, differential remelting, and moderate granitization. In this respect the Rosenbusch-Niggli school has met with defeat. But not everybody is willing to follow Backlund, Holmes, Perrin, Roubault, and other contemporary geologists who question the magmatic mode of origin of granitic rocks in general.

Sederholm explained many of the Finnish granites as formed by *palingenesis* * or rebirth, whereby a granitic juice or liquor was formed, the so-called *ichor* † (corresponding to the *colonnes filtrantes* of Termier), that was capable of granitizing large surrounding areas.

Rocks formed by granitization contain abundant remnants of the pre-existing rocks, that is, irregular dark or gray patches in a more or less digested state: so-called cloudy relics or ghost-like remnants. ‡ Partly granitized rocks, that is, mixed gneisses containing beds, layers, and schlieren of the older rocks into which the granitic liquor had insinuated itself, were called migmatites. § They should be compared with the aorites || (unausgereifte Migmatite) of Erdmannsdörffer.

Eskola describes the ichor as a *pore solution* that at great depths fills the pores of all rocks and is mobilized by orogenic movements, squeezed out, and intruded as granite into the upper strata of the fold mountains.

Reinhard speaks of *migma*, that is, mobilized migmatic material which consists of a mixture of solid and liquid and may intrude like magma.

Wegmann regards the ichor as an *intergranular film*, and he thinks that the material transport takes place in this film along the grain boundaries. It may be mobilized by orogenic movements, but it is also able to rise by its own force. It is specificially lighter than its

* From πάλιν = again, and γένεσις = birth.

† ιχώρ = blood of the gods, lymph, plasma.

‡ Such inclusions have been called *skialiths* (σκιά = shadow) by Goodspeed. (See Fig. IV-59.)

§ μῖγμα = mixture.

|| ἀωρία = not ripe.

environment and will, therefore, driven by its own buoyancy, rise diapire-like, as salt domes rise through the encasing sedimentary strata (diapire granites) (Fig. II-5).

In the surrounding schistose rocks the folding axes in the nearest vicinity of such granite masses are usually steeply inclined or vertical, the granite occupying the axial culminations. Generally speaking, the traces of movements, that is, the platy and linear structures, are mostly directed upwards, proving that "something" has moved in that direction.

Today the physical nature of the ichor is much debated. *Holmes* and *Backlund,* for example, called it *emanation.* The emanation is of pulsating character and represents a migration of ions facilitated by *lattice disorder, substitution, polymorphic inversions, lattice deformations,* etc.

Sediments of granitic composition (arkoses) are most easily granitized. If the chemical differences are large, the original rock will long resist granitization. This relationship is illustrated by the following (relict) series in granitization (in order of increasing ability to survive): (1) mica schist, (2) quartzite, (3) limestone, (4) basite, (5) ore.

The continuous succession of the granitizing processes may be presented as follows:

						extrusion (acid lava)
					↗	
				rheomorphism (granite II)	→	intrusion (granite III)
				↑		
sedimentation (sediment)	→	metamorphism (metamorphite)	→	metasomatism (gneiss)	→	granitization (granite I)

The ichor may also be regarded simply as a cloud of free molecules and atoms loosened from the crystalline bonds, that is, as a kind of gas present everywhere in the rock. It possesses, of course, a high power of penetration and is able to move along the grain boundaries and through the crystalline lattices. Whenever physico-chemical conditions are right it acts as a strongly granitizing agent. Some authors claim that *all* granites are formed by reaction in the solid state (Perrin and Roubault).

The present status of the name ichor is unsatisfactory because, if used at all, it may be used for everything from a volatile-rich highly fluid magma to a highly disperse aggregate of slowly migrating ions.

All field evidence suggests that regional metamorphism and granitization are closely related; both are characteristic of the orogenic belts. Granitization works from below; it rises as a "front" and is preceded by a wave of metamorphism. It is therefore suggested that granitic melts are present in thick orogenic belts of the earth's crust at all times, at depths exceeding 20 kilometers. Ichor emanates from this locale and soaks into the overlying rocks which become studded with porphyroblasts of alkali feldspars and with spots of endemic pegmatites (see Fig. IV-54).

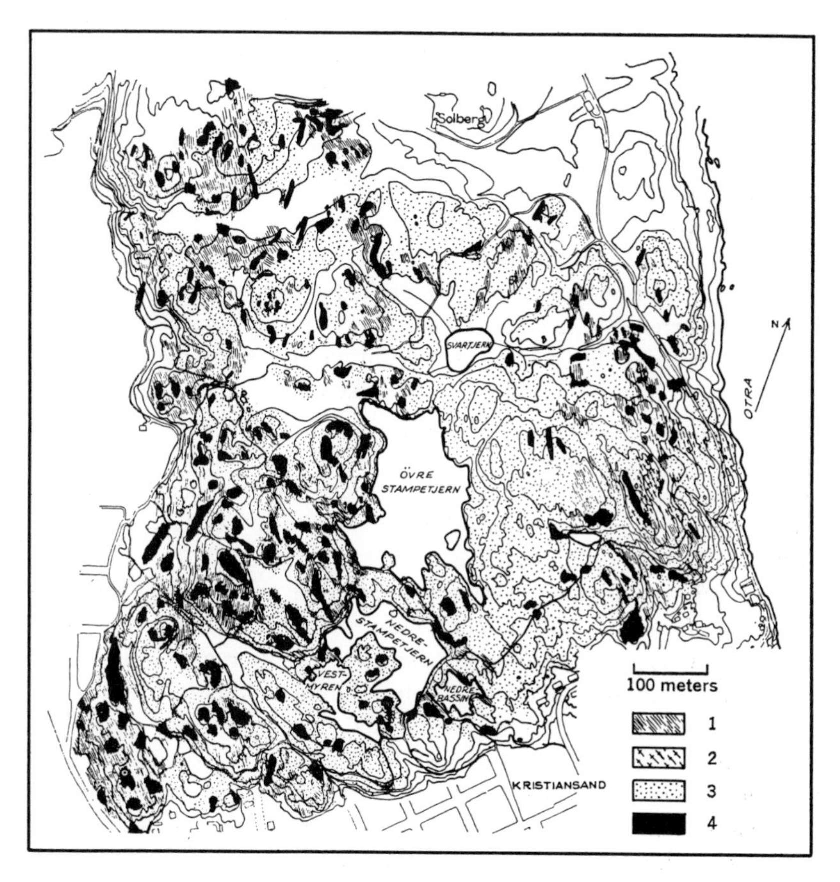

FIG. IV-54. Migmatite area at Kristiansand, southern Norway. (1) Amphibolite. (2) Augen gneiss. (3) Gneiss granite. (4) Massive granite and pegmatite. (Barth, 1933.) The solid black areas are pegmatites looking so much like blots that the gneiss complex may be likened to blotting paper that has sucked up all the pegmatitic juices. This blotting action was actually much more general than the map might suggest, for there are thousands of additional patches of pegmatite throughout the area that are too small to be shown on the map.

The larger pegmatite dikes are discrete bodies exhibiting sharp contacts to the gneiss.

In many places the pegmatites are not associated with any large granite body or any other rock of parental relation. They have no ducts or feeders, they lie isolated as endocrinic glands in a body. They show no textural or structural criteria indicative of movement in a once-molten magma. On the contrary, instead of showing platy foliation parallel to the walls, the pegmatites usually show all minerals oriented perpendicular to the walls. (See Fig. IV-35.)

Aplite is frequently associated with pegmatite, although the genetical relations are obscure. In their composition some of the granite pegmatites and aplites are potassic, and cannot, therefore, be accepted as residual solutions of any usual igneous rock. They are often related to augen gneisses (see page 296) into which they may merge with gradual transitions. Possibly the *pegmatitization* represents a special kind of granitization taking place at greater depths.

One of the most prominent features of pre-Cambrian areas, which are usually identified with deeply eroded orogenic zones, is the extensive occurrence of granitic rocks. Furthermore, granites are more common in the older pre-Cambrian formations than in its younger divisions. In its most ancient portions, like the so-called granite-gneiss region of Eastern Fennoscandia, they are almost universal. Likewise in all mountains, granite is found to be more extensively developed the deeper the mountains have been eroded. Consequently, granites have become more and more common as erosion has exposed deeper horizontal sections of the earth's crust.

On this basis Eskola (1932) concluded that granitic magmas must have been formed mainly in connection with orogenic movements by the pressing out, or squeezing, of the lowest melting materials, partly from more basic rocks not yet entirely solidified and partly from rocks partially refused in the deep regions of the geosynclines.

This idea may now be modified by tentatively assuming a comparable process, which, according to Misch (1949), consists of selective mobilization of ions and molecules producing an alkali-rich, mobile phase, some kind of emanation, which by means of infiltration, permeation, and diffusion rises into the geosynclinal prism and effects metasomatic granitization. Excess material, such as water and carbon dioxide, on the one hand, and iron, magnesium, and calcium, on the other, will move away from the sediments, upward and downward, respectively. These are the diffusions *à double sens* thought so important by Perrin and Roubault, silicium and alkali ions forming the granitic front, iron, magnesium and calcium forming the basic front (basic behind) of which Bowen said: ". . . a basic affront to the geologic fraternity." Retorted Reynolds: "This is accusing Nature of an intentional break of politeness."

Arguments in favor of metasomatic rather than igneous origin are, for example:

1. The room problem, which is linked with the abundance of inclusions and the usual lack of disturbance of the country rock adjacent to large batholiths. In some granites traces of sedimentary strata lie in exactly the position indicated by extension of the beds, either in the original form or as endomorphic modifications of the granite.
2. The common occurrence of small masses, irregular veins, schlieren (*trainées diffuses*), and isolated granitic bodies without visible connection to any magma source (ductless bodies). They are commonly "explained" as products of exudation or *Ausblütung.*
3. The preponderance of granite over basic rocks in the geosynclinal belts.
4. The scarcity of rhyolites, dacites, and other effusive equivalents of granite. If enormous masses of granitic igneous magma rose to shallow levels over tens of thousands of square miles, why did it never break through to the surface, drowning the geosynclinal area in floods of rhyolites and dacites?
5. Finally, the continuous gradation of granite to gneiss—a fact of profound importance—should not be forgotten.

Granitization involves both chemical action and introduction of heat conveyed by the mobile agent that granitizes the geosynclinal sediments. Additional heat may be generated *in situ* by exothermic reactions, but the primary source of heat is at great depths, 20 to 30 km or more, that is, at the place of the source of the granitizing elements (alkalies, silica, etc.) which ascend and react metasomatically with the overlying sediments (see Fig. III-81). Additional granitizing material is again generated *in situ* by redistribution (that is, mobilization, transfer, and reprecipitation) within the sediments. As distinct from *metasomatism,* the redistribution may be thought of as a *metabolism* of the rock.

There is an enormous literature on granite and granitization. Over 300 references are quoted by Mehnert (1959) in his useful and thus far most complete survey of the present stand of the granite problem.

11 · MIGMATITES

I conclude that migmatisation is the prime cause of regional metamorphism.

H. H. Read, 1944.

The name, meaning *mixed rock,* was introduced by Sederholm in 1907. According to his first notion, migmatites represent gneissous rocks, although neither ortho- nor paragneisses, but gneisses developed by mixing of sedimentary and igneous material by a special kind of injection metamorphism. It is "necessary to use a designation for these hybrid rocks which really characterizes their appearance and origin. They look like mixed rocks, and they originate by the mixture of older rocks and a later erupted granitic magma, and therefore the name migmatite is the most appropriate." (J. J. Sederholm, 1926.)

Later studies have forced most students to believe that the provenance of the "magma" portions of a migmatite is not as essential to the definition. This "magmatic" portion can arise in place and not be connected with a granitic magma; migmatites may be special products of metamorphic differentiation.

The formation of migmatites is closely related to sialic orogenic magmatism and granitization as expounded in an earlier section (page 221). In the root parts of the folded mountains (Unterkruste of Wegmann) a pore fluid or ichor is generated by differential melting (palingenesis, anatexis), and the solid residues and the contiguous sediments were stewed in this magmatic-anatectic liquid, which reacted with the sediments and metasomatically transformed them into migmatic gneisses. A *migma* is formed that corresponds to the magma of the higher levels.

Such types of rocks cannot be described adequately in conservative petrographical terms. To explain and to define the processes and their formation and special stages of these processes, various terms have been introduced: migmatism, metasomatism, replacement, imbibition, syntexis, anatexis, diatexes, palingenesis, contamination, hybridization, ultrametamorphism, injection metamorphism, hydrothermal metamorphism.

The following additional terms have been proposed by K. H. Scheumann:

Both venite and arterite designate veined gneisses, often of identical aspect and composition; but according to definition venite denotes that the material in the veins is derived by secretion from the rock itself, arterite implies that the vein-material was injected from a magma. In practical cases it is usually impossible to make this distinction, and the word *phlebite* (φλεψ = vein) should be used.

A magma-like portion of a migmatite should be called *metatect*, and the process of its emplacement *metatexis*. The border cases of the metatectic series may be called *ectect* (*ectexis*) in the case of pure secretion and elaboration, and *entect* (*entexis*) in the case of direct injection of a foreign melt. The metatect may also be referred to as *metasome* in contradistinction to *paleosome* or the relatively immobile (older) parts (Restgeweben). See Figs. IV-55 and IV-56.

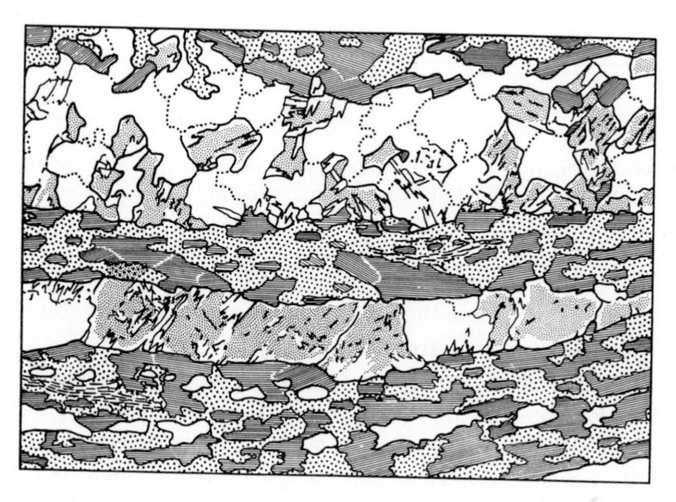

Fig. IV-55. Microphoto (×10) of metatectic gneiss showing metamorphic differentiation into alternating dark and light layers, respectively, of biotite-cordierite-sillimanite and quartz-feldspar. The light layers represent the metatect (metasome), the dark layers the paleosome. (From Black Forests, Germany, after K. R. Mehnert et al., *Neues Jahrb. f. Min.*, 1945/48.)

Metablastesis is the process of recrystallization and growth (blastesis) of preferred minerals or mineral groups (crystalloblastesis) until eventually the total mineral assemblage becomes texturally homogeneous and of massive structure. The neo-crystallization may imply only a relocation of the constituents of the rock itself or may depend on an influx of new matter by diffusion.

A series of rocks of "ultrametamorphic" origin has been classified by Backlund as follows:

Venites–(Augen gneisses)–Arterites–Migmatites–Palingenites–Diapirites

↘ ↘ ↘

(Coarse porphyries) (Syntectites) (Anatectites)

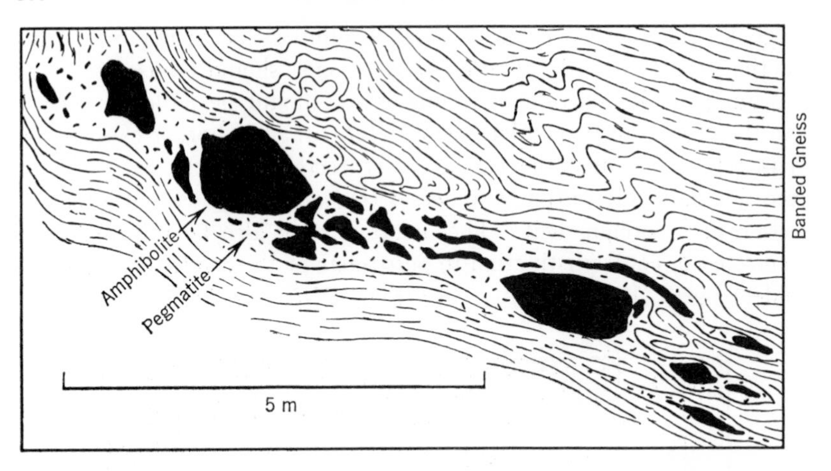

FIG. IV-56. Migmatite, Iveland, southern Norway. Pegmatite is metatectic and contains fragments of amphibolite in agmatite-like fashion. Pegmatite also represents the metasome; amphibolite is the paleosome.

In this series, the mobility and temperature of formation of the several members increase from left to right.

The fundamental difference in tectonics and in petrogenetic conditions of rocks from the higher levels of a mountain chain as compared with those from the subcrust has been emphasized repeatedly by Wegmann. He has made it clear that in great depths there is no sharp distinction between magmatic and non-magmatic rocks. This distinction was introduced for rocks of the upper parts of the crust and for the transitional zone, but it does not exist for rocks of great depths. Figures IV-57 and IV-58, slightly modified from Wegmann's paper, serve to explain schematically the process of migmatization in the deeper zones of a geosyncline. They illustrate that migmatization from below has to be regarded as a rejuvenescence just as is the sedimentation from the top.

It is consistent to reason that the term *migmatite* should be used not only for rocks displaying macroscopically a mixture of paleosome and metasome components but, generally, for *ultrametamorphic* rocks that are neither metamorphic sediments nor metamorphic igneous rocks but a mixture of both. That is, they are ultrametamorphites in which, owing to metasomatic processes, the mixing of the igneous and the sedimentary components has been so intimate that every mineral molecule of the rock has received contributions from both sources.

The modern French school has followed up this line of thought and

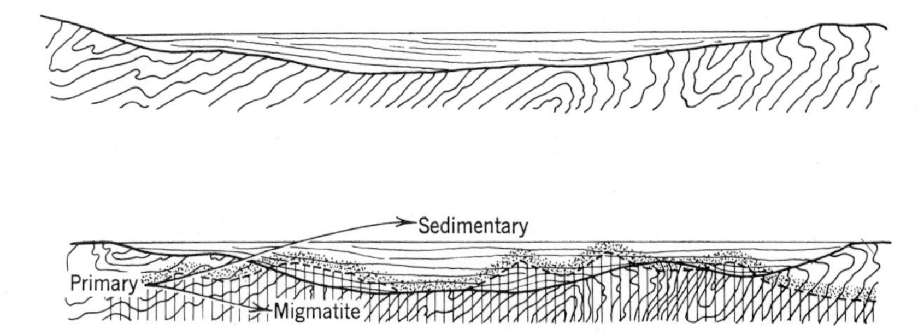

FIG. IV-57. Process of migmatization in the deeper zones of a geosyncline. *Above:* Geosyncline filled with sedimentary material. *Below:* Hot solutions from below are intruded into the geosyncline (vertically hatched area = zone of migmatism). Over this zone (represented by dotted area) is the zone of regional metamorphism.

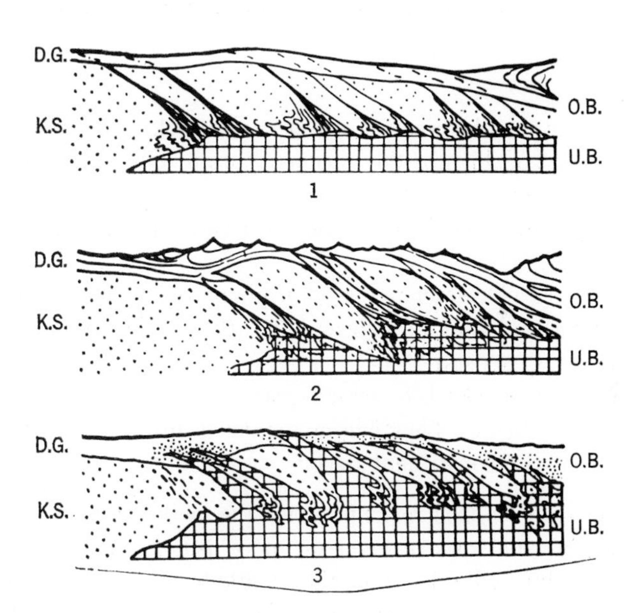

FIG. IV-58. (1) The subcrust of a geosyncline is activated and begins to move. The deep flow zones extend upward as zones of schistosity which higher up become zones of mylonite. (2) The whole subcrust is displaced upward. (3) The activated material is injected between the old masses, and relative to them it is a magma. It moves upward, filling the arches in the superstructure, which displays overthrusts, whereas the transitional zone shows great nappes and the infrastructure steep flowage folds. (After Wegmann, 1935.)

D.G. = Cover. K.S. = Crystalline Socle. O.B. = Superstructure (Oberbau). U.B. = Infrastructure (Unterbau).

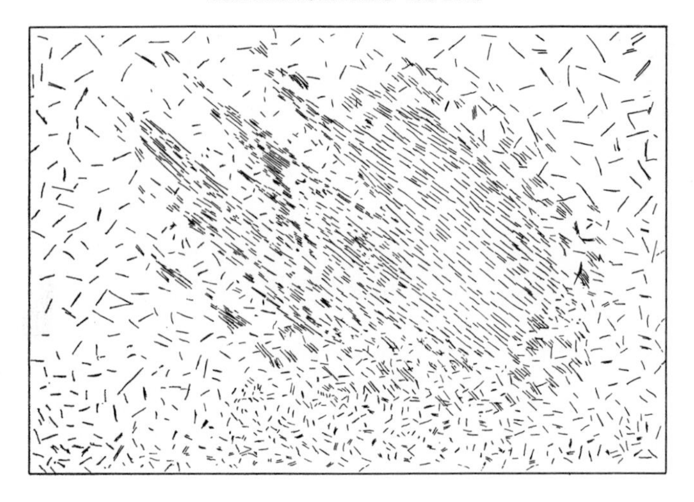

FIG. IV-59. Schist relic pre-Cambrian anatexite, southern Norway. (Half natural size.)

has worked out a scheme suitable for field work and also reflecting the genetic relationships of the various types of migmatite (Jung and Roques, 1938; Roques, 1941).

All metamorphic rocks are divided into two large groups: ectinites and migmatites.

Ectinites * are characterized as non-migmatic, which is about the same as saying that they have been left rather intact by granitization processes (slates, schists).

Migmatites are rocks into which granite material was introduced during metamorphism. They can be arranged into a series of increasing blending of the constituent parts:

1. *Diadysites* †—mixed rocks in which the metatectic granitic material appears in a network of veins or, more usually, in a multitude of isolated lenticular bodies in which the granite has an oriented structure.
2. *Embrechites*—mixed rocks in which the structural features of the usual crystalline schists are still preserved, although often partly obliterated by metablastesis: augen gneisses, banded gneisses (phlebites), etc. (*Injektionsmetamorphose* of V. M. Goldschmidt.)

* ἐκτείνω = tension, because the dynamic phenomena play a great part in the recrystallization of the rocks.

† διάδυω = pass through.

3. *Anatexites*—the regular schistose structure becomes fainter and the composition more granite-like. Skialiths or ghost-like remnants are abundant (nebulitic anatexites). By transitions these rocks grade into anatexe granites.

There is at present much confusion. As Dietrich and Mehnert (1960) justly remark in their proposal for the nomenclature of migmatites and associated rocks presented before the International Geological Congress in Copenhagen, 1960: One and the same outcrop could be called: agmatite, anatectic breccia, arterite, chorismite, diabrochite, ectexite, extinitic breccia, embrechite, entexite, granitized breccia, injection breccia, intrusion breccia, metatexite, migmatite, rheomorphic breccia and venite.

According to Wegmann, the occurrence of migmatites is of two types, one in the marginal zone of granites which have been intruded as magma, and the other in a zone where the country rocks have been transformed into granitic rocks. Since all sorts of sedimentation textures and tectonic structures are preserved in granite, it follows that the migmatization has passed like a wave through the country rocks, leaving a granitic and gneiss zone behind as a witness of its passage.

A distinction must be drawn between happenings in the non-migmatitic superstructure (Oberbau) and the migmatitic infrastructure (Unterbau) lying below. The two zones fold in disharmonic fashion, and the migmatite zone comes to fill the arches in the superstructure. Between the two zones is a transitional zone which is the site of regional metamorphism; it may be narrow as in southern Finland or wide as in the Scandinavian Caledonides. During the pressing forward of the migmatite front, the wave of "regional metamorphism" of the transitional zone moves forward in sympathy. In fold belts with great cross-shortening, the superstructure is the site of overthrusts, the transitional zone of great nappes, and the infrastructure of steep flowage folds. Further, since both the upper part of the infrastructure and the transitional zone were both once in the superstructure, it is not unexpected that relics of superstructural characters are found in them.

In his classical paper on granite and gneiss, Sederholm (1907) writes (translated from Swedish):

Orderly arrangement is, indeed, as fine a thing in science as in social life. But it must not grow into a screen of neglect. And it is a neglect of facts, if one tries to compress all rocks into two distinct groups—igneous and sedimentary—each with its accompaniment of metamorphic derivatives. For the two groups are not as distinct as has been assumed by some masters of petrography, who have loved order. The igneous and sedimentary rocks grade into each other on the surface of the earth, in the untransported alteration products of igneous rocks. And they grade into each other in the depths of the earth, where the high temperatures iron out all differences. The continuity is complete, the cycle in inorganic nature is without a break.

Says Read (1944):

There are three great classes of rocks, Neptunic, Volcanic, Plutonic. There are a few odds-and-ends left over that can be tucked in at leisure. I present the following summary of the position:

THE THREE CLASSES OF ROCKS

1. *Neptunic.* The sedimentary rocks, dominantly marine.
2. *Volcanic.* Effusive and associated intrusives; dominantly basic, magmatic, igneous; non-orogenic; comprises the orthodox volcanic rocks plus related minor intrusion, sills, laccolites, etc., as of gabbro, dolerite, andesite, trachyte, etc.; crystallization-differentiation may run in this class.
3. *Plutonic.* Of two associated kinds, the metamorphic rocks, and the ultrametamorphic, migmatitic, metasomatic, granitic rocks.

12 · SELECTED REFERENCES

General

Becke, F., *Über Mineralbestand und Struktur der kristallinen Schiefer,* Denkschrift Wiener Akademie, 1903.

Deer, W. A., R. A. Howie, and J. Zusman, *Rock Forming Minerals,* London, 1962.

Eskola, P., *Die metamorphen Gesteine, in Entstehung der Gesteine,* Berlin, 1939.

Fyfe, W. S., F. J. Turner, and J. Verhoogen, Metamorphic reactions and metamorphic facies, *Geol. Soc. Am. Mem.* **73,** 1958.

Grubenmann, U., *Die Kristallinen Schiefer,* Berlin, 1904–1906, 1910.

Grubenmann, U., and P. Niggli, *Die Gesteinsmetamorphose,* Berlin, 1924.

Harker, A., *Metamorphism,* London, 1932.

Hise, C. R. Van, *A Treatise on Metamorphism,* U.S. Geological Survey Monograph 47, 1904.

Lafitte, P., *Introduction a l'etude des roches métamorphiques et des gites métallifères,* Paris, 1957.

Leith, C. K., and W. J. Mead, *Metamorphic Geology,* New York, 1915.

Ramberg, H., *The Origin of Metamorphic and Metasomatic Rocks,* University of Chicago Press, 1952.

Turner, F. J., Mineral and structural evolution of metamorphic rocks, *Geol. Soc. Am. Mem.* **30,** 1948.

Turner, F. J., and J. Verhoogen, *Igneous and Metamorphic Petrology,* 2nd ed., New York (McGraw-Hill), 1959.

There is a very large and compendious literature on granite and granitization. Instead of attempting to compile this literature here, it is sufficient to mention the great essay review by Mehnert, 1959 (v.i.), who discusses and gives references to all important papers up to 1959.

Structural Geology and Petrofabrics

Billings, M. P., *Structural Geology,* New York, 1942.

Bucher, W. H., *The Deformation of the Earth's Crust,* Princeton, 1933.

Cloos, E., *The Application of Recent Structural Methods in the Crystalline Rocks of Maryland,* Maryland Geological Survey, 1937.

Cloos, E., Lineation, *Geol. Soc. Am. Mem.* **18,** 1946.

Fairbairn, H. W., and F. Chayes, *Structural Petrology of Deformed Rocks,* 2nd ed., Cambridge, Mass., 1949.

Holmes, A., *Principle of Physical Geology,* New York, 1946.

Knopf, E. B., and E. Ingerson, Structural Petrology, *Geol. Soc. Am. Mem.* **6,** 1938.

Nevin, C. M., *Principles of Structural Geology,* New York, 1949.

Sander, B., *Gefügekunde der Gesteine,* Vienna, 1930. *Einführung in die Gefügekunde der geologischen Körper,* Vienna, 1948.

Schmidt, W., *Tektonik und Verformungslehre,* Berlin, 1932.

Willis, B., and R. Willis, *Geologic Structures,* 3rd ed., New York and London, 1934.

Articles

Balk, R., and T. F. W. Barth, Structural and petrological studies in Dutchess County. *Bull. Geol. Soc. Am.* **47,** 1936.

Berthelsen, A., Structural studies in the Pre-Cambrian of W. Greenland, *Medd. Grönland* **135,** 1950; **123,** 1960.

Chinner, G. A., Pelitic gneisses with varying ferrous/ferric ratios from Glen Clova, *J. Petr.* **1,** 1960.

Coombs, D. S., Lower grade mineral facies in New Zealand, *XXI Geol. Congr. Norden, Part XIII,* 1960.

Dietrich, R. V., and K. R. Mehnert, Proposals for the nomenclature of migmatites and associated rocks, *XXI Intern. Geol. Congr., Norden, Part XXVI,* 1960.

Engel, A. E. J., and C. G. Engel, Grenville series in the N.W. Adirondack Mts., *Bull. Geol. Soc. Am.* **64,** 1953.

Eskola, P., The mineral facies of rocks, *Norsk Geol. Tidsskr.* **6,** 1921.

Fyfe, W. S., Hydrothermal synthesis and determination of equilibrium in the subsolidus region, *J. Geol.* **68,** 1960.

Goldschmidt, V. M., Die Kontaktmetamorphose des Kristiania-Gebietes, *Norske Videnskaps-Akad. Skr.* **1,** 1911.

———, Die Injektionsmetamorphose im Stavanger-Gebiete, *Norske Videnskaps-Akad. Skr.* **10,** 1920.

Mehnert, K. R., Der gegenwärtige Stand des Garnitproblems. *Fortschr. Mineral.* **37,** 1959.

———, Zur Geochemie der Alkalien im tiefen Grundgebirge, *Beitr. Min. Petr.* **7,** 1960.

Michot, P., Phénomènes géologiques dans la catazone profonde, *Geol. Rundschau* **46,** 1957.

Misch, P., Metasomatic granitization of batholithic dimensions, *Am. Jour. Sci.* **247,** 1949.

Ringwood, A. E., The constitution of the mantle, *Geochim. et Cosmochim. Acta* **13,** 303, 1958; **15,** 18, 195, 1959; **16,** 191, 1959.

Sederholm, J. J., On synantetic minerals, *Bull. comm. géol. Finlande* **48,** 1916.

Sederholm, J. J., On migmatites I, II, III, Idem. No. **58, 77, 107,** 1923, 1926, 1934.

Shaw, D. M., Geochemistry of pelitic rocks, *Bull. Geol. Soc. Am.* **65, 67,** 1954, 1956.

Wegmann, E. C., Zur deutung der migmatite, *Geol. Rundschau* **26,** 1935.

Winkler, H. G. F., and H. von Platen, Experimentelle gesteinsmetamorphose I, II, III, IV, V. *Geochim. et Cosmochim. Acta* **13,** 1958, **15,** 1959, **18,** 1960, **24,** 1961.

Yoder, H. S., The $MgO—Al_2O_3—SiO_2—H_2O$ system and the related metamorphic facies, *Am. J. Sci.,* Bowen Vol., 1952.

Part V · Geochemical Cycles

πανταρεί. (*All things flow.*)
Héraclitus.

The solid rocks of the crust are not stable and petrified as most think, but full of movements and action, physically and chemically. Physically the crust cracks and faults, heaves and sinks, shrinks and expands. Chemically the crust is likewise full of changes. Melting, crystallization, remelting, and assimilation occur; and liquids and emanations soak through, and react with, the solid rocks. And even the hard minerals themselves, without any help from fluids, are in a state of ceaseless change. Drifts of atoms loosened from the crystalline lattice will diffuse away from places of high chemical activity, migrate, and consolidate at places of low chemical activity.

Only by integration of data from seismic, electric, magnetic, thermal, chemical, mechanical, and radioactive surveys can the problems of earth deformation and chemical alteration be approached. And only by physico-chemical reasoning, by calculating chemical energies, by applying thermodynamical equations to transitions in minerals and to changes in rocks, can we hope to discover the quantitative rules that govern the reactions in the earth's crust, and to form a model picture of the mechanics and causes of crystallization and mineral formation.

Processes involving the shaping of the earth and the development of the crust and the surface of the earth have their beginnings in pregeological time. According to cosmogonists, the young earth had neither atmosphere nor hydrosphere; all gaseous and volatile matter at the surface had been lost to cosmos (see page 4).

Some of the original gases and volatiles were incarcerated, however, in the deeper parts of the earth, and together with other compounds (ions) of high fictive volume they began a slow migration upward, driven by forces explained by equations 25–28 in the Appendix.

The discharge of light substances from the interior is a geochemical process of great importance. It has been going on since the beginning of time, and represents a steady addition to the matter already present in the lithosphere, hydrosphere, and atmosphere. A certain exchange takes place between the upper and the lower parts of the crust of the earth.

One of the objects of geochemistry is to trace the migration of each element and explain the chemical reactions taking place at various stages. As emphasized first by Vernadsky, much of the migration of the elements can be described in terms of geochemical cycles.

In addition to the cyclic migration, we must not forget, however, the new supply brought in from below and, correspondingly, the continued loss of hydrogen to cosmos and the probable loss of heavy elements or compounds with small fictive volumes from the lithosphere to the deeper parts of the earth.

Water, nitrogen, and the noble gases, partly oxygen, as well as large quantities of chlorine, fluorine, boron, sulfur, and selenium are products of degassing of the lithosphere.

A small fraction of some of the metals, for example, mercury and perhaps sodium, are likewise possible products of degassing and are eventually being introduced into the external geochemical cycle mainly by hydrothermal processes. Thus the concentric structure of the earth is augmented, and the following elements are slowly accumulating in the outer earth shells:

H, N, Ne, A, Kr, Xe, O, Cl, F, B, S, Se, (Hg), (Na)

These are the marks of age imparted upon the earth. All planets undergo changes with the passing of time; the world ages. But for most geological processes, the aging of the earth is of no direct consequence; the processes of aging are too slow to interfere seriously with the much faster cyclic processes of weathering, transportation, deposition, followed by downbuckling, folding, orogenesis, until the original sediments again become incorporated into the consolidated rocks of the continents; then a new cycle will start, and the rocks are again exposed to the agencies of weathering. The full geological cycle can be divided into an external and an internal part.

The external geochemical migration derives its energy from the sun; it consists of rock weathering, transportation, and sedimentation. In this part of the cycle, the sea occupies a central position:

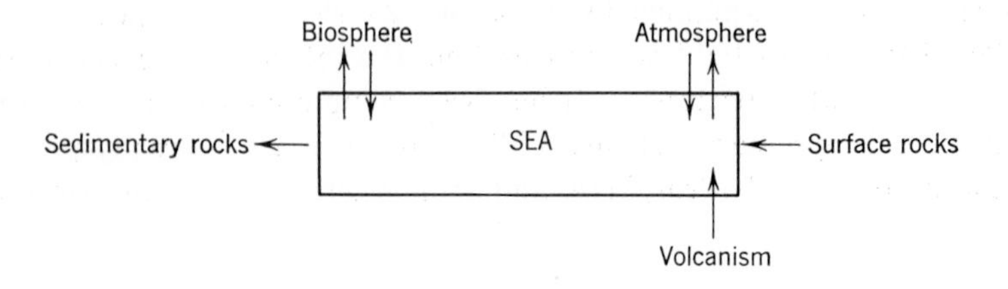

The weathered rock material is delivered into the sea as sand and mud, in colloidal suspension, and in true solution. The suspended material is soon precipitated, but the dissolved matter remains in the sea for some time.

"All rivers run into the sea, yet the sea is not full" (Ecclesiastes, 1:7).

Are the cycles endless? V. M. Goldschmidt and all geochemists following him did not believe in "endless" cycles, but thought that sodium accumulated in the sea; this notion was inherited from the Rosenbusch school of thought whose ideas have prevailed in modern geochemistry: igneous rocks are regarded as primary, they come from the depths of the earth, from "die ewige Teufe," i.e., they are "juvenile," whereas sedimentary (and metamorphic) rocks are secondary and ultimately derived from igneous material.

Clarke and Washington adhered to this theory: The average composition of the crust is, in effect, that of igneous rocks; crustal abundances of the elements can be computed, therefore, by averaging the chemical analyses of the igneous rocks. The overall average of the 5159 analyses of igneous rocks compiled by Washington is still the accepted basis for nearly all geochemical calculations.

V. M. Goldschmidt, a scientist of unusual capacity and the founder of modern geochemistry, was likewise deeply rooted in Rosenbusch's school; so was N. L. Bowen, an equally great man and the founder of modern petrology.

It still is a formal or intensional truth that the average composition of the sedimentary rocks must correspond to that of the igneous rocks; for ultimately all sediments derive from igneous material.

As a consequence, Goldschmidt maintained *in principle:* *

$$\text{Age of the Sea} = \frac{\text{Total amount of Na in the sea}}{\text{Na supplied per year}} \qquad \text{(V-1)}$$

* He made a modification by assuming that 33 per cent of the sodium was locked up in evaporites and other sediments. It should also be added that he did not think that the present supply of Na (as observed in river water, etc.) should be used.

It is highly interesting to note that early in the eighteenth century Edmund Halley suggested that the age of the ocean might be found from the amount of salt which had accumulated in it and the present rate at which it was being added from rivers; but at that time the data were not available from which to make the calculation. He wrote: "It were to be wished that the ancient Greek and Latin authors had delivered down to us the degree of saltness of the sea as it was about 2000 years ago, for then it cannot be doubted that the difference between what is now found and what then was, would become very sensible" (Halley, 1715).

Assuming that the sodium missing in the sediments is to be found in the sea, he derived the second famous equation:

Total weight of weathered rocks

$$W = \frac{\text{Total amount of Na in the sea}}{\text{Na\% of igneous rocks} - \text{Na\% of sediments}} \tag{V-2}$$

Goldschmidt made a correction in accordance with the fact that the amount of sediments (shales) is slightly less than that of the igneous rocks (v.i.). The concentration of sodium in igneous rocks and in sediments are about 3 per cent and 1 per cent respectively. The total amount of sodium in the ocean corresponds to a mass of 2.9 kg per cm^2 of the earth's surface. From equation V-2 we thus obtain,

$$W_{\text{total}} = \frac{2.9}{3\% - 1\%} = 145 \text{ kg/cm}^2$$

By introducing the small corrections mentioned above, Goldschmidt actually arrived at 160 kg/cm^2.

It is hard to understand why sodium should occupy such a special position in the geochemical migration. Neither the chemical properties of sodium nor geological evidence suggest that anything particular should happen to sodium. In fact, there is every reason to think that sodium behaves like other elements—K, Ca, Mg, Fe, etc.—which migrate through the sea in recurring cycles.

Sodium is another element in the cycles, and for sodium, also, there must be approximate balance between supply and removal. Through weathering, solution, and transport sodium is delivered to the sea, through precipitation and sedimentation it is again extracted from the sea.

Thus Barth (1961) maintained that in principle it is impossible to determine the age of the sea by this method. But if perfect balance is assumed between input and output, it becomes possible to calculate how long each element remains in the sea before it is removed. The period of passage, τ, may be defined as the average time expressed in years necessary for an element i to pass through the sea in the geochemical cycle. Thus

$$\tau_i = \frac{\text{total amount of an element in the sea}}{\text{amount supplied per year}} \tag{V-3}$$

An approximate solution for sodium is furnished by the figures of Table V-3 (see page 375):

$$\tau_{\text{N}} = \frac{2.9}{2.4 \times 10^{-8}} = 120 \text{ million years}$$

This is the period of passage, or, statistically, the average period of residence of a Na ion in the sea.

Equation V-3 gives a totally different meaning to the Goldschmidt equation here quoted as V-1. Obviously Goldschmidt's second equation will also be influenced. It is easily seen that the weight of weathered rock per year W_y has the following relation:

$$W_y = \frac{\text{Na supplied per year}}{\text{Na\% in source rock} - \text{Na\% in sediments}} \qquad \text{(V-4)}$$

By combining equations (V-3) and (V-4):

$$\tau \cdot W_y = \frac{\text{total amount of Na in the sea}}{\text{Na\% in source rock} - \text{Na\% in sediments}} \qquad \text{(V-5)}$$

Equations (V-1), (V-2)—and (V-3), (V-5) form two sets of equations of the same form, but of different meaning.

The expression:

$$\frac{\text{total amount of Na in sea}}{\text{Na supplied per year}}$$

gives, according to Goldschmidt, in principle the age of the sea. According to Barth, it gives, statistically, the period of passage through the sea of a Na ion.

The expression:

$$\frac{\text{total amount of Na in the sea}}{\text{Na\% in source rock} - \text{Na\% in sediment}}$$

is, according to Goldschmidt, proportional to the total weathering in the earth. According to Barth, it is the amount of weathering which takes place in 120 million years (120 million = the period of passage of a Na ion).

The equations of Goldschmidt apply only to Na which, according to him, is exceptional. Barth's equations are of general application, and can be used in conjunction with other elements, in particular Mg, Ca, and K, which go into true solution in river waters and stay comparatively long in the sea. Thus values for W_y (= the amount of sediments formed in a time unit, e.g. in a million years) may be computed from equation (V-4) by inserting the numerical values of either Mg, Ca, Na, or K; the result should be the same. Table V-5 shows this to be true—all sets of computation give the same order of magnitude of the amount of sediments (on an average about 2.4 kg per cm^2 per one million years). This again corresponds to an erosion to a depth of 8 m in a million years. Putting the age of the earth at somewhat more than 3×10^9 years, the total erosion will correspond to a depth of 30 km all over the globe; and if we restrict the weathering to the continents, the erosion would correspond to more than 60 km.

This sounds like very much; and it is, indeed, very much more than that calculated by contemporary geochemists. But most estimates based on geological and stratigraphical evidence are of the same order of magnitude. They range upward of 30 km referred to the surface of the whole earth.

There is no reason to reject these estimates. Thus the total effect of weathering corresponds to a great erosional depth. Actually, however, the erosion has never dug so deeply into the crust: the sediments return to the continents and are recycled. The old students of pre-Cambrian geology always searched for a beginning. At one time it was believed that the banded gneisses represented the first crystalline crust of the earth. This belief had to be abandoned. Time seemed to be without bounds—it was futile to seek confines. Sederholm, the great explorer of the Fennoscandian pre-Cambrian, was seeking a floor upon which the sediments were deposited. He found no floor. The oldest rocks are sediments. Said Hutton 200 years ago: "In the economy of the world I can find no traces of a beginning, no prospect of an end."

Thus we are forced to the conclusion that our good earth is so old that the continental crust repeatedly has been put through the geochemical cycles. The cycles are endless as far as the present geology is concerned. Most rocks of the continents have once been sediments; there is very little "juvenile" material. The rocks have been modified by "plutonism," by metamorphism, and metasomatism; they have been, partly at least, remelted and have been in the form of magmas and lavas; but some time back they were derived from sediments. This is further explained by the internal part of the geochemical cycle.

Computations and Tables. In Table V-1 some areas and volumes are listed, and Table V-2 shows the concentration of the various elements in terms of milligrams per kilogram of sea water. These figures,

TABLE V-1

AREAS AND VOLUMES

Ocean	361,059,200 km^2
Shelf and slope	66,200,000 km^2
Surface of earth	510,100,934 km^2
Volume of the ocean	1.37×10^9 km^3; density = 1.028
Total dissolved solids	5×10^{16} ton = 500 Gg *

* In these calculations the geogram, as introduced by Conway, is a convenient unit: 1 Gg = 10^{20} g = 10^{14} tons. (The symbol Gg is also used for gigagram = 10^9 g).

TABLE V-2

ELEMENTS PRESENT IN SOLUTION IN SEA WATER

(Dissolved Gases Not Included)

(After Sverdrup et al., 1942 with amendments of Sugawara 1955)

Element	mg/kg	Element	mg/kg
Chlorine	18,980	Iron	0.002–0.02
Sodium	10,561	Manganese	0.001–0.01
Magnesiun	1,272	Copper	0.001–0.01
Sulfur	884	Arsenic	0.001–0.002
Calcium	400	Zinc	0.005
Potassium	380	Lead	0.004
Bromine	65	Selenium	0.004
Carbon	28	Cesium	0.002
Strontium	8	Uranium	0.0015
Boron	4.6	Molybdenum	0.0005
Silicon	0.02–4.0	Thorium	0.0005
Fluorine	1.4	Cerium	0.0004
Nitrogen	0.01–0.7	Silver	0.0003
Aluminum	0.5	Lanthanum	0.0003
Rubidium	0.2	Yttrium	0.0003
Lithium	0.1	Nickel	0.0001
Phosphorus	0.001–0.10	Scandium	0.00004
Barium	0.05	Mercury	0.00003
Iodine	0.04	Gold	0.000006
Vanadium	0.002	Radium	$0.2\text{–}3 \times 10^{-10}$

TABLE V-3

THE MOST IMPORTANT IONS OF THE HYDROSPHERE

	D Dissolved Ions Delivered to the Sea kg/cm²/10⁶ years				M kg/cm²/- 10⁶ years	τ 10⁶ years
	1	2	3	4	5	6
Mg	0.018	0.016	0.014	0.015	0.35	23.
Ca	0.109	0.108	0.061	0.085	0.11	1.3
Na	0.031	0.014	0.036	0.024	2.90	120.
K	0.011	0.011	0.010	0.011	0.105	10.
Sum	...	0.149	0.121	0.135	...	...
Other ions	0.366	0.331	...	...	...	...

1. Recalculated from Clarke's (1924) data of the total discharge and the dissolved ions in the rivers of the world.
2. Recalculated from Conway's (1942) data of rain corrected river waters.
3. Theoretically delivery of rivers of the world *if* the amount of chemical weathering of the composite source rock is 2.5 kg/cm² (see Table V-5).
4. Weighted average for rain corrected river waters.
5. Mass of ions present in the sea.
6. Period of passage, $\tau = \frac{M}{D}$, using the values of columns 4 and 5.

multiplied by the factor 0.014, give the total amount of the several elements in the sea expressed in geograms, and the factor 2.75×10^{-4} converts them into kg/cm² of the earth's surface.

Let W be a unit mass of the source rock. By weathering W splits into three parts: (1) D dissolves in surface waters and is delivered to the sea, (2) S is carried in suspension by surface waters, sediments, and builds up shales and schists, and (3) Q remains as residue (sandstone, quartzite).

Consequently, $W = D + S + Q$. According to Wickman, $S = 0.845W$, and $Q = 0.088W$, thus

$$W = \frac{D}{0.067} \tag{V-6}$$

TABLE V-4 AVERAGE COMPOSITION OF IGNEOUS ROCKS (AFTER CLARKE) COMPARED WITH THAT OF SEDIMENTS (AFTER CORRENS)

	Igneous Rocks	Sediments
SiO_2	59.12	55.64
TiO_2	1.05	0.69
Al_2O_3	15.34	14.44
Fe_2O_3	3.08	6.87
FeO	3.80	....
MnO	0.12	0.12
MgO	3.49	2.93
CaO	5.08	4.69
Na_2O	3.82	1.21
K_2O	3.13	2.87
P_2O_5	0.30	0.17
H_2O^+	1.15	5.54
CO_2	0.10	3.86
S	0.05 (SO_3)	0.32
C	....	0.65

TABLE V-5 SELECTED VALUES FOR THE CONTENTS OF Mg,Ca,Na,K IN THE COMPOSITE SOURCE ROCK AND IN AVERAGE SHALE

(See Barth, 1962)

	c Per Cent	$s' = 0.845 \times s$ Per Cent	W kg/cm^2/10^6 y
Mg	1.80	1.19	2.6
Ca	3.00	0.34	3.2
Na	2.30	0.85	1.7
K	2.60	2.16	2.5

c Concentration (per cent) in composite source rock.
s' Concentration (per cent) in shale $\times$ 0.845 (see equation V-7).
W Calculated amount of weathering, $W = \dfrac{D_i}{c_i - s_i'}$ in kg/cm^2 per million years.

All calculated values for W are of the same order of magnitude, on an average 2.5 kg/cm^2 per one million years. This amount should be compared with the amount computed from equation V-6, $W = 2.0$ kg/cm^2 per one million years. Putting the age of the earth somewhat higher than 3×10^9 years, the total amount of weathered rocks becomes about 8000 kg/cm^2, corresponding to a depth of erosion of 30 km all over the earth (or about 60 km only on the continents).

The elements Mg, Ca, Na, K do not occur in quartzite. The concentration of one of these elements, i, in the primary source rock and in shale is c_i and s_i respectively. The quantity of the element dissolved in surface waters is D_i. For each of the four elements the following relation holds:

$$W \cdot c_i = D_i + S \cdot s_i$$

substituting for S,

$$W = \frac{D_i}{c_i - 0.845 \cdot s_i} = \frac{D_i}{c_i - s_i'} \tag{V-7}$$

The total weathering per time unit, W_y, can be computed from equation V-6, D being the total amount of dissolved ions delivered to the sea in one year. W_y can also be computed from equation V-7, and here the four elements, Mg, Ca, Na, K, whose concentrations in river water, primary source rock, and schist are, D_i, c_i, and s_i, respectively, provide us with four sets of values that should all give the same result.

According to Clarke's estimates, the total amount of dissolved substances supplied by the rivers to the sea is 2.74×10^9 tons per year. A surprisingly large fraction of this amount does not, however, take part in the true geochemical cycles. Practically all chlorine and about 55 per cent of the sodium in river water are supposed to be "cyclic." Sodium chloride is blown from the sea over the continents, and by rain and river waters again returned to the sea; this fraction must be subtracted from the river water in computing the amounts delivered by the lithosphere to the sea. Thus are obtained, according to Conway, the figures in the column of "rain corrected" river water in Table V-3.

The internal geochemical migration. The earth is an active globe geologically. As soon as the sediments have been laid down in a geosyncline, they come under the sway of the endogenic forces. The processes of diagenesis begin to work: compaction and cementation, also recrystallization and replacement to some extent, the result being a conversion of loose, incoherent sediments into hard rocks.

Through geosynclinal downwarping and deep burying, temperature and pressure increase, and true metamorphism begins. The mineral assemblages which were adjusted to low temperatures and pressure are no longer stable; reactions set in. Volatile constituents and, for reasons already discussed, atoms and molecules of low specific gravity will pick up a high tendency to escape. Thus sediments will lose water, carbon dioxide, and a number of rarer components which are returned to the hydrosphere and atmosphere.

Through extended recrystallization, replacement, and metasomatism, the chemical constituents of the sediments become completely rearranged. Depending upon temperature, pressure, and stress conditions, loss and gain and exchange of matter will take place, resulting in a complete chemical reconstitution of the several rocks. They lose their identity as sediments; their chemical components disperse in various directions.

At high temperature, differential melting sets in (anatexis, rheomorphism), and, if high pressures also prevail, the zone of anamigmatism is reached, where the sialic material (rich in potassium, sodium, aluminum, silicon) is squeezed out and returned to the upper crust.

At any stage in this sequence of alteration the rocks, through orogenic movements and denudation, may be returned to the surface and may again be subjected to the weathering agents.

A graphical survey of the endogenic geochemical migration is given by the following diagram:

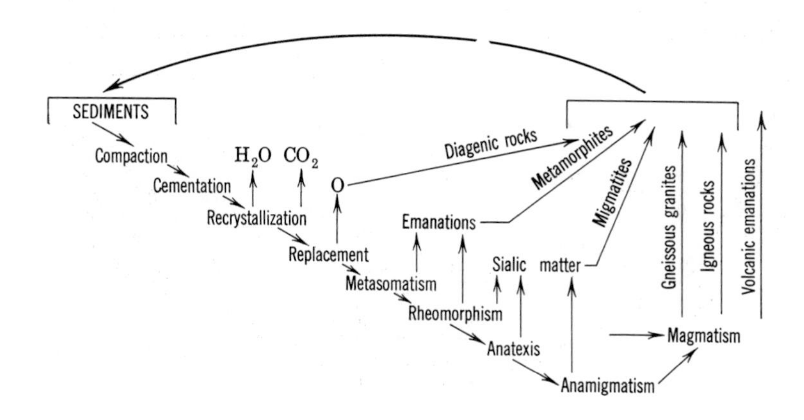

Each chemical species travels with its own intrinsic speed; on an average the light molecules, radicals, or atoms take a shorter path and therefore use less time on their endogenic journey than the heavier substances. In this fact more than in the effect of differentiation in magmas lie the ultimate causes of the chemical diverseness of the metamorphic and igneous rocks.

Conclusion: *"Let us leave a few problems for our children to solve; otherwise they might be so bored."*

In Part III of this book, igneous activity was regarded as a complex of processes which, generally speaking, resulted in differentiation. Starting with a homogeneous magma, the diversity of rock types was derived. But in many ways this is contrary to common sense. Most people would use the expression "melting pot" as a figure of speech conveying the idea of homogenization, not diversification. By melting, assimilation, and stirring, the igneous processes will lead to homogenization. If the igneous processes were allowed to go on without interruption, there would be very little rock variation in spite of the many silicate systems investigated in the laboratories. An example of the point is the petrography of Hawaii. Great quantities of basaltic

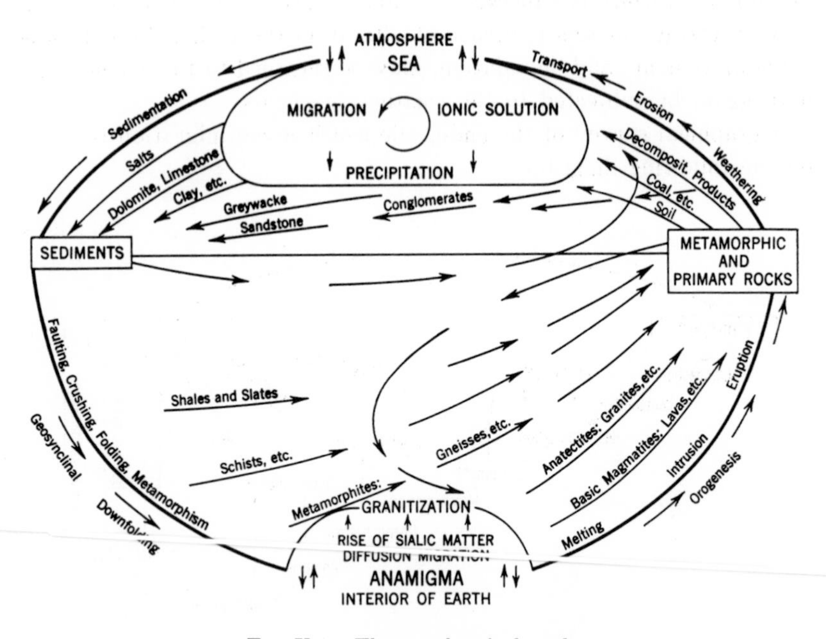

Fig. V-1. The geochemical cycle.

rocks are extruded, igneous activity is intense, and the processes of magmatic differentiation are fully active; chemical analyses and thin section examinations clearly show that crystal fractionation, mostly following the rules of Bowen, operates continuously. And what is the result? Just homogeneous basaltic rocks covering nearly the totality of the islands. As *rarae aves* small amounts of trachyte or melilite basalt, for example, may be found; but volumewise such differentiates are pitiably small. The melting pot does its work, and in the absence of sedimentary processes interfering, the result is nearly only basaltic rock of remarkable homogeneity.

The point is that in the continents the igneous processes are not allowed to keep the show for themselves; they are interrupted by sedimentary processes, and sedimentation results in differentiation. Thus it may be said, although it sounds like a paradox:

The diversification of igneous rocks is caused by sedimentary processes.

SELECTED REFERENCES

Barth, T. F. W., Abundance of the elements, areal averages, and geochemical cycles, *Geochim. et Cosmochim. Acta* **23,** 1, 1961.

———, Die Menge der Kontinentalsedimente und ihre Beziehung zu den Eruptivgesteinen, *N. Jb. Miner.,* Mh. 1962.

Conway, E. J., Mean geochemical data in relation to oceanic evolution, *Proc. Roy. Irish Acad.* **48,** 1942–1943.

Goldschmidt, V. M., *Geochemistry,* London, 1954.

Mason, B., *Principles of Geochemistry,* 2nd ed., New York, 1958.

Nieuwenkamp, W., Géochimie classique et transformiste, *Bull. Soc. Géol. France* **6,** 1956.

Vernadski, V. I., *La Géochimie,* Paris, 1924.

Wickman, F. E., The total amount of sediments and the composition of the average igneous rock, *Geochim. et Cosmochim. Acta* **5,** 97, 1953.

Appendix · Thermodynamics as a Help to the Study of Rocks

*Es gibt nichts praktischeres als die Theorie.**
Boltzmann, Vorlesungen, Leipzig, 1895.

In the investigation of the rocks of the crust, the exact methods of physics and chemistry are employed where they are appropriate. But as distinct from mathematics and theoretical physics, natural phenomena cannot be treated with schematic accuracy; all calculated conditions are to be regarded as idealized cases that correspond approximately to the physical facts.

Thermodynamics is a means by which chemical and physical changes can be quantitatively treated and often predicted. In the studies of the crystallization of magmas, polymorphic transformations, mineral reactions—in short, in magmatism, metamorphism, and metasomatism—thermodynamics correlates and interprets the phenomena in a quantitative way; it shows that the driving forces of all reactions are energy differentials of some sort, and that the end results is a tendency for all matter to transfer into a state of lowest free energy. This is perhaps the most important thing to remember. The difficulty is that this state of lowest free energy cannot be determined once and for all; it changes with temperature, pressure, composition, and other factors. Therefore each case must be specially considered, and each combination of these variables must be taken into consideration.

THE FIRST LAW OF THERMODYNAMICS

The fact that a mineral assemblage changes into another shows that the new assemblage has lower free energy than the old. Under the conditions of the metamorphic environment a new set of minerals, U, V, W, . . . , has a lower free energy than the old set of minerals, A, B, C, ... It is important to note that the energy drop represents the physico-chemical force which drives the metamorphic process, and that its potential comes from a difference in the free energy of the old set of crystals and the free energy of a new set.

* There is nothing more practical than theory.

The energy relations, therefore, are the controlling factors in petrology, and a quick survey of these relations can be given only in terms of some of the simple, fundamental concepts of thermodynamics.

The total energy of all sorts contained in a system, for instance, in a mineral (namely, the kinetic and potential energies of all the atoms of the system), is called the *internal energy* and is denoted by U.† It depends only on the state of the system, that is, in a given mineral, on temperature, pressure, and position in the field of gravity. It cannot be determined in absolute values; only differences have significance.

In an infinitesimal change of the system, the energy which has entered the system as heat flow is called dQ, and the energy which has left the system as mechanical work is called $-dW$. Then the law of conservation of energy demands that

$$dU = dQ - dW \tag{1}$$

This is a mathematical statement of *the first law of thermodynamics*.

The mechanical work, dW, is usually measured by a change of volume acting against a hydrostatic pressure,

$$dW = P\,dV \tag{2}$$

consequently

$$dU = dQ - P\,dV \tag{3}$$

$$U_2 - U_1 = Q - P(V_2 - V_1)$$

$$U_2 - U_1 = Q_v \qquad (v \text{ constant}) \tag{3'}$$

The *enthalpy* or heat content of a system is denoted by H and defined as:

$$H = U + PV$$

$$dH = dU + P\,dV + V\,dP \tag{4}$$

If the process takes place at constant pressure, $dP = 0$, we obtain in combination with equation 3,

$$dH = dU + P\,dV = dQ$$

$$H_2 - H_1 = Q_p \qquad (p \text{ constant}) \tag{4'}$$

The internal energy content and the enthalpy content have definite values for a mineral or a mineral assemblage at any definite state of pressure, temperature, and aggregation. They are usually expressed in calories per mole.

† In many texts the symbol E is used.

But the energy contents cannot be determined in absolute values—they are only known as differences. The usual (but arbitrary) zero point for any element is the so-called *standard state* (pure solid, pure liquid, or gas at one atmosphere pressure at room temperature, 298°K) and is indicated by the symbol ΔH°.

The *heat of formation*—for example, of quartz—from the elements can be measured calorimetrically, and is found to be 205 cal per mole, at room temperature and pressure. This is equivalent of saying that the enthalpy of quartz, $\Delta H^\circ 298 = -205$ cal per mole, provided that the standard state enthalpies are taken as zero for both $Si_{(c)}$ and $O_{2(g)}$. ((c) and (g) stand for crystalline and gaseous, respectively.)

The *heat capacity* is the heat required to raise a unit mass of the material one degree. The *molar* heat capacity is measured in cal per mole per degree. The heat capacity at constant volume has the symbol c_v; at constant pressure the symbol c_p is used.

If the material is raised from temperature 1 to temperature 2 it follows from equations 3′ and 4′ that the increase in energy content equals the heat absorbed, consequently,

$$U_2 - U_1 = \int_1^2 c_v \, dT \backsim c_v(T_2 - T_1) \qquad (v \text{ constant}) \tag{5}$$

$$H_2 - H_1 = \int_1^2 c_p \, dT \backsim c_p(T_2 - T_1) \qquad (p \text{ constant}) \tag{6}$$

Similarly the heat of reaction, ΔH, will change with the temperature; this is the Kirchhoff equation,

$$\Delta H_{T_2} - \Delta H_{T_1} = \int_{T_1}^{T_2} \Delta c_p \, dT \qquad (p \text{ constant})$$

where Δc_p is the heat capacity of the products formed, less the heat capacity of the reactants.

For minerals and rocks $c_v \backsim c_p \backsim 0.2$ cal/g $\cong 0.5$ cal/cm^3. In an example on page 304 heat capacities have been used to calculate the time it takes to heat a rock mass by radioactive energy.

THE SECOND LAW OF THERMODYNAMICS

It follows from equation 1 that the change from state 1 into state 2 can be expressed by the following integral:

$$\int_1^2 dU = \int_1^2 dQ - \int_1^2 dW$$

This integral is independent of the path going from 1 to 2, but the separate integrals, $\int_1^2 dQ$ and $\int_1^2 dW$, are not independent of the path but may be entirely different for different processes, since mechanical and heat energy are interchangeable. Therefore we have to introduce entropy.

Entropy is a thermodynamic function of great importance: It is implied in the second law of thermodynamics that a certain fraction of the heat entering the system (namely, the quantity dQ/T) is not available for mechanical work, but remains bound as heat. The integral of this quantity, which is known as the *entropy* and defined as follows:

$$dS = \frac{dQ}{T} \tag{7}$$

is, therefore, independent of the path and a function only of the state of a system. The quantity TS is sometimes called *the bound energy*. In any process, the change in TS is given by $T\,dS + S\,dT$. If the process is reversible and isothermal, $dT = 0$, so that $d(TS) = T\,dS = dQ$. Combining equations 1 and 7 gives

$$T\,dS = dU + dW \tag{8}$$

which is a mathematical statement of *the second law of thermodynamics*.

Equation 8 applies only to reversible processes, that is, processes which do not lose energy to the environment; in irreversible processes, $T\,dS > dU + dW$.

Thus entropy is a measure of that portion of the total energy which is unavailable for external work; it is bound within the matter as energy of the random arrangement and thermal motion of the atoms. Another way of defining entropy is, therefore, to associate it with the randomness of the system. At absolute zero temperature a perfectly ordered crystal has zero entropy, and with increasing temperature a certain disorder is induced—the higher the disorder, the larger the entropy.

The entropy at a higher temperature, T, can be calculated from the relation

$$S_T = \int_0^T \frac{c}{T}\,dT \qquad \text{(at constant pressure)}$$

The entropy first rises slowly since the heat capacity, c, goes to zero at the absolute zero. As the temperature rises, the increase assumes a logarithmic form because c becomes constant. At high temperatures we have, approximately,

$$S = \int c \frac{dT}{T} \backsim c \ln T + \text{constant}$$

On heating, many minerals suffer polymorphic inversions. There is a discontinuous change of entropy at the transformation point, called the entropy of transformation, denoted by ΔS, and determined by

$$\Delta S = \frac{\Delta H}{T}$$

where ΔH and T are the latent heat and the temperature of transformation.

Some minerals undergo several polymorphic inversions, and for each inversion the accompanying change of entropy should be added. Thus:

$$S_{T_n} = \int_0^{T_1} \frac{c_1}{T_1} dT + \frac{\Delta H_1}{T_1} + \int_{T_1}^{T_2} \frac{c_2}{T_2} dT + \frac{\Delta H_2}{T_2} + \cdots + \int_{T_{n-1}}^{T_n} \frac{c_n}{T_n} dT \tag{9}$$

For other transitions, that is, melting or evaporation, the same formula can be used.

By combining equations 2 and 8, we obtain

$$dU = T\,dS - P\,dV \tag{10}$$

and from equations 4 and 10 we obtain,

$$dH = T\,dS + V\,dP \tag{11}$$

The bound energy, TS, represents in a sense the energy bound as heat; the remaining part of the internal energy, $U - TS$, may be referred to as the mechanical part of the energy. It is denoted by A and called the *Helmholtz free energy:*

$$A = U - TS$$

$$dA = dU - T\,dS - S\,dT$$

By equation 10 this is

$$dA = -P\,dV - S\,dT$$

or

$$\left(\frac{\partial A}{\partial V}\right)_T = -P; \left(\frac{\partial A}{\partial T}\right)_V = -S \tag{12}$$

If $dT = 0$, or if the process is taking place at constant temperature, the work done is equal to the decrease in Helmholtz free energy. Thus the Helmholtz free energy is a measure of the maximum work which can be done by a system in an isothermal change.

Instead of using volume and temperature as independent variables, one may wish to use pressure and temperature. Then it is more convenient to use the *Gibbs free energy*, denoted by G (or by F), and defined by the equation

$$G = H - TS = U + PV - TS = A + PV \tag{13}$$

By equation 11 we obtain for a reversible process

$$dG = V\,dP - S\,dT \tag{14}$$

or

$$\left(\frac{\partial G}{\partial P}\right)_T = V; \left(\frac{\partial G}{\partial T}\right)_P = -S \tag{15}$$

Absolute values of free energies cannot be determined, only differences have significance. As is done for heat contents it is customary to take the free energy of elements in their standard state at 298°K and one atmosphere pressure equal to zero. The standard molar free energy of any compound is then the free energy increase, ΔG°, occurring in the formation of one mole of the compound from the elements.

To determine ΔG for a new temperature we may use equation 13

$$\Delta G = \Delta H - T\,\Delta S,$$

and determine ΔH and ΔS for the new temperature as described by equations 6 and 9. If the change in heat capacity with temperature is negligible, then ΔH and ΔS will be constant and

$$\Delta G_T \backsim \Delta H_{298} - T\,\Delta S_{298} \tag{16}$$

This is often a useful approximation for work at high temperatures.

An exact relation is

$$d\left(\frac{\Delta G}{T}\right) = -\frac{\Delta H\,dT}{T^2} \qquad (p \text{ constant}) \tag{17}$$

Clapeyron's Equation

Clapeyron's equation describes the relation between the pressure and a temperature of transition when two phases (1) and (2) of the same compound coexist in equilibrium.

Equation 13 holds for both phases, hence

$$\Delta G = \Delta V\, dP - \Delta S\, dT \tag{18}$$

where $\Delta G = G_2 - G_1$, $\Delta V = V_2 - V_1$, and $\Delta S = S_2 - S_1$.

If P and T change in such a way that both phases remain in equilibrium, i.e., if $\Delta G = 0$, then we have:

$$\frac{dP}{dT} = \frac{\Delta S}{\Delta V} = \frac{\Delta H}{T\,\Delta V} \tag{19}$$

which is called Clapeyron's equation (presented in 1834, reproduced in *Pogg. Ann.* **59**, 1843).

Note that

$$\frac{dT}{dP} = \frac{T\,\Delta V}{\Delta H}, \quad \frac{\dfrac{dT}{T}}{dP} = \frac{\Delta V}{\Delta H}, \quad \frac{d \ln T}{dP} = \frac{\Delta V}{\Delta H}$$

If one of the phases is a gas, the volume of the other can be neglected, and $\Delta V = V_{\text{gas}} = RT/P$, where R is the gas constant. The heat of evaporation is usually denoted by ℓ. This gives us:

$$\frac{d \ln P}{dT} = \frac{\ell}{RT^2}, \quad \text{or} \quad \frac{d \ln P}{d\dfrac{1}{T}} = -\frac{\ell}{R} \tag{20}$$

which is the law of Clausius-Clapeyron (*Pogg. Ann.* **79**, 1850).

The integral of equation 20 becomes:

$$\ln P = \text{constant} - \frac{\ell}{RT}; \quad \log \frac{P_2}{P_1} = \frac{\ell}{R \ln 10}\left(\frac{1}{T_2} - \frac{1}{T_1}\right)$$

If $\log P$ is plotted against $1/T$, the numerical value of ℓ can be computed from the slope of the line.

The thermodynamic variables so far encountered are P, V, T, S, U, H, A, G. The number of partial derivatives which can be found from these is $8 \times 7 \times 6 = 336$.

Note the following equations:

$$\text{Thermal expansion} = \frac{1}{V}\left(\frac{\partial V}{\partial T}\right)_P$$

$$\text{Isothermal compressibility} = -\frac{1}{V}\left(\frac{\partial V}{\partial P}\right)_T$$

$$\text{Adiabatic compressibility} = -\frac{1}{V}\left(\frac{\partial V}{\partial P}\right)_S$$

$$\text{Heat capacity at constant volume } c_v = \left(\frac{\partial U}{\partial T}\right)_V = T\left(\frac{\partial S}{\partial T}\right)_V$$

$$\text{Heat capacity at constant pressure } c_p = T\left(\frac{\partial S}{\partial T}\right)_P = \left(\frac{\partial H}{\partial T}\right)_P$$

$$= \left(\frac{\partial U}{\partial T}\right)_P + P\left(\frac{\partial V}{\partial T}\right)_P$$

ON POLYMORPHIC TRANSITIONS

The most important phase transformations in minerals are melting and polymorphic changes; they occur reversibly at constant temperature and pressure; consequently, the free energy must be the same for both phases in equilibrium. On the other hand, at a temperature and pressure which do not correspond to equilibrium between the two phases, the free energy will be different for the two phases. The stable phase has the lowest free energy, and all other polymorphic forms tend to transform into the one stable form characterized by the minimum free energy.

Consider the Hemholtz free energy and its relation to temperature.

$$A = U - TS$$

At absolute zero, the TS vanishes, and the free energy becomes equal to the internal energy of the crystal, U. Thus, at very low temperatures, the internal energy dominates the free energy, and the polymorphic form with the least internal energy tends to be the stable one. But with increasing temperature the TS term becomes increasingly important. It may happen (Fig. A-1), therefore, that because of the possibility of larger entropy in a second structural arrangement, its TS term so reduces its free energy that, in spite of greater U, the free energy drops below that of the first polymorphic form. If this occurs before the first form disintegrates by melting, the second form becomes the stable one, and the first form tends to transform into it. The temperature at which the free energies become just equal is the transition temperature.

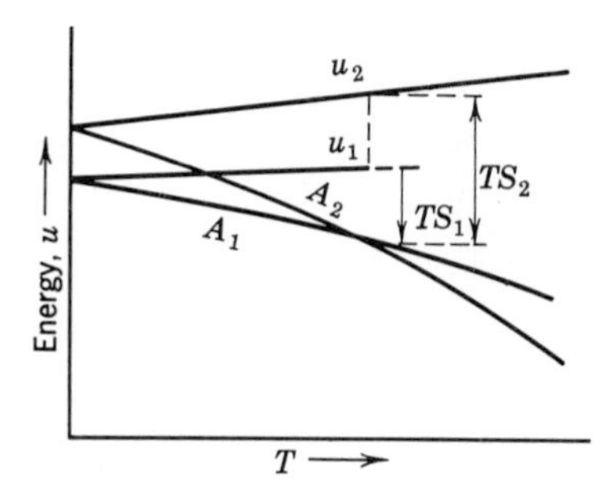

FIG. A-1. Diagram illustrating the relation $A = U - TS$, showing how a high temperature form with high internal energy (u_2), because of its high entropy (S_2), corresponds to the state of lowest free energy (A_2). (After M. J. Buerger, 1948.)

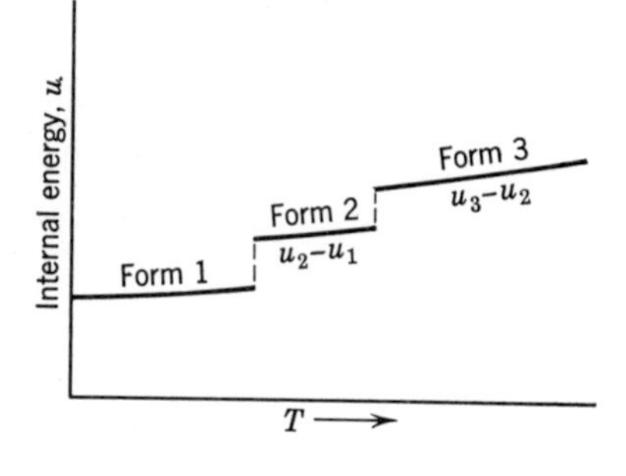

FIG. A-2. In a series of polymorphic forms 1, 2, and 3, stable at increasing temperatures, the internal energy increases discontinuously from one form to the other. (After M. J. Buerger, 1948.)

Note that the second form has a higher internal energy, U, than the first form, and that the difference between the U of the second form and the U of the first form must be supplied to the latter to make the transition occur. This difference is the latent heat of the transition, and, from what has been said, it must be positive for the first transformation. More generally, it can be shown on thermodynamic grounds that it must always be positive in the direction of increasing temperature. For this reason, if we plot the internal energies of a series of polymorphic forms, a series of sloping steps (Fig. A-2) results. The significance of this is that a series of polymorphic forms stable at increasing temperatures must have increasing internal energies.

Analogously, the entropies of a series of polymorphic forms increase for the forms stable at increasing temperatures. Because the entropy affects the volume over which the atoms may be disordered, there is a tendency for the forms of higher entropy to have greater open spaces available for thermal motion. Although this tendency does not necessarily affect the openness of the entire structure, it often does, so that high-temperature forms tend to be less dense than low-temperature forms.

Thermal energy in a crystal manifests itself in vibrational disorder, that is, the constituent atoms (ions) of the space lattice show thermal vibration around their center of equilibrium. With increasing temperature, this vibration may shift into rotation (rotational disorder), that is, single atoms or groups of atoms rotate, occupying no fixed lattice positions. Finally the temperature may engender substitutional disorder. In many silicates, especially at elevated temperatures, Al and Si are disordered. They occupy together the same set of lattice sites over which they are statistically distributed. See under feldspar, page 252.

THE ROLE OF PRESSURE

In the crystal structure temperature and pressure tend to impose opposite conditions: whereas temperature tends to require an open structure, high pressure tends to require a compact one. This can be seen by plotting the free energy against pressure. The relation is shown by equation 15:

$$\left(\frac{\partial G}{\partial P}\right)_T = V$$

For many minerals the volume does not change much with pressure. Plotting G against pressure will, therefore, give almost straight lines. The slope of the lines is proportional to V; consequently, minerals of small volume, that is, minerals with compact structures, will have the lowest free energy at high pressures. It should be observed that in a transition from a low-pressure modification to a high-pressure modification the latent heat of transition may be either positive or negative.

There are many examples of transformations induced by high pressure; quartz → coesite is one of the most famous, see pages 75 ff. A general difficulty in studying high-pressure modifications is that they often quickly reinvert when the pressure is relieved. It is possible, therefore, that many silicates possess high-pressure modifications that are as yet unknown to science.

PARTIAL FREE ENERGY

The free energy may be applied to a pure substance or to a specified compound (i) in a solution. The solution may be either solid or liquid, and the symbol F_i is often used to represent the *partial molal free energy* or the *chemical potential* per mole of the species i.

The *activity* of a chemical species is denoted by a_i and defined by the relation

$$F_i = RT \ln a_i + F_i^0 \tag{21}$$

where F_i^0 is the free energy in the "standard state," R the gas constant, and T the absolute temperature.

Note that the numerical value of F involves an arbitrary molecular weight or formula weight. For n moles of the components 1, 2, 3, . . . i in the system

$$G = n_1F_1 + n_2F_2 + n_3F_3 + \cdots n_iF_i$$

For constant temperature and pressure,

$$dG = \Sigma F_i \, dn_i$$

where $F_i = M_i\mu_i$. M_i is the molecular weight of the chemical species, and μ_i is the *chemical potential per gram.*

The activity of a chemical species is proportional to its vapor tension. In dilute solutions it is proportional also to its concentration. Just as the free energy has to be the same for two phases in equilibrium, so must the chemical potential and, in most cases, the activity and vapor tension be the same.

The change of activity with pressure follows the Gibbs free energy, in analogy with equation 15:

$$\left(\frac{\partial \ln a}{\partial P}\right)_T = \frac{V}{RT}$$

The *law of mass action,* propounded by Guldberg and Waage, is a quantitative statement relating concentration (or velocity of a reaction) to the activity of its reactions. Consider the hypothetical reaction

$$2A \rightleftharpoons C + D$$

Two molecules of A react to form one molecule of C and one of D. The rate of the reaction is determined by the chance of collision between any two molecules. It is easy to show that this chance is proportional to the concentration of A molecules squared. This can be expressed as

$$\text{Rate} \rightarrow \; = k_1 \times (A)^2$$

where (A) represents the concentration of A.

It is also easy to show that the reverse reaction, i.e., $C + D$ combining to form 2 A-molecules, has a probability proportional to the product of the concentrations of C and D. Thus

$$\text{Rate} \leftarrow \; = k_2 \times (C) \times (D)$$

For equilibrium the rates of the two reversible reactions must be the same. Consequently:

$$\frac{k_1}{k_2} = \frac{(C) \times (D)}{(A)^2} = K$$

For the general reaction,

$$nA + mB = pC + rD$$

where n, m, p, r are small whole numbers, the following relation holds:

$$\frac{(C)^p \times (D)^r}{(A)^n \times (B)^m} = K$$

where (A), (B), (C), and (D) represent the activities of A, B, C, and D, respectively, and K is the equilibrium constant, which is constant at constant temperature and pressure.

This is a generalized statement of the law of mass action or the law of chemical equilibrium.

The activity of pure liquids or pure solids is taken as unity at all moderate pressures and temperatures.

For any reaction a relation like equation 21 will apply

$$\Delta F = \Delta F^\circ + RT \ln \frac{(C)^p \times (D)^r}{(A)^n \times (B)^m}$$

If the activities in this equation are the equilibrium values, then $\Delta F = 0$, and

$$\Delta F^\circ = -RT \ln K \tag{22}$$

By combining equations 17 and 22 we obtain

$$\left(\frac{\partial \ln K}{\partial T}\right)_P = \frac{\Delta H^\circ}{RT^2} \qquad (p \text{ constant}) \tag{23}$$

Since, according to equation 15, $dF/dP = V$, it follows from equation 22 that

$$\left(\frac{\partial \ln K}{\partial P}\right)_T = \frac{\Delta V}{RT} \qquad (T \text{ constant}) \tag{24}$$

A combination of equations 23 and 24 yields the famous Clausius-Clapeyron equation

$$\frac{dT}{dP} = \frac{T \, \Delta V}{\Delta H}$$ (see page 387)

Great caution must be exercised by the application of these equations to mineral reactions in rocks; for rocks are usually "open" for "mobile" phases (like H_2O, CO_2) in the surrounding, and if any of the surrounding phases interact, the temperature of the reaction will change. For example, the reaction

$$\text{Quartz} \rightleftharpoons \text{Tridymite}$$

usually takes place in a vapor phase, some of the constituents of which enter into the tridymite lattice:

$$SiO_2 + H_2O \rightleftharpoons [SiO_2(H_2O)]$$

$$\text{Quartz} + \text{Vapor} \rightleftharpoons \text{stuffed tridymite}$$ (see page 74)

Stuffed tridymite is not pure silica but contaminated; therefore the temperature of the reaction is depressed. Thermodynamically this is similar to the increase in the boiling point or the depression of the melting point by contamination of a liquid (page 78), but it has often been neglected in petrology and it is the merit of Rosenfeld (1961) to have pointed out the geological importance of this phenomenon by setting up his "contamination-reaction-rules."

FREE ENERGY IN THE GRAVITATIONAL FIELD

The free energy has one property dissimilar to activity: it varies with the position in the gravitational field. Chemists usually do not pay much attention to this fact, because they are not often concerned with reactions over great vertical distances. But geologists are, and to them it becomes of great importance. Gibbs himself was aware of this factor. He writes that "the difference of the values of the intrinsic potential for any component at two different levels is equal to the work done by the force of gravity when a unit of matter falls from the higher to the lower level." Consequently,

$$\left(\frac{\partial F}{\partial h}\right)_T = g \cdot M_i \qquad (25)$$

M_i is the molecular weight (atomic weight) of the species, and h is the vertical distance, which is positive upward; g is the gravity constant.

This is a very important relation. It shows that equilibrium is affected simply by moving a chemical species up or down in the field of gravity. In petrology we often are interested in sections of large vertical extensions. The position in the gravitational field is, therefore, of great consequence.

Clearly, equation 25 is valid only at constant temperature. In a vertical section in the crust of the earth there is, however, a temperature gradient which impairs the petrological usefulness of the equation. However, it can be shown that a temperature gradient is not able to change the trend indicated by the equation.

It should be noted that equilibrium is obtained in a system of large vertical extension if the free energy of a specified component is the same all through the system. But the activity, or the vapor tension, must not be the same; these quantities are independent of the position and must, under equilibrium conditions, display different values in the different depth levels.

Let V_i be the specific volume of a species, i. It is related to the molecular volume, V_{mol}, by

$$V_{\text{mol}} = V_i \cdot M_i$$

Equation 15 thus becomes

$$\frac{dF}{dP} = V_i \cdot M_i \tag{26}$$

Great depths in the crust of the earth are now considered, that is, the vertical distance, h, of equation 25 is large.

The relation of depth, h, pressure, P, and specific volume of the overlying load, V_{load}, is

$$h = PV_{\text{load}} \cdot \frac{1}{g} \tag{27}$$

Let $\Delta F_{(h)}$ and $\Delta F_{(P)}$ denote the free energy increments of a chemical species, i, for a depth h and for a pressure P, respectively.

According to equations 25, 26, and 27,

$$\begin{aligned} \Delta F_{(h)} &= -hgM_i = -PV_{\text{load}} \cdot M_i \\ \Delta F_{(P)} &= \qquad\qquad\quad +PV_iM_i \end{aligned}$$

The change of free energy with both depth and pressure is, therefore:

$$\Delta F_{(h)(P)} = PM_i(V_i - V_{\text{load}}) \tag{28}$$

In order for the chemical species, i, to be in equilibrium, the change in free energy must be zero; this happens if $V_i = V_{\text{load}}$. Consequently, only if the chemical species is overlain by itself (or a rock of exactly the same specific gravity) will it be in equilibrium.

The lighter the chemical species concerned, the larger the increase in the $\Delta F_{(h)(P)}$–function.. The physical effect of high ΔF-values is a high *tendency of escape*, the lighter the constituent the faster it will migrate toward the upper parts of the earth crust. In view of the small density of, for example, the oxygen ion, it would at great depths be "squeezed" toward the outside of the globe. Likewise water-bearing and CO_2-bearing minerals are not stable at greater depths, but lose water that migrate upward. Analogously, if the chemical species concerned is denser than the overburden, it will migrate downward.

If we disregard the chemical forces the relation is very simple: the heaviest matter will concentrate in the core of the earth, and toward the crust concentric shells of successively lighter material will form. This simple arrangement will be modified, however, for certain elements by the chemical bonding energies. But it does not alter the fact that if one starts with a homogeneous earth, a vertical transport of the various components will take place with some rising and some sinking in an attempt to attain the equilibrium conditions required in the gravitational field. Many compounds, for example, iron oxides, will not be stable at great depths—calculations show that somewhere around 1000 km depth (depending upon the temperature) all oxides of iron will become unstable, the chemical bonds will break, and iron will concentrate at lower depths, while oxygen will concentrate in the opposite direction.

Thus homogeneous rock is not stable in the gravitational field. The very presence of the field makes the homogeneous rock unstable, initiates chemical migrations, and results in differentiation.

The phenomenon of a compositional gradient resulting from a steady thermal gradient is known as the *Soret* effect.

In systems open to water the Clausius-Clapeyron equation then becomes

$$\frac{dT}{dP_S} = \frac{\Delta V_S + V^*_{\mathrm{H_2O}} \cdot \Delta n\mathrm{H_2O}}{\Delta S_S + S^*_{\mathrm{H_2O}} \cdot \Delta n\mathrm{H_2O}}$$

where subscript S applies to the immobile phases of the solid rock, $V^*_{\mathrm{H_2O}}$ is the fictive volume of water and $S^*_{\mathrm{H_2O}}$ is the entropy of water including the entropy of transport.

THE PHASE RULE

Any mixture of, for example, MgO and SiO_2 will fuse and form a melt which may be homogeneous or inhomogeneous. Upon cooling, the melt will crystallize, but it is difficult to predict the chemical nature of the solid products thus formed. Experiments have shown that the following compounds may occur: (1) periclase (MgO), (2) forsterite (Mg_2SiO_4), (3) clinoenstatite ($MgSiO_3$), and (4) cristobalite (SiO_2). There are four additional minerals that, according to their composition, belong here, but do not occur as primary products of crystallization in the system: (5) enstatite, (6) tridymite, (7) quartz, and (8) coesite. The theory of the *heterogeneous equilibria* tells us when we should expect to find only one of the possible minerals, and when several minerals and phases will occur together. The following definitions are important.

The *phases* of a heterogeneous system are the parts of the system that are mechanically separable and physically different, for instance, the solid, liquid, and gaseous parts. The number of phases which can exist side by side may vary greatly in various systems. In all cases, however, there can be but one gas phase, because gases are miscible with one another in all proportions. Liquids also are often miscible in all proportions, but occasionally two (or several) liquid phases may coexist. Any number of crystalline phases may be present together.

The *components* of the system are the smallest number of chemical species necessary to make up all the phases. The components are not synonymous with the chemical elements, that is, with the constituents, of the system. Uncertainties occasionally arise as to what components should be selected.

The *degrees of freedom* of a system are determined by the relations of the variables. In a system of constant chemical composition we usually consider but two variables, temperature (T) and pressure (P). We shall not take into account changes of equilibrium due to the action of electrical, magnetic, or capillary forces, or gravity. Let us consider a system in which all phases are in equilibrium with each other under constant T and P. According to whether none, one, or both of the variables can be changed without disturbing the equilibrium, the system is said to possess none, one, or two degrees of freedom.

An example is afforded by the system of H_2O. Three phases, ice-water-vapor, can coexist only at a definite value of temperature and pressure. The system is *invariant;* it possesses no degree of freedom, for none of the two variables can be changed without affecting the equilibrium.

If one of the variables is arbitrarily chosen, the system has become *monovariant*. It possesses one degree of freedom, but only two phases can now coexist: ice—water, or ice—vapor, or water—vapor.

If both variables are chosen arbitrarily the system is *divariant*, with two degrees of freedom. But the number of coexisting phases is now reduced to one.

For a system in equilibrium the following relation holds between the number of coexisting phases p, components c, and degrees of freedom f:

$$p = c + 2 - f \tag{29}$$

This is the phase rule of Willard Gibbs (1874).

The deduction of this rule is easy from a consideration of the Gibbs free energy, or better, the related quantities: partial molal free energy or chemical potential of the components of the system (see equation 21).

The components distribute themselves over all the phases of the system. The concentration of a specified component may differ enormously from one phase to another, but no phase can be without some contribution from all components. This follows immediately from the fact that under equilibrium conditions, the chemical potential (or the activity) of each component must be the same in all the phases of the system. The variables in the system are P, T, and the concentration of $(c - 1)$ of the components in each of the p phases, thus all together $2 + p(c - 1)$ variables. Under equilibrium conditions, the chemical potential of each of the components must have the same value in all phases. This fact presents us with $(p - 1)$ relations for each component: all together $c(p - 1)$ relations. Now we have $2 + p(c - 1)$ variables and $c(p - 1)$ equations. The difference between these two numbers gives us the number of the degrees of freedom:

$$f = 2 + p(c - 1) - c(p - 1) = c + 2 - p,$$

which is identical to equation 29.

The phase rule is useful in petrology, for example, in the study of the crystallization of magmas. It is likewise useful in the study of solid rocks, for it places a limitation on the number of minerals which can occur in equilibrium in a given rock. For such studies it should be remembered that the maximum number of phases can be attained only in an invariant system, that is, with both T and P fixed invariantly, corresponding to a single point in the P-T diagram. During a process of mineral formation P and T are not constant. On the contrary, the processes usually take place over a large P-T interval, corresponding to a region in the P-T diagram. Thus both P and T are variable, corresponding to two degrees of freedom. Under these conditions equation 1 reduces to

$$p \leqq c \tag{30}$$

which has become known as the *mineralogical phase rule,* propounded by V. M. Goldschmidt (1912). According to this rule, the number of different minerals in a rock should not exceed the number of components; because of the phenomenon of solid solution the number of minerals may be less.

If a third phase existed in the pores during the formation of the rock, but was later squeezed out another reduction is imposed on the maximum number of minerals; now

$$p \leqq c - 1$$

This case has been discussed by Korzhinskii (1950, 1957) who has shown that, generally, if n "mobile" components are present, the relation becomes:

$$p \leqq c - n \tag{31}$$

i.e., in an "open" system the number of different minerals must not exceed the number of total components minus the number of mobile components.

H_2O and CO_2 are always mobile, but also other components, e.g., soda, or even lime may under special conditions act as mobile components.

This is particularly significant in the study of the chemistry of the metamorphic rocks and has actually often been assumed or implied for mineral assemblages in that the water, or carbon dioxide content of the assemblage, has been neglected, frequently on the assumption that water or carbon dioxide was present in *excess* during the metamorphism.

Korzhinskii distinguishes between (1) *isolated systems* which are separated from the environment by rigid walls impervious to flow of heat and matter; (2) *closed systems* which cannot exchange matter but can exchange heat; (3) *open systems* which can exchange heat and, under certain conditions, also matter with the surroundings.

J. B. Thompson (1955) of Harvard University independently concerned himself with open systems and endeavored to show that if a rock with a given content of fixed components is open to a mobile component, m, the most stable state of the rock (= lowest free energy expressed by the L-function) will be dependent not only on P and T but also upon the value of the partial free energy of the mobile component, F_m, in the immediate environment:

$$L = G + \Delta n F_m = \Sigma F_i n_i$$

$$d\,\Delta L = 0 = d\,\Delta G + \Delta n\, dF_m$$

where Δn represents the change of the number of moles of the mobile component in the environment. If there is no mobile component, L becomes identical to the Gibbs free energy.

It is readily seen that the partial derivatives of L with respect to both P and T will also depend on a function of F_m (in addition to V and $-S$, see equation 15). This emphasizes the importance of mobile constituent and the great caution which must be exercised in the computation of free energy relations of natural rock systems (see pages 248, 257–258).

SURFACE ENERGY

Surface energies are extremely important in geology. A better understanding of the surface effects would give us a much better understanding of all sorts of movements and migrations, both mechanical and ionic, in rocks along grain boundaries, in intergranular films, in the filling of interstices, etc.

The total free energy of a surface is proportional to its area A. It is well-known that the vapor pressure of small drops is higher than the normal vapor pressure. Therefore, highly curved particles have an excess free energy over plane-surface particles of the same material.

$$\Delta F = RT \ln \frac{p'}{p} = \gamma M V\left(\frac{2}{r}\right) \tag{32}$$

Here p' is the vapor pressure of a particle with radius r, the normal pressure is p, specific volume V, and molecular weight M. Consequently, small mineral grains involve large total energies; therefore, small grains are unstable in relation to larger grains and will eventually be devoured by the larger grains (German: "Sammelkristallisation"). The force required to extend the surface, the surface tension, γ, can be measured and is numerically equal to the surface energy:

$$\frac{dF}{dA} = \gamma$$

In Fig. A-3 the contact angle Θ is a measure of the relative wettabilities, because it depends on the relative surface energies of the interfaces and will adjust itself until equilibrium is achieved with the lowest surface energy. The resultant of the tension of the vapor-liquid interface acting on the plane surface must be equal to the difference between the interfacial tensions of the two fluids against the solid, that is

$$\gamma_{13} - \gamma_{23} = \gamma_{12} \cdot \cos \Theta$$

If $\gamma_{13} < \gamma_{23}$, then $\Theta > 90°$, as in Fig. A-3. If $\gamma_{13} > \gamma_{23}$, then $\Theta < 90°$; this can be observed for example in a drop of mercury on a glass plate. If γ_{12} is less than $\gamma_{13} - \gamma_{23}$ the equation cannot be satisfied; when this occurs, phase 2 spreads indefinitely over the surface of the solid and entirely displaces phase 1. This is why certain silicate solutions may spread over large surfaces often insinuating themselves into small (capillary) openings and cracks, which become completely filled.

It is interesting that the theory can be extended to explain the microstructure of rocks as exhibited in solid-solid interfaces, as has been done for metals (C. S. Smith, 1948). In the absence of any crystal boundaries in either phase, the boundary of one phase when completely surrounded by another must eventually become a sphere, for this is the smallest surface to contain a fixed volume. If there are two crystals of one phase meeting with one crystal of a second

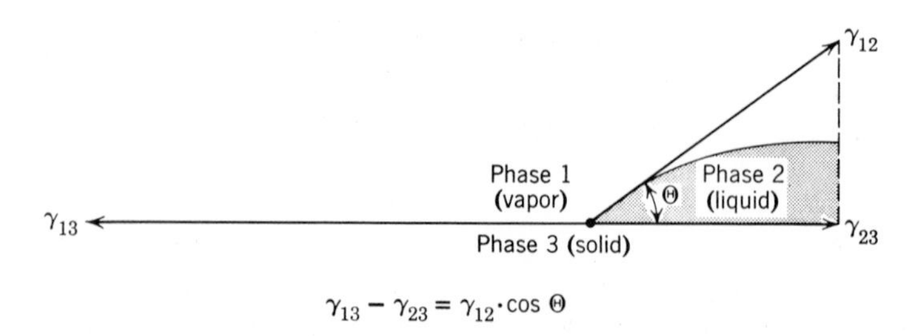

Fig. A-3. Relations between the surface energies and the dihedral angle at which a drop of liquid meets the surface on which it rests.

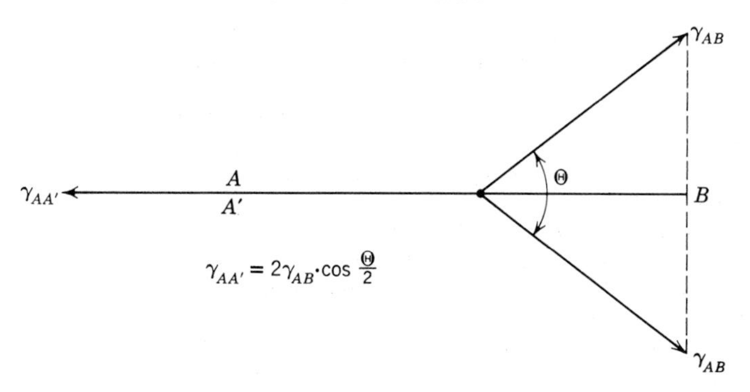

FIG. A-4. Equilibrium between a grain boundary *AB* and two equal interphase boundaries AA′.

phase, the one grain boundary will establish geometric equilibrium with the two interphase boundaries, and the angles will be determined by the relative values of the interfacial tensions, see Fig. A-4. Assuming that the two interphase boundaries are not subject to crystallographic influences, then

$$\gamma_{AA} = 2\gamma_{AB} \cdot \cos\frac{\Theta}{2}$$

For $\gamma_{AB} \gtreqless \gamma_{AA}$ we have $\Theta \gtreqless 120°$. But if $\gamma_{AB} < \frac{1}{2}\gamma_{AA}$ there is no value of Θ that can satisfy the equation, and the second phase will penetrate along the boundary indefinitely. Figure A-5 shows in an idealized form the shapes, for various values of Θ, that a small volume of a second mineral must have if it appears at the corner of three grains.

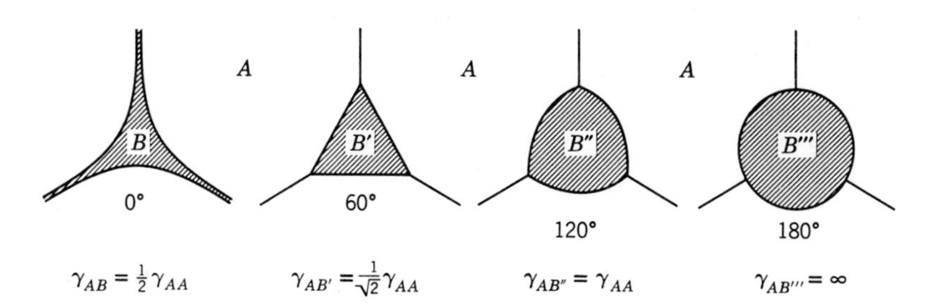

FIG. A-5. Effect of relative values of grain boundary tension on shape of a mineral *B* formed at line of intersection between three grains. (After C. S. Smith, 1948.)

SELECTED REFERENCES

Buerger, M. J., The role of temperature in mineralogy, *Am. Mineralogist* **33,** 1948.

Bridgman, *A Condensed Collection of Thermodynamic Formulas,* Cambridge, Mass., 1926.

Carroll and Lehrman, *J. Chem. Educ.* **24,** 1947.

Findlay, A., *The Phase Rule,* 8th ed. (Dover), New York, 1945.

Goldschmidt, V. M., Die Gesetze der gesteinsmetamorphose, *Norske Vid.-Selsk. Skr.* **1,** Oslo, 1912.

Kelley, K. K., Contributions to the Data on Theoretical Metallurgy. XIII. High-temp. heat content, heat-capacity, and entropy data for the elements and inorganic compounds. *Bull. Bureau of Mines, 50 Anniversary* **584,** 1960.

Korzhinskii, D. S., Phase rule and geochemical mobility of elements, *XVIII International Geol. Congr.* (Great Britain), Part d, 50, 1948.

———, *Physicochemical Basis of the Analysis of the Paragenesis of Minerals,* USSR Acad. Sci. Press, Moscow, 1957. Translation by Consultants Bureau, 1959.

Kretz, Ralph, Some applications of thermodynamics to coexisting minerals of variable composition, *J. Geol.* **69,** 1961.

Kubaschewski, O., and E. L. Evans, *Metallurgical Thermochemistry,* London (Butterworth-Springer), 1951.

Margenau and Murphy, *The Mathematics of Physics and Chemistry,* New York, 1943.

Marsh, J. S., *Principles of Phase Diagrams,* New York (McGraw-Hill), 1935.

Ricci, J. E., *The Phase Rule and Heterogeneous Equilibrium,* New York (Van Nostrand and Macmillan), 1951.

Rosenfeld, J. L., The contamination-reaction rules, *Am. J. Sci.* **259,** 1, 1961.

Rossini, F. D., and F. R. Bichowsky, The Thermochemistry of the Chemical Substances, New York (Reinhard), 1936.

Rossini, F. D., and others, Selected Values of Chemical Thermodynamic Properties, *National Bureau Standards Circ.* **500,** 1950.

Slater, J. C., *Introduction to Chemical Physics,* New York, 1939.

Smith, C. S., *Grain, Phases, and Interfaces,* AIME Techn. Publ. No. 2387, 1948.

Thompson, J. B., Thermodynamic basis for the mineral facies concept, *Am. J. Sci.* **253,** 65, 1955.

———, Graphical analysis of mineral assemblages in pelitic schists, *Am. Mineralogist* **42,** 842, 1957.

———, Local equilibrium in metasomatic processes, in *Recherches in Geochemistry,* edited by P. H. Abelson, New York (Wiley), 1959, p. 427.

CONSTANTS

$R = 8.31$ j degree^{-1} mole^{-1} = 1.99 cal degree^{-1} mole^{-1}.

1 j = 0.239 cal = 10^7 erg = 9.87 ml atm = 0.102 kg m.

1 cal = 4.19 j = 0.427 kg m.

1 atm = 760 torr = 1.013 bar = 1.033 kg cm^{-2}.

1 Å = 10^{-8} cm; g = 980.7 cm sec^{-1}; 0°C = 273°K.

Avogadro's number, $N = 6.0228 \times 10^{23}$ molecules mole^{-1}.

Index